T0295440

Robotics: Design, Kinematics and Motion Planning

Robotics: Design, Kinematics and Motion Planning

Edited by Maggie Ross

CLANRYE INTERNATIONAL
www.clanryeinternational.com

Clanrye International,
750 Third Avenue, 9th Floor,
New York, NY 10017, USA

Copyright © 2023 Clanrye International

This book contains information obtained from authentic and highly regarded sources. Copyright for all individual chapters remain with the respective authors as indicated. All chapters are published with permission under the Creative Commons Attribution License or equivalent. A wide variety of references are listed. Permission and sources are indicated; for detailed attributions, please refer to the permissions page and list of contributors. Reasonable efforts have been made to publish reliable data and information, but the authors, editors and publisher cannot assume any responsibility for the validity of all materials or the consequences of their use.

Trademark Notice: Registered trademark of products or corporate names are used only for explanation and identification without intent to infringe.

ISBN: 978-1-64726-676-9

Cataloging-in-Publication Data

Robotics : design, kinematics and motion planning / edited by Maggie Ross.
     p. cm.
Includes bibliographical references and index.
ISBN 978-1-64726-676-9
1. Robotics. 2. Robots--Design and construction. 3. Automation.
4. Machine theory. I. Ross, Maggie.
TJ211 .R63 2023
629.892--dc23

For information on all Clanrye International publications
visit our website at www.clanryeinternational.com

Contents

Preface

The world is advancing at a fast pace like never before. Therefore, the need is to keep up with the latest developments. This book was an idea that came to fruition when the specialists in the area realized the need to coordinate together and document essential themes in the subject. That's when I was requested to be the editor. Editing this book has been an honour as it brings together diverse authors researching on different streams of the field. The book collates essential materials contributed by veterans in the area which can be utilized by students and researchers alike.

Robotics is a branch of engineering involved in the conception, design, manufacture and operation of robots. The plan or convention which is used for constructing a robot or a robotic system is known as robot design. Robot kinematics refers to the application of geometry in order to study the movement of multi-degree of freedom kinematic chains that make up the framework of robotic systems. It investigates the relationship between the connectivity and dimensions of kinematic chains. Furthermore, it studies the relation between acceleration, location and velocity of each link in the robotic system for planning and controlling movement, and for computing torques and actuator forces. Robot kinematics also includes singularity avoidance, collision avoidance, motion planning, redundancy and kinematic synthesis of robots. Motion planning refers to the process of breaking down a desired movement of the robot into discrete motions that are subject to certain movement constraints, and optimize the required movement. This book unfolds the innovative aspects of robot design, kinematics and motion planning, which will be crucial for the progress of robotics in the future. It will serve as a valuable source of reference for graduate and post graduate students.

Each chapter is a sole-standing publication that reflects each author's interpretation. Thus, the book displays a multi-facetted picture of our current understanding of application, resources and aspects of the field. I would like to thank the contributors of this book and my family for their endless support.

Editor

A Multi-Switching Tracking Control Scheme for Autonomous Mobile Robot in Unknown Obstacle Environments

Jianhua Li, Jianfeng Sun *[ID] and Guolong Chen[ID]

School of Mechanical & Electrical Engineering, Lanzhou University of Technology, Lanzhou 730050, China;
li_jhlz@lut.edu.cn (J.L.); cgl20061273@126.com (G.C.)
* Correspondence: jianfeng.sun.lut@gmail.com

Abstract: The obstacle avoidance control of mobile robots has been widely investigated for numerous practical applications. In this study, a control scheme is presented to deal with the problem of trajectory tracking while considering obstacle avoidance. The control scheme is simplified into two controllers. First, an existing trajectory tracking controller is used to track. Next, to avoid the possible obstacles in the environment, an obstacle avoidance controller, which is used to determine the fastest collision avoidance direction to follow the boundary of the obstacle at a constant distance, is proposed based on vector relationships between the robot and an obstacle. Two controllers combined via a switch strategy are switched to perform the task of trajectory tracking or obstacle avoidance. The stability of each controller in the control scheme is guaranteed by a Lyapunov function. Finally, several simulations are conducted to evaluate the proposed control scheme. The simulation results indicate that the proposed scheme can be applied to the mobile robot to ensure its safe movement in unknown obstacle environments.

Keywords: trajectory tracking; obstacle avoidance; switch strategy; mobile robots

1. Introduction

In recent years, the wheeled mobile robot (WMR) has received much attention due to its many practical applications, which is widely used in various aspects, such as search and rescue [1], multi-robotic formation [2,3], industrial applications [4,5], military operations [6,7], and so on. These applications require the mobile robot to move autonomously and carry out a variety of automated tasks which include trajectory tracking, obstacle avoidance, formation control, etc. For the control of mobile robots, the major challenge is to develop effective controllers to deal with various tasks. Among those tasks, the problems of obstacle avoidance and trajectory tracking are especially important for autonomous movement.

Obstacle avoidance is a necessary function in robotics technology. It aims to ensure the robot would not collide with obstacles in unknown environments. The obstacle avoidance problem has been investigated by many researchers. In [8], a path planning algorithm was designed using the sensor fusion of a camera and a laser radar to generate a collision-free path. In addition, artificial potential field (APF) [9,10] methods were presented for obstacle avoidance. They use a potential field function to generate an obstacle-free trajectory by creating an attractive force for the goal and a repulsive force around the obstacle to avoid collision. In [11,12], The genetic algorithms (GA) inspired by evolutionary theory were devised to generate optimal paths from one start point to the target location within given resented to find a feasible path for the multi-objective path planning problem. In [14,15], some common optimization technologies like particle swarm optimization (PSO) and ant colony optimization (ACO) were also presented to resolve the obstacle avoidance problem in terms of multi-objective optimization. However, the main drawbacks of those heuristic or evolutionary methods based on optimization techniques are that they have possible local minimum problem in computation, and their computations are complicated. In order to solve those limitations, researchers studied the obstacle avoidance problem based on control theory, and obstacle avoidance methods using the geometric relationship between the robot and an obstacle were proposed to avoid possible obstacle [16–19]. Whereas the above-mentioned works only consider obstacle avoidance for mobile robots without taking trajectory tracking into account.

Robots follow a desired trajectory generated by a virtual mobile robot based on its kinematic model and initial posture, which can navigate the mobile robot to the desired position. This tracking problem has been studied by many researchers, and there are a lot of control strategies are proposed for mobile robots, such as model predict control (MPC) [20–22], sliding mode control (SMC) [23–26], fuzzy control [27–29], adaptive control [30–32], and intelligent control [33,34], etc. However, most of them studied the trajectory tracking problem under the assumption that the movement of the robot is in an obstacle-free environment. Hence, the design of the controller lacks the consideration of the possible collisions in the environment.

In the above discussion, the obstacle avoidance problem is addressed by pathing planning algorithm or obstacle avoidance controller. Besides those mentioned literatures, some studies presented some unified controllers by considering the obstacle avoidance function in trajectory tracking controller to handle the problems of tracking and obstacle avoidance over the past few years [35–38], few studies have focused on the combination of multiple controllers [39]. Furthermore, it is extremely hard to address the problem of trajectory tracking with obstacles utilizing only one controller for the difficulties in design and the high computational cost. Therefore, the combination of practical, low-computational cost, and effective controllers is important to ensure the movement of the mobile robot.

This paper presents a control scheme for a mobile robot to navigate it from a start position to a desired destination. The mobile robot tracks a pre-planned trajectory by using the reference posture and reference velocities as input signals for its control system. Owing to unknown obstacles in the environment, an obstacle avoidance controller is presented to escape the obstacle. In this paper, the problem of trajectory tracking in unknown obstacle environments is simplified by dividing the objective of control scheme into two controllers, trajectory tracking and obstacle avoidance controllers. Both controllers are designed separately based on their own error dynamic model to execute the corresponding task. In this control scheme, a switch strategy is introduced to combine two controllers, and the current controller executed by the control system of mobile robot is switched between the trajectory tracking controller and the obstacle avoidance controller, which means that only one controller works at specific condition to guarantee the performance of each task In this paper, an existing trajectory tracking controller is used to track the pre-planned trajectory. Once an obstacle is detected by the mobile robot, the safe boundary and risk area of the obstacle are generated by the obstacle controller. When the obstacle avoidance condition is satisfied, a blending vector used to determine the fastest obstacle avoidance direction and follow the boundary of the obstacle to avoid the obstacle. After the completion of the obstacle avoidance, the trajectory tracking controller is activated to track the pre-planned trajectory. The advantage of this control scheme is to divide the complex tracking problem into two simple and low-computational controllers combined via a switch strategy. The stability of the proposed control scheme is proved by a Lyapunov function. In addition, the effectiveness of the proposed control scheme is evaluated by the simulation results.

The rest of this paper is organized as follows. In Section 2, the problem statement related to control scheme is stated. In Section 3, an obstacle avoidance control method is presented to avoid the possible collision. In Section 4, a switch strategy used to combine the tracking and obstacle avoidance controllers is introduced. In Section 5, several simulations are given to validate the effectiveness of the proposed control scheme. In Section 6, brief conclusions and future studies are discussed.

2. Problem Statement

The robot discussed in this paper is a two-wheeled mobile robot. The kinematic model of the robot is described in detailed in [40,41]. The environment where the mobile robot works is always filled with unknown obstacles. That is, static obstacles or dynamic obstacles, and their position information is prior or measured by the sensors attached to the mobile robot during its movement. In this paper, the obstacle avoidance problem can be defined as designing a feasible path from the perspective of control theory, meaning that will not collide with the obstacles. In the control domain of the mobile robot, the task of the mobile robot requires navigating the robot from a start point to a desired position. Trajectory tracking typically plays an important role in the navigation task. In order to implement the trajectory tracking, we use a classic nonlinear control rule to track the trajectory, and then combined with our proposed obstacle avoidance method. In this paper, to deal with the problem that the mobile robot encounters an obstacle when tracking, an obstacle avoidance controller is presented. The basic idea of the controller is to drive the mobile robot to a direction determined by a blending vector to follow the boundary of the obstacle at a constant distance, and then escape the obstacle to track the trajectory.

3. Methodology

3.1. Trajectory Tracking Control

In this section, an existing trajectory tracking controller is introduced, which aims to find appropriate control inputs of the mobile robot and then make tracking errors to zero when the tracking time goes to infinite.

The trajectory tracking problem that the mobile robot tracks the reference trajectory generated by a moving virtual mobile robot is depicted in Figure 1. Let $q_r = \begin{bmatrix} x_r & y_r & \theta_r \end{bmatrix}^T$ denote the posture of the virtual mobile robot, and the current posture of the mobile robot is expressed as $q = \begin{bmatrix} x & y & \theta \end{bmatrix}^T$. The tracking errors e_1, e_2, and e_3 between the mobile robot and the virtual mobile robot denoted in the body coordinate system $\{O_1\}$ are given by

$$\begin{bmatrix} e_1 \\ e_2 \\ e_3 \end{bmatrix} = \begin{bmatrix} \cos\theta & \sin\theta & 0 \\ -\sin\theta & \cos\theta & 0 \\ 0 & 0 & 1 \end{bmatrix} \begin{bmatrix} x_r - x \\ y_r - y \\ \theta_r - \theta \end{bmatrix} \tag{1}$$

A classic control rule for the trajectory tracking is given [42]

$$v = v_r \cos e_3 + k_1 e_1 \tag{2}$$

$$\omega = \omega_r + k_2 v_r e_2 + k_3 v_r \sin e_3 \tag{3}$$

where k_1, k_2, and k_3 are positive constants. the control inputs described in Equations (2) and (3) are used, the tracking errors e_1, e_2, and e_3 will converge to zero.

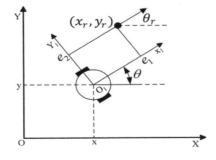

Figure 1. Diagram of trajectory tracking.

3.2. Obstacle Avoidance Control

3.2.1. Analysis of Obstacle Avoidance Problem

In this section, a mobile robot is considered with the mission of avoiding collision between the robot and circular obstacles in a two-dimensional plane. Let $M(x, y)$ and $O(x_0, y_0)$ be the positions of the robot and the obstacle, respectively. In order to facilitate the design and analysis of the obstacle avoidance controller, several definitions are introduced in the following.

Definition 1. *An obstacle can be considered be detected by the mobile robot if the following expression is satisfied:*

$$D_r \leq D_{det} \tag{4}$$

where D_{det} represents the maximum measurement distance of the range sensor attached to the mobile robot. D_r denotes the relative distance between the robot and an obstacle, and the D_r is calculated by the expression

$$D_r = \|u_{mo}\| = \sqrt{(x_0 - x) + (y_0 - y)} \tag{5}$$

where

$$u_{mo} = \begin{bmatrix} x_0 - x \\ y_0 - y \end{bmatrix}$$

represents a vector pointing from the position of the mobile robot to the position of the obstacle, and $\|\cdot\|$ represents the Euclidean norm of a vector.

Definition 2. *In practical situations, to avoid the possible collision occurred when the mobile robot tracks the trajectory, a safe distance relative to the obstacle surface is required to be defined to form the safe boundary of the obstacle, which can be denoted as*

$$D_s \geq R_m \tag{6}$$

where D_s denotes a constant safe distance from the obstacle, and R_m represents the radius of the mobile robot.

Definition 3. *The collision occurs if the relative distance and the safe distance satisfy the following relationship*

$$D_r < R_m \tag{7}$$

Definition 4. *The robot would collide an obstacle if it is going on tracking the reference trajectory, in this case, the robot needs to perform an obstacle avoidance controller to avoid collision when the relative distance D_r satisfies the condition*

$$D_s < D_r \leq D_{act} \tag{8}$$

where D_{act} denotes a distance value to activate the obstacle avoidance controller under the condition that the obstacle avoidance is not completed. The value of D_{act} is a little larger than D_s, it gives a chance for the robot to activate the obstacle avoidance controller instead of activating the obstacle avoidance controller at the distance D_s rapidly.

To drive the mobile robot to keep a constant distance from the obstacle, a vector is defined to steer the robot in the direction of the vector, which can be described as

$$u_p = u_{mo} - D_s \frac{1}{\|u_{mo}\|} u_{mo}$$
$$= \begin{bmatrix} x_0 - x \\ y_0 - y \end{bmatrix} - D_s \frac{1}{\sqrt{(x_0-x)^2 + (y_0-y)^2}} \begin{bmatrix} x_0 - x \\ y_0 - y \end{bmatrix} \tag{9}$$

The vector u_p is used to maintain a constant distance to the obstacle when the robot is following the boundary of the obstacle. It is a vector pointing towards the obstacle when the relative distance $D_r > D_s$. The vector u_p will be zero vector when the robot follows the boundary of the obstacle at a constant distance D_s. It is a vector pointing away from the obstacle when $D_r < D_s$.

At the same time, we also expect the robot to drive in the direction that is parallel to the boundary of the obstacle, another vector u_f is determined by the expression

$$u_f = Ru_{mo} \tag{10}$$

where

$$R = \begin{bmatrix} \cos\alpha & -\sin\alpha \\ \sin\alpha & \cos\alpha \end{bmatrix}$$

Notice that the rotation matrix R is used to transform the vector u_{mo} to the vector u_f, where the value of α is $\pi/2$ or $-\pi/2$, which can be determined by the following equations

$$\varphi = atan2(y_0 - y, x_0 - x) \tag{11}$$

$$\alpha = \begin{cases} \pi/2, & \theta \geq \varphi \\ -\pi/2, & \theta < \varphi \end{cases} \tag{12}$$

where φ denotes the angle between the robot and the obstacle. θ is the current heading orientation of the mobile robot.

The Equations (11) and (12) are used to determine the fastest direction of obstacle avoidance to follow the boundary of the obstacle when the mobile robot activates the obstacle avoidance controller. As shown in Figure 2b, when $\theta > \varphi$, $\alpha = \pi/2$, the vector u_f can be obtained by rotating the vector u_{mo} by α radians counterclockwise. The value of α remains constant in the stage of obstacle avoidance, and the robot moves towards its left side to follow the boundary of obstacle.

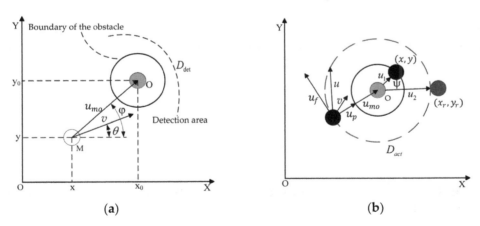

(a)   (b)

Figure 2. Geometrical relations between the robot and the obstacle: (**a**) The detection of an obstacle; (**b**) The completion of obstacle avoidance.

Combining the two vectors u_p and u_f, a blending vector used to maintain the constant distance and follow the boundary of the obstacle is defined as

$$u = u_p + u_f \tag{13}$$

Therefore, the desired direction of the motion of the robot can be calculated based on the vector u as

$$\beta = atan2(u_y, u_x) \tag{14}$$

where β represents the angle of the vector u and the positive direction of the x-axis. The u_y and u_x represent the vector components on the x and y axes, respectively.

During the process of obstacle avoidance, the vectors u_p and u_f are varying with time. The time derivative of the u_p can be represented as

$$\dot{u}_p = -\begin{bmatrix} \dot{x} \\ \dot{y} \end{bmatrix} + D_s \frac{\dot{x}(x_0-x)+\dot{y}(y_0-y)}{\sqrt{(x_0-x)^2+(y_0-y)^2}}\begin{bmatrix} x_0-x \\ y_0-y \end{bmatrix} + \frac{D_s}{\|u_{mo}\|}\begin{bmatrix} \dot{x} \\ \dot{y} \end{bmatrix}$$
$$= \left(\frac{D_s}{\|u_{mo}\|}-1\right)\begin{bmatrix} \dot{x} \\ \dot{y} \end{bmatrix} + D_s(\dot{x}\cos\varphi+\dot{y}\sin\varphi)\begin{bmatrix} x_0-x \\ y_0-y \end{bmatrix} \qquad (15)$$
$$= \left(\frac{D_s}{\|u_{mo}\|}-1\right)\begin{bmatrix} \dot{x} \\ \dot{y} \end{bmatrix} + D_s(\dot{x}\cos\varphi+\dot{y}\sin\varphi)u_{mo}$$

The time derivative of the vector u_f is represented by the equation

$$\dot{u}_f = -R\begin{bmatrix} \dot{x} \\ \dot{y} \end{bmatrix} \qquad (16)$$

The time derivative of the blending vector u can be expressed as

$$\dot{u} = \dot{u}_p + \dot{u}_f \qquad (17)$$

Combining Equations (15) and (16), Equation (17) can be rewritten as

$$\dot{u} = \begin{bmatrix} \dot{u}_x \\ \dot{u}_y \end{bmatrix} = \left[-R + \left(\frac{D_s}{\|u_{mo}\|}-1\right)I\right]\begin{bmatrix} \dot{x} \\ \dot{y} \end{bmatrix} + D_s(\dot{x}\cos\varphi+\dot{y}\sin\varphi)u_{mo} \qquad (18)$$

where I represents an identity matrix.

In order to analysis the completion of the obstacle avoidance, two vectors are defined as

$$u_1 = \begin{bmatrix} x-x_0 \\ y-y_0 \end{bmatrix}$$

$$u_2 = \begin{bmatrix} x_r-x_0 \\ y_r-y_0 \end{bmatrix}$$

where vectors u_1 and u_2 are the vectors pointing from the obstacle to the mobile robot and the virtual mobile robot, respectively.

Definition 5. *If the virtual mobile robot is outside of the region formed by the value of D_{act}, and the projection of vector u_1 onto the vector u_2 has the same direction of the vector u_2, obstacle avoidance can be considered be completed. The completion of obstacle avoidance can be represented by the expressions*

$$\|u_2\| > D_{act} \qquad (19)$$

$$u_1 \cdot u_2 = \|u_1\|\|u_2\|\cos\psi > 0 \qquad (20)$$

where ψ is the angle of the vector u_1 and the vector u_2. From Equations (19) and (20), the mobile robot can be viewed as having a clear shot to the virtual mobile robot, and then starts the trajectory tracking.

3.2.2. Obstacle Avoidance Control Design

The obstacle avoidance controller is designed to drive the robot to track the desired direction β, which means that the error e between the desired angle β and the current heading orientation θ converges to zero. The angle e is described as

$$e = \beta - \theta \tag{21}$$

By differentiating e, the error dynamic model can be expressed as

$$\dot{e} = \dot{\beta} - \dot{\theta} = \frac{\dot{u}_y \cos\beta - \dot{u}_x \sin\beta}{\| u \|} - \omega \tag{22}$$

A control law for the obstacle avoidance controller is proposed as

$$v = v_r \cos e_3 + k_1 e_1 \tag{23}$$

$$\omega = \frac{\dot{u}_y \cos\beta - \dot{u}_x \sin\beta}{\| u \|} + k_2 e \tag{24}$$

where k_1 and k_2 are positive control gains.

Notice that the linear velocity of the mobile robot is the same as Equation (2). Combining Equations (14), (18) and (21), the angular velocity of the mobile robot for obstacle avoidance controller can be calculated. Equation (24) can be ultimately converted into a formula without derivative terms, meaning that will not produce possible noise in the obstacle avoidance control.

Substituting Equation (24) into (22), the $\dot{e}$ can be rewritten as

$$\dot{e} = -k_2 e \tag{25}$$

Theorem 1. *The error e will converge to zero if the control law proposed for obstacle avoidance is chosen as the control inputs of the mobile robot.*

Poof of Theorem 1. Considering a scalar-valued Lyapunov function candidate as

$$V_1 = \frac{1}{2} e^2 \tag{26}$$

Combining Equation (25), the time derivative of the Lyapunov function candidate V_1 can be expressed as

$$\dot{V}_1 \quad = \dot{e}e$$
$$= -k_2 e^2 \leq 0$$

Then, V_1 becomes a Lyapunov function, and the controller is asymptotically stable around $e = 0$.
□

4. Switch Strategy

In this control system, two kinds of dynamics—tracking error dynamics and obstacle avoidance dynamics—are utilized. The previous section has utilized a classic trajectory tracking controller and designed an obstacle avoidance controller. Considering the switch between two controllers, a switch strategy is introduced to combine two controllers.

A transition between two controllers is shown in Figure 3, where u_{TT} and u_{AO} represent the proposed control laws for the trajectory tracking and obstacle avoidance, respectively. The control system of the mobile robot switches different controllers according to the conditions mentioned in the previous section. Considering the practical situation, we define the conditions that represented in the Figure 3 as

$$p_1 : \begin{cases} \|\boldsymbol{u_2}\| > D_{act} \\ \boldsymbol{u_1 \cdot u_2} > 0 \end{cases}$$

where p_1 represents the condition that the completion of obstacle avoidance, and it is also used to switch the obstacle avoidance controller to trajectory tracking controller.

$$p_2 : \begin{cases} \|\boldsymbol{u_1}\| < D_{act} \\ p_1 \; is \; not \; satisfied \end{cases}$$

where p_2 represents the switch condition for control system to activate the obstacle avoidance controller.

$$p_3 : \begin{cases} v_r = 0 \\ \omega_r = 0 \end{cases}$$

where p_3 represents a condition that the control inputs of the mobile are zero, meaning that the robot will stop moving if the reference velocities of the virtual mobile robot are zero.

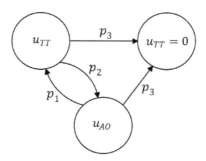

Figure 3. Switch strategy of the proposed control system.

The above-mentioned three conditions are used to generate the appropriate control inputs to switch the controllers. At every time epoch, there is only one controller is activated. we can use q_0 and q_1 to represent the activated state of the trajectory tracking controller and obstacle avoidance controller, respectively. The determination of values of two variables q_0 and q_1 is the same, which depends on the state of the current controller executed by the control system. As an example, the value of the q_0 is 0 or 1, where 1 represents the controller is under activated state, otherwise, the value of q_0 is 0.

The control scheme is composed of the two controllers, in order to analysis the stability of this hybrid control system. A total Lyapunov function is defined as

$$V = q_0 V_0 + q_1 V_1$$

where V_0 is the Lyapunov function in [42], and V_1 is the Lyapunov function of the obstacle avoidance controller. The time derivative of the total Lyapunov function is expressed as

$$\dot{V} = q_0 \dot{V}_0 + q_1 \dot{V}_1 \tag{27}$$

From Equation (27), it can be derived that the derivative of V is $\dot{V}_0$ or $\dot{V}_1$. Thus, the stability of control system is guaranteed.

5. Simulation Results and Discussion

Simulations based on the proposed control scheme illustrated in Figure 4 are performed for trajectory tracking in obstacle environments. The simulations are composed of two examples, and two kinds of reference trajectories are utilized. The reference trajectory of the first example is a straight line. The reference trajectory of the second example is circular.

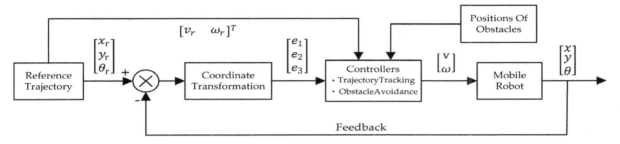

Figure 4. Structure of the proposed control scheme for the mobile robot.

Table 1 shows the parameters used in two simulations.

Table 1. Parameters used in two simulations.

Parameter	Description	Value
R_m	Radius of the mobile robot	0.2 m
R_{ob}	Radius of the obstacle	0.1 m
D_{det}	Detected distance of the sensor	3.5 m
D_{act}	Distance of activating the obstacle avoidance controller	0.5 m
D_s	Safe distance from the obstacle	0.35 m
k_1, k_2, k_3	Trajectory tracking control gains	3, 12, 6
k_1, k_2	Obstacle avoidance control gains	3, 6

The first example is simulated with a straight line. The initial posture of virtual mobile robot is $\begin{bmatrix} 0.1 & 2.6 & 0.7 \end{bmatrix}^T$, and the reference trajectory is generated by $v_r = 0.5$ m/s and $w_r = 0$ rad/s. The mobile robot starts to track the virtual mobile robot at the posture $\begin{bmatrix} 0 & 2.5 & 0.5 \end{bmatrix}^T$. The positions of the obstacles are O_1 (3, 5.2) and O_2 (4, 5.8), respectively.

Figure 5 shows the simulation results of the first example. At $t_1 = 0.82$ s, the mobile robot was tracking trajectory and the obstacle O_1 was detected in its working place. The tracking errors e_1, e_2 and e_3 gradually converged to zero, as shown in Figure 5b,c. At $t_2 = 3.10$ s, the obstacle O_2 was detected. In this case, two obstacles were under the range of the sensors. At $t_3 = 6.82$ s, the mobile robot moved into the region formed by the D_{act}, and the obstacle avoidance controller was activated to drive the robot to its right side to follow the boundary of the obstacle at a constant distance (Figure 5a). The relative distance D_{r1} between the robot and the obstacle O_1 kept constant during time period $t_3 < t \leq t_4$ shown in Figure 5f. At $t_4 = 8.76$ s, the reference posture of the virtual mobile robot was outside the region formed by D_{act}, and the projection of two vectors u_1 and u_2 have same direction in x-axis. Obstacle avoidance can be considered as being completed, and the controller was switched from the obstacle avoidance controller to the trajectory tracking controller. The mobile robot gone on tracking the reference trajectory. After this moment, the relative distance D_{r2} between the mobile robot and the obstacle O_2 gradually decreased. At $t_5 = 9.36$ s, $D_{r2} \leq D_s$, the trajectory tracking controller was stopped and the obstacle avoidance controller was activated. As shown in Figure 5a,f, the robot moved towards its left side to follow the boundary of the obstacle and kept a constant distance D_s to the obstacle O_2 during time period $t_5 < t \leq t_6$. At $t_6 = 11.10$ s, the condition of completion of obstacle avoidance was satisfied, the trajectory tracking controller was performed, the relative distance D_{r2} gradually increased, and the tracking errors gradually decreased to zero (Figure 5b,c), which meant that the robot had escaped the obstacle and started the tracking. During the whole moving process, the relative distances D_{r1} and D_{r2} always satisfied the conditions $D_{r1} \geq D_s$ and $D_{r2} \geq D_s$. Therefore, there is no collision occurred between the robot and the obstacles. The control inputs of the mobile robot in whole simulation are shown in Figure 5d,e.

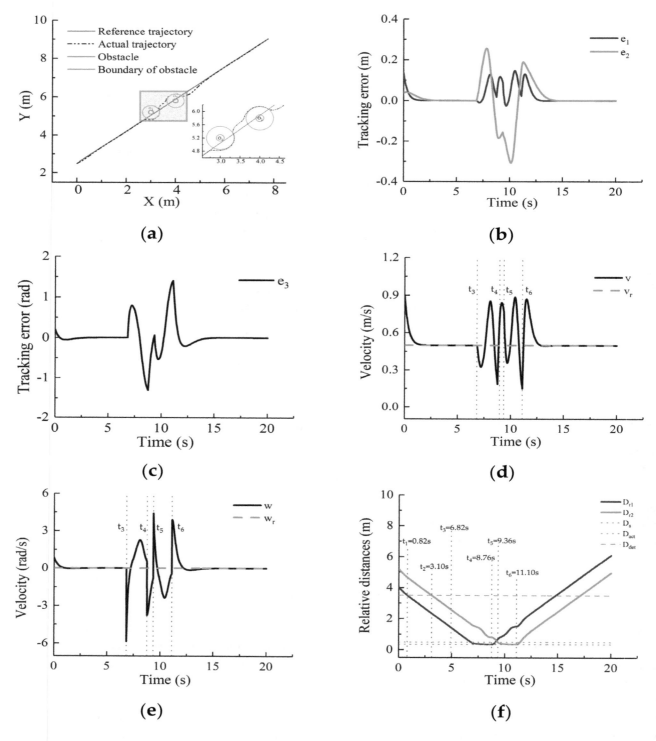

Figure 5. Simulation results of the first example: (**a**) Trajectory tracking and obstacle avoidance results; (**b**) position tracking error; (**c**) angular tracking error; (**d**) control input v and reference velocity v_r; e) control input ω and reference velocity ω_r; (**f**) relative distances between the robot and obstacles.

To further validate the effectiveness of the proposed control strategy, the second example is performed with a circular trajectory. The initial posture of the reference trajectory is $\begin{bmatrix} 4.5 & 6.5 & 0.8 \end{bmatrix}^T$, and the reference trajectory is generated by $v_r = 0.48$ m/s and $w_r = -0.3$ rad/s. The initial posture of the mobile robot is $\begin{bmatrix} 4.3 & 6.4 & 0.7 \end{bmatrix}^T$. The positions of the obstacles are $O_1(6.9, 6.2)$ and $O_2(4.9, 3.8)$, respectively.

Figure 6 shows the results of the second example. Before $t_1 = 4.94$ s, the mobile tracked the desired trajectory and detected the obstacles O_1 and O_2. At $t_1 = 4.94$ s, the relative distance D_{r1} between the robot and the obstacle O_1 satisfied the condition $D_{r1} \leq D_{act}$. The control system activated the obstacle avoidance controller to drive the robot to its left side to follow the boundary of the obstacle O_1, and the mobile robot kept a constant distance D_s to the obstacle O_1 before the completion of obstacle avoidance (Figure 6a,f). After $t_2 = 7.06$ s, the controller executed by the control system was switched to the trajectory tracking controller. As shown in Figure 6b,c, the tracking errors gradually converged to zero during $t_2 < t \leq t_3$. At $t_3 = 13.68$ s, the obstacle avoidance controller was activated. The mobile robot moved towards its right side to follow the boundary of the obstacle O_2 during $t_3 < t \leq t_4$. At $t_4 = 15.58$ s, the obstacle avoidance with the obstacle O_2 was completed, and the trajectory tracking controller was activated to track the reference trajectory. As shown in Figure 6f, the relative distances between the robot and two obstacles represented by D_{r1} and D_{r2} always satisfied the conditions $D_{r1} \geq D_s$ and $D_{r2} \geq D_s$. This shows that no collision occurred in unknown obstacle environments. The control inputs of the mobile robot are given in Figure 6d,e, and tracking errors are shown in Figure 6b,c.

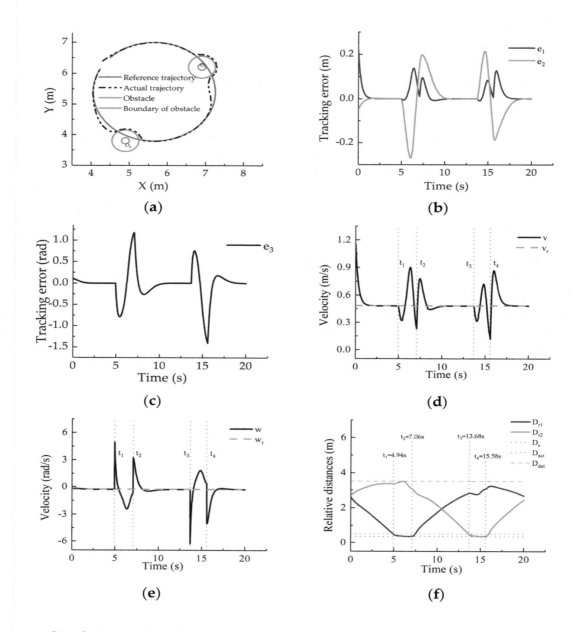

Figure 6. Simulation results of the second example: (**a**) trajectory tracking and obstacle avoidance results; (**b**) position tracking error; (**c**) angular tracking error; (**d**) control input v and reference velocity v_r; (**e**) control input ω and reference velocity ω_r; (**f**) relative distances between the robot and obstacles.

From the results of two simulation scenarios, it can be concluded that the mobile robot can track the trajectory before encountering obstacles in the environment. Once an obstacle is detected by the sensors attached to the mobile robot, the safe boundary and risk area of an obstacle are generated by the obstacle avoidance controller. When the robot moves into the region form by an activated distance, the obstacle avoidance controller is activated to determine the fastest obstacle avoidance direction to follow the obstacle boundary at a constant safe distance, and then after the completion of the obstacle avoidance, a transition between trajectory tracking and obstacle avoidance is triggered. The mobile robot goes on to track, and the tracking errors gradually decrease to zero. This combination of two controllers can well address the problem of trajectory tracking in obstacle environments. Therefore, the proposed scheme can be applied to the mobile robot to track the trajectory while considering obstacles.

6. Conclusions

In this paper, a control scheme consisting of the trajectory tracking and obstacle avoidance controllers is proposed to address the trajectory tracking problem in unknown obstacle environments. The trajectory tracking controller is employed to track pre-planned trajectory based on its tracking error dynamics. To deal with the problem of the possible obstacles in the environment, a blending vector is introduced to control the mobile robot toward the fastest obstacle avoidance direction to follow the boundary of an obstacle at a constant distance to escape the obstacle. Finally, two controllers combined by a switch strategy are switched to calculate the control inputs to track the trajectory or avoid the obstacle.

The results obtained from simulations indicate that the proposed control scheme can effectively ensure the safe movement of the mobile robot. In brief, the proposed control scheme provides a new simple method with application values for solving the tracking problem in unknown obstacle environments. However, this article does not consider dynamic obstacles and external disturbances existing in the environment. For future work, we will focus on the extension of the proposed control scheme to dynamical environments filled with disturbances.

Author Contributions: J.L. supervised the work, reviewed and edited the paper. J.S. designed the control algorithms, analyzed the data, and wrote this paper. G.C. simulated the methodology. All authors have read and agreed to the published version of the manuscript.

Acknowledgments: The authors would like to thank the support of National Key R&D Program of China (grant no. 2018YFB1309005), the New Equipment Non-ferrous Metallurgy Engineering Research Center of the Ministry of Education, Lanzhou University of Technology.

References

1. Zhao, J.C.; Gao, J.Y.; Zhao, F.Z.; Liu, Y. A Search-and-Rescue Robot System for Remotely Sensing the Underground Coal Mine Environment. *Sensors* **2017**, *17*, 2426. [CrossRef]

2. Kowalczyk, W. Formation Control and Distributed Goal Assignment for Multi-Agent Non-Holonomic Systems. *Appl. Sci.* **2019**, *9*, 23. [CrossRef]

3. Consolini, L.; Morbidi, F.; Prattichizzo, D.; Tosques, M. Leader-follower formation control of nonholonomic mobile robots with input constraints. *Automatica* **2008**, *44*, 1343–1349. [CrossRef]

4. Galasso, F.; Rizzini, D.L.; Oleari, F.; Caselli, S. Efficient calibration of four wheel industrial AGVs. *Robot. Comput. Integr. Manuf.* **2019**, *57*, 116–128. [CrossRef]

5. Villagra, J.; Herrero-Perez, D. A Comparison of Control Techniques for Robust Docking Maneuvers of an AGV. *IEEE Trans. Control Syst. Technol.* **2012**, *20*, 1116–1123. [CrossRef]

6. Bhat, S.; Meenakshi, M. Military Robot Path Control Using RF Communication. *Proc. First Int. Conf. Intell. Comput. Commun.* **2017**, *458*, 697–704. [CrossRef]

7. Adamczyk, M.; Bulandra, K.; Moczulski, W. Autonomous mobile robotic system for supporting counterterrorist and surveillance operations. In *Counterterrorism, Crime Fighting, Forensics, and Surveillance Technologies*; Bouma, H., CarlysleDavies, F., Stokes, R.J., Yitzhaky, Y., Eds.; Spie-Int Soc Optical Engineering: Bellingham, WA, USA, 2017; Volume 10441.

8. Ali, M.A.H.; Mailah, M. Path Planning and Control of Mobile Robot in Road Environments Using Sensor Fusion and Active Force Control. *IEEE Trans. Veh. Technol.* **2019**, *68*, 2176–2195. [CrossRef]

9. Rostami, S.M.H.; Sangaiah, A.K.; Wang, J.; Liu, X.Z. Obstacle avoidance of mobile robots using modified artificial potential field algorithm. *Eurasip J. Wirel. Commun. Netw.* **2019**, *19*. [CrossRef]

10. Orozco-Rosas, U.; Montiel, O.; Sepulveda, R. Mobile robot path planning using membrane evolutionary artificial potential field. *Appl. Soft Comput.* **2019**, *77*, 236–251. [CrossRef]

11. Xue, Y. Mobile Robot Path Planning with a Non-Dominated Sorting Genetic Algorithm. *Appl. Sci.* **2018**, *8*, 27. [CrossRef]

12. Elhoseny, M.; Tharwat, A.; Hassanien, A.E. Bezier Curve Based Path Planning in a Dynamic Field using Modified Genetic Algorithm. *J. Comput. Sci.* **2018**, *25*, 339–350. [CrossRef]

13. Nazarahari, M.; Khanmirza, E.; Doostie, S. Multi-objective multi-robot path planning in continuous environment using an enhanced genetic algorithm. *Expert Syst. Appl.* **2019**, *115*, 106–120. [CrossRef]

14. Antonakis, A.; Nikolaidis, T.; Pilidis, P. Multi-Objective Climb Path Optimization for Aircraft/Engine Integration Using Particle Swarm Optimization. *Appl. Sci.* **2017**, *7*, 22. [CrossRef]

15. Yang, H.; Qi, J.; Miao, Y.C.; Sun, H.X.; Li, J.H. A New Robot Navigation Algorithm Based on a Double-Layer Ant Algorithm and Trajectory Optimization. *IEEE Trans. Ind. Electron.* **2019**, *66*, 8557–8566. [CrossRef]

16. Lin, T.C.; Chen, C.C.; Lin, C.J. Wall-following and Navigation Control of Mobile Robot Using Reinforcement Learning Based on Dynamic Group Artificial Bee Colony. *J. Intell. Robot. Syst.* **2018**, *92*, 343–357. [CrossRef]

17. Matveev, A.S.; Wang, C.; Saykin, A.V. Real-time navigation of mobile robots in problems of border patrolling and avoiding collisions with moving and deforming obstacles. *Robot. Auton. Syst.* **2012**, *60*, 769–788. [CrossRef]

18. Wang, C.; Savkin, A.V.; Garratt, M. A strategy for safe 3D navigation of non-holonomic robots among moving obstacles. *Robotica* **2018**, *36*, 275–297. [CrossRef]

19. Thanh, H.; Phi, N.N.; Hong, S.K. Simple nonlinear control of quadcopter for collision avoidance based on geometric approach in static environment. *Int. J. Adv. Robot. Syst.* **2018**, *15*, 17. [CrossRef]

20. Maurovic, I.; Baotic, M.; Petrovic, I. Explicit Model Predictive Control for Trajectory Tracking with Mobile Robots. In Proceedings of the 2011 IEEE/ASME International Conference on Advanced Intelligent Mechatronics, Budapest, Hungary, 3–7 July 2011; IEEE: New York, NY, USA, 2011; pp. 712–717.

21. Klancar, G.; Krjanc, I. Tracking-error model-based predictive control for mobile robots in real time. *Robot. Auton. Syst.* **2007**, *55*, 460–469. [CrossRef]

22. Kumar, P.; Anoohya, B.B.; Padhi, R. Model Predictive Static Programming for Optimal Command Tracking: A Fast Model Predictive Control Paradigm. *J. Dyn. Syst. Meas. Control. Trans. ASME* **2019**, *141*, 12. [CrossRef]

23. Woo, C.; Lee, M.; Yoon, T. Robust Trajectory Tracking Control of a Mecanum Wheeled Mobile Robot Using Impedance Control and Integral Sliding Mode Control. *J. Korea Robot. Soc.* **2018**, *13*, 256–264. [CrossRef]

24. Sahloul, S.; Benhalima, D.; Rekik, C. Tracking trajectory of a mobile robot using sliding mode control. In Proceedings of the 2018 15th International Multi-Conference on Systems, Signals And Devices, Hammamet, Tunisia, 19–22 March 2018; IEEE: Piscataway, NJ, USA, 2018; pp. 1386–1390.

25. Goswami, N.K.; Padhy, P.K. Sliding mode controller design for trajectory tracking of a non-holonomic mobile robot with disturbance. *Comput. Electr. Eng.* **2018**, *72*, 307–323. [CrossRef]

26. Mu, J.Q.; Yan, X.G.; Spurgeon, S.K.; Mao, Z.H. Generalized Regular Form Based SMC for Nonlinear Systems With Application to a WMR. *IEEE Trans. Ind. Electron.* **2017**, *64*, 6714–6723. [CrossRef]

27. Falsafi, M.H.; Alipour, K.; Tarvirdizadeh, B. Tracking-Error Fuzzy-Based Control for Nonholonomic Wheeled Robots. *Arab. J. Sci. Eng.* **2019**, *44*, 881–892. [CrossRef]

28. Abbas, M.A.; Milman, R.; Eklund, J.M. Obstacle Avoidance in Real Time With Nonlinear Model Predictive Control of Autonomous Vehicles. *Can. J. Electr. Comput. Eng. Rev. Can. Genie Electr. Inform.* **2017**, *40*, 12–22. [CrossRef]

29. Castillo, O.; Martinez-Marroquin, R.; Melin, P.; Valdez, F.; Soria, J. Comparative study of bio-inspired algorithms applied to the optimization of type-1 and type-2 fuzzy controllers for an autonomous mobile robot. *Inf. Sci.* **2012**, *192*, 19–38. [CrossRef]

30. Shu, P.F.; Oya, M.; Zhao, J.J. A new adaptive tracking control scheme of wheeled mobile robot without longitudinal velocity measurement. *Int. J. Robust Nonlinear Control* **2018**, *28*, 1789–1807. [CrossRef]

31. Cui, M.Y.; Liu, H.Z.; Liu, W.; Qin, Y. An Adaptive Unscented Kalman Filter-based Controller for Simultaneous Obstacle Avoidance and Tracking of Wheeled Mobile Robots with Unknown Slipping Parameters. *J. Intell. Robot. Syst.* **2018**, *92*, 489–504. [CrossRef]

32. Wang, Y.; Shuoyu, W.; Tan, R.; Jiang, Y. Adaptive control method for path tracking of wheeled mobile robot considering parameter changes. *Int. J. Adv. Mechatron. Syst.* **2012**, *4*, 41–49. [CrossRef]

33. Rahmani, B.; Belkheiri, M. Adaptive neural network output feedback control for flexible multi-link robotic manipulators. *Int. J. Control* **2019**, *92*, 2324–2338. [CrossRef]

34. Abdelhakim, G.; Abdelouahab, H. A New Approach for Controlling a Trajectory Tracking Using Intelligent Methods. *J. Electr. Eng. Technol.* **2019**, *14*, 1347–1356. [CrossRef]

35.　Yoo, S.J. Adaptive tracking and obstacle avoidance for a class of mobile robots in the presence of unknown skidding and slipping. *IET Control Theory Appl.* **2011**, *5*, 1597–1608. [CrossRef]

36.　Yang, H.J.; Fan, X.Z.; Shi, P.; Hua, C.C. Nonlinear Control for Tracking and Obstacle Avoidance of a Wheeled Mobile Robot With Nonholonomic Constraint. *IEEE Trans. Control Syst. Technol.* **2016**, *24*, 741–746. [CrossRef]

37.　Kowalczyk, W.; Michalek, M.; Kozlowski, K. Trajectory tracking control with obstacle avoidance capability for unicycle-like mobile robot. *Bull. Pol. Acad. Sci. Tech. Sci.* **2012**, *60*, 537–546. [CrossRef]

38.　Ji, J.; Khajepour, A.; Melek, W.W.; Huang, Y. Path planning and tracking for vehicle collision avoidance based on model predictive control with multiconstraints. *IEEE Trans. Veh. Technol.* **2017**, *66*, 952–964. [CrossRef]

39.　Ha, L.N.N.T.; Bui, D.H.P.; Hong, S.K. Nonlinear Control for Autonomous Trajectory Tracking while Considering Collision Avoidance of UAVs Based on Geometric Relations. *Energies* **2019**, *12*, 1551. [CrossRef]

40.　Tzafestas, S.G. Mobile Robot Control and Navigation: A Global Overview. *J. Intell. Robot. Syst.* **2018**, *91*, 35–58. [CrossRef]

41.　Thanh, H.L.N.N.; Hong, S.K. Completion of Collision Avoidance Control Algorithm for Multicopters Based on Geometrical Constraints. *IEEE Access* **2018**, *6*, 27111–27126. [CrossRef]

42.　Kanayama, Y.; Kimura, Y.; Miyazaki, F.; Noguchi, T. A stable tracking control method for an autonomous mobile robot. In Proceedings of the IEEE International Conference on Robotics and Automation, Cincinnati, OH, USA, 13–18 May 1990; Volume 381, pp. 384–389.

Kinematic Modeling of a Combined System of Multiple Mecanum-Wheeled Robots with Velocity Compensation

Yunwang Li [1,2,*], **Shirong Ge** [1], **Sumei Dai** [2,3], **Lala Zhao** [1], **Xucong Yan** [1], **Yuwei Zheng** [1] and **Yong Shi** [2]

[1] School of Mechatronic Engineering, China University of Mining and Technology, Xuzhou 221116, China
[2] Department of Mechanical Engineering, Stevens Institute of Technology, Hoboken, NJ 07030, USA
[3] School of Mechanical and Electrical Engineering, Xuzhou University of Technology, Xuzhou 221018, China
* Correspondence: yunwangli@cumt.edu.cn

Abstract: In industry, combination configurations composed of multiple Mecanum-wheeled mobile robots are adopted to transport large-scale objects. In this paper, a kinematic model with velocity compensation of the combined mobile system is created, aimed to provide a theoretical kinematic basis for accurate motion control. Motion simulations of a single four-Mecanum-wheeled virtual robot prototype on RecurDyn and motion tests of a robot physical prototype are carried out, and the motions of a variety of combined mobile configurations are also simulated. Motion simulation and test results prove that the kinematic models of single- and multiple-robot combination systems are correct, and the inverse kinematic correction model with velocity compensation matrix is feasible. Through simulations or experiments, the velocity compensation coefficients of the robots can be measured and the velocity compensation matrix can be created. This modified inverse kinematic model can effectively reduce the errors of robot motion caused by wheel slippage and improve the motion accuracy of the mobile robot system.

Keywords: kinematic model; robot combination system; Mecanum-wheeled robot; velocity compensation; cooperative motion

1. Introduction

In recent years, intelligent and flexible manufacturing has motivated the development of autonomous mobile robots for workpiece and equipment handling and transportation [1,2], and in particular, automated guided vehicle (AGV) technology is widely studied and applied [3–7]. The omni-directional mobile AGV with Mecanum-wheeled robot platform, which has good driving force and easy control performance, can improve the utilization of workshop space and the efficiency of workshop transportation, and is very appropriate for transporting heavy goods in complex industrial environments [8,9]. The AGV that adopts a four-Mecanum-wheeled mobile robot platform with symmetrical structure is the most basic form and most widely used in industry [10–12]. In the field of large-scale equipment manufacturing, such as electric multiple unit (EMU) and aircraft, the objects are very heavy and large, and the carrying capacity and size of the four-Mecanum-wheeled AGV cannot meet the transportation requirements. In order to solve the problem, two schemes can be adopted: (1) Using a heavy-duty omnidirectional mobile platform with more Mecanum wheels, such as 8- or 12-wheeled platforms. These robot platforms can also be used in tandem to carry larger loads. For example, German company CLAAS (which has been acquired by MBB Industries AG, and is now Aumann Beelen GmbH) developed an omnidirectional mobile heavy-duty mobile handling robot called MC-Drive, and the MC-Drive TP200 robot has been used to carry aircraft at Airbus manufacturing plants [13,14]. The KUKA omniMove UTV-2 set is a heavy-duty mobile platform with 12 Mecanum wheels that can be used to carry large objects [15]. (2) Using a combination of multiple Mecanum-wheeled robot platforms, which can be considered as a whole with cooperative omnidirectional motion and transport. Usually, mobile platforms used for cooperative transportation are symmetrically arranged. For example, a railcar body can be carried cooperatively by four KUKA omniMove mobile platforms at the Siemens plant in Krefeld, Germany [16]. The four mobile platforms are symmetrically arranged at four corners of the railcar body. Omni-directional mobile platforms with 8, 12, 16, or 32 Mecanum wheels can also be considered as specific combinations of multiple basic mobile platforms.

The mature basic theory of four-Mecanum-wheeled robots is the basis of research on multi-robot systems. Muir [17–19] carried out basic research on Mecanum-wheeled robots and developed a kinematic and dynamic model and control on a four-Mecanum-wheeled robot. Campion et al. [20] studied structural properties and classification of kinematic and dynamic models of mobile wheeled robots and derived a motion constraint equation of a Mecanum wheel that can be used in kinematic research of multiple Mecanum-wheeled robots. Robots with four or more Mecanum wheels are overactuated systems with one or more motion constraints (wheel velocities are linearly correlated). Every additional wheel, and every additional robot, adds new motion constraints. So, the motion constraints of a multiple-Mecanum-wheeled robot system can be developed [21], which is also valid for multi-robot systems. The problem of transporting objects with a combined system is also a typical cooperative object transport problem in multi-robot systems, which is a growing research interest in recent years. Tuci et al. [22] provided a comprehensive summary for the scientific literature of cooperative object transport. Alonso-Mora et al. [23] presented a method that exploits deformability during manipulation of soft objects by robot teams including three KUKA YouBot robots. Alonso-Mora et al. [24] presented a constrained optimization method for multi-robot formation control in dynamic environments. Habibi et al. [25,26] presented a scalable distributed path planning algorithm for transporting large objects through unknown environments using a group of homogeneous robots. Lippi et al. [27] studied the modeling and planning problems of a system composed of multiple ground and aerial robots involved in a transportation task. Verginis et al. [28] studied the problem of cooperative transportation of objects rigidly grasped by N robotic agents and presented the communication-based decentralized cooperative object transportation using nonlinear model predictive control. Tsai et al. [29] presented a decentralized cooperative transportation control method with obstacle avoidance using fuzzy wavelet neural networks and a consensus algorithm for a group of Mecanum-wheeled omnidirectional robots with uncertainties, in order to move a large payload together. Wang et al. [30] proposed a distributed force and torque controller for a group of robots to collectively transport objects with both translation and rotation control and proved that follower robots can synchronize both their forces and torques with a leader robot that guides the group, and thus contribute positively to the transport. Paniagua-Contro et al. [31] presented an extension of leader–follower behaviors for the case of a combined set of kinematic models of omnidirectional and differential-drive wheeled mobile robots.

Slippage between the wheels and the ground is a disadvantage of Mecanum-wheeled robots, causing them to lose velocity and affecting their positioning accuracy. There are currently some studies on reducing the impact of slippage on the accuracy of robot motion. Chu [32] proposed a method to eliminate position and orientation errors; in this method, multiple ultrasonic distance sensors were used to measure the position and orientation of the mobile robot, and a position compensation algorithm was developed to minimize the position error between the current position and the desired position. Kulkarni et al. [33] proposed a technique to negate wheel slippage by using an additional sensor, a gyroscope, and a nested closed loop control structure to compensate for the increased processing required for slip negation. Tian et al. [34] used a method of back propagation (BP) network for nonlinear motion compensation to reduce the influence of slippage and improve the accuracy of 8 ×

8 omnidirectional platform motion. Udomsaksenee et al. [35] proposed a global control method for Mecanum-wheeled vehicles with slip compensation.

The latest research on cooperative transportation mainly focuses on path planning and navigation algorithms in working environments, and control methods in the cooperative transportation process. Research on the Mecanum-wheeled robot platform is mainly focused on single-Mecanum-wheeled robots [8,9,36–39], while there is less research on the kinematic of a combination system and motion compensation of multiple-Mecanum-wheeled robot platforms. However, studying the kinematic and characteristics of the combined omnidirectional mobile system is the basis for studying the motion control, path planning, and navigation of the combination system. In this research work, we study the combined mobile system of multiple Mecanum-wheeled robots as a whole, not as a cooperative multi-robot system, and the main research interest is the kinematic of this combined system and velocity compensation for the kinematic model, aiming to provide a theoretical basis for motion control of the combined system.

This paper is organized as follows: In Section 2, on the basis of studying the kinematic constraints of a single Mecanum wheel in an omnidirectional mobile system, kinematic models of a four-Mecanum-wheeled robot platform with symmetrical structure and a combined robot system composed of multiple Mecanum-wheeled robots are established, and a motion compensation model is proposed. In Section 3, in order to verify the correctness and feasibility of the kinematic and motion compensation models, the typical motion modes of a virtual prototype of a four-Mecanum-wheeled robot are simulated on RecurDyn. In Section 4, translation motion tests are carried out using a four-Mecanum-wheeled robot physical prototype, and the test results with and without motion compensation are compared. The tests in Section 4 verify the correctness of the motion and motion compensation models and the feasibility of the simulation method in Section 3. In Section 5, the motions of a variety of combined mobile system of a multiple Mecanum-wheeled robot are simulated on RecurDyn, and the kinematic and motion compensation models of a multi-robot system are verified.

2. Kinematic Model of Mecanum-Wheeled Mobile Robot

2.1. Kinematic Constraint Model of a Single Mecanum Wheel

The kinematic constraints of the i-th Mecanum wheel of robot system $O-XYZ$ consisting of n Mecanum wheels are shown in Figure 1 [40–42].

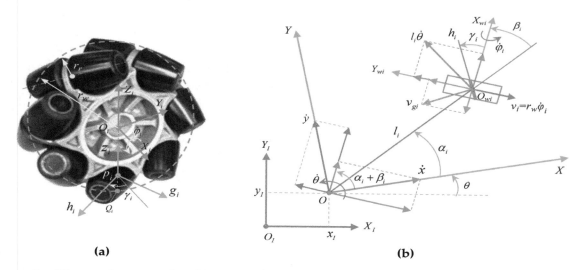

(a)   (b)

Figure 1. Kinematic constraint diagram of a robotic Mecanum wheel: (**a**) structural principle; (**b**) kinematic constraint.

In Figure 1, the Cartesian coordinate systems of the i-th Mecanum wheel and the roller of the wheel are $O_{wi}-X_{wi}Y_{wi}Z_{wi}$ and $P_i-g_ih_iz_i$; r_w and r_r are the radius of the wheel and the roller, respectively;

(l_i, α_i) is used to describe the relative installation orientation of origin O of the body coordinate system and center O_i of the wheel; and the angle between the X_{wi} axis and l_i is β_i, which is defined as the installation attitude angle of the local coordinate system of the wheel. The velocity of the motion center is v_o in the current state, and the angle between v_o and the X axis is θ_o; $\dot{\theta}$ is the rotation angular velocity of the system when moving in the plane. The angle between the projection of X_{wi} and h_i on the plane is the tilt angle γ_i $(0° < |\gamma_i| < 90°)$ of the roller; and v_{gi} is the velocity of the roller touching the ground on the i-th Mecanum wheel.

It is assumed that the movement between the roller and the ground is pure rolling, the contact point of the roller with the ground does not slip, and the instantaneous velocity is 0. According to the constraints of rolling and sliding, the following formulas can be obtained [20,43].

$$\begin{cases} \dot{x}\sin(\alpha_i + \beta_i) - \dot{y}\cos(\alpha_i + \beta_i) - l_i\dot{\theta}\cos\beta_i = r_w\dot{\varphi}_i - v_{gi}\cos\gamma_i \\ \dot{x}\cos(\alpha_i + \beta_i) + \dot{y}\sin(\alpha_i + \beta_i) + l_i\dot{\theta}\sin\beta_i = -v_{gi}\sin\gamma_i \end{cases} \tag{1}$$

Because the rollers rotate passively, the velocity of roller v_{gi} is an uncontrollable variable, which is usually not taken into account. By eliminating v_{gi} from Equation (1), we obtain the following [21]:

$$\dot{x}\cos(\alpha_i + \beta_i + \gamma_i) + \dot{y}\sin(\alpha_i + \beta_i + \gamma_i) + l_i\dot{\theta}\sin(\beta_i + \gamma_i) = -r_w\dot{\varphi}_i\sin\gamma_i \tag{2}$$

Let $\dot{\zeta} = \begin{bmatrix} \dot{x} & \dot{y} & \dot{\theta} \end{bmatrix}^T$, then the form of the matrix of Equation (2) is as shown in Equation (3), that is, the inverse kinematic equation of any (i-th) Mecanum wheel.

$$\dot{\varphi}_i = \frac{-1}{r_w\sin\gamma_i}\begin{bmatrix} \cos(\alpha_i + \beta_i + \gamma_i) & \sin(\alpha_i + \beta_i + \gamma_i) & l_i\sin(\beta_i + \gamma_i) \end{bmatrix}\dot{\zeta} \tag{3}$$

The motion state of the robot in local coordinate system $O - XYZ$ can be mapped to global coordinate system $O_I - X_IY_IZ_I$, which is expressed as:

$$\dot{\varphi}_i = \frac{-1}{r_w\sin\gamma_i}\begin{bmatrix} \cos(\alpha_i + \beta_i + \gamma_i) & \sin(\alpha_i + \beta_i + \gamma_i) & l_i\sin(\beta_i + \gamma_i) \end{bmatrix}R(\theta)\dot{\zeta}_I \tag{4}$$

where $\dot{\zeta}_I = \begin{bmatrix} \dot{x}_I & \dot{y}_I & \dot{\theta}_I \end{bmatrix}^T$, $R(\theta) = \begin{bmatrix} \cos\theta & \sin\theta & 0 \\ -\sin\theta & \cos\theta & 0 \\ 0 & 0 & 1 \end{bmatrix}$.

2.2. Kinematic Model of Four-Mecanum-Wheeled Mobile Robot with Symmetrical Structure

In this section, a four-Mecanum-wheeled robot with symmetrical structure is taken as the research object, as shown in Figure 2, and the coordination motion relationship of each wheel is discussed [44–47]. Choosing the geometrically symmetrical center as system origin O, the rectangular coordinate system XOY fixed with the mobile platform is established. The coordinate systems of Mecanum wheels $X_{wi}O_{wi}Y_{wi}$ $(i = 1, 2, 3, 4)$ are established, taking the wheel centers as the system origins. $\begin{bmatrix} \dot{x} & \dot{y} & \dot{\theta} \end{bmatrix}^T$ is defined as the generalized velocity of the mobile platform.

According to the geometric characteristics of the robot, for any Mecanum wheel O_{wi}, $\alpha_i + \beta_i = 0$. According to Equation (3), the following formula can be obtained.

$$\dot{\varphi}_i = \frac{-1}{r_w\sin\gamma_i}\begin{bmatrix} \cos\gamma_i & \sin\gamma_i & l\sin(\beta_i + \gamma_i) \end{bmatrix}\dot{\zeta} = -\frac{1}{r_w}\begin{bmatrix} \cot\gamma_i & 1 & W - H\cot\gamma_i \end{bmatrix}\begin{bmatrix} \dot{x} \\ \dot{y} \\ \dot{\theta} \end{bmatrix} \tag{5}$$

where roller angle $\gamma_1 = \gamma_3 = -\gamma = -45°$, $\gamma_3 = \gamma_4 = \gamma = 45°$, then the inverse kinematic equation for the four-Mecanum-wheeled robot is [17–19,21]:

$$\begin{bmatrix} \dot{\varphi}_1 \\ \dot{\varphi}_2 \\ \dot{\varphi}_3 \\ \dot{\varphi}_4 \end{bmatrix} = -\frac{1}{r_w} \begin{bmatrix} -\cot\gamma & 1 & W + H\cot\gamma \\ \cot\gamma & 1 & -W - H\cot\gamma \\ -\cot\gamma & 1 & -W - H\cot\gamma \\ \cot\gamma & 1 & W + H\cot\gamma \end{bmatrix} \begin{bmatrix} \dot{x} \\ \dot{y} \\ \dot{\theta} \end{bmatrix} = -\frac{1}{r_w} \begin{bmatrix} -1 & 1 & W + H \\ 1 & 1 & -(W + H) \\ -1 & 1 & -(W + H) \\ 1 & 1 & W + H \end{bmatrix} \begin{bmatrix} \dot{x} \\ \dot{y} \\ \dot{\theta} \end{bmatrix} \qquad (6)$$

The inverse kinematic Jacobian matrix J is expressed as:

$$J = -\frac{1}{r_w} \begin{bmatrix} -1 & 1 & W + H \\ 1 & 1 & -(W + H) \\ -1 & 1 & -(W + H) \\ 1 & 1 & W + H \end{bmatrix} \qquad (7)$$

In the global coordinate system $X_I O_I Z_I$, the inverse kinematic equation of the four-Mecanum-wheeled robot is

$$\begin{bmatrix} \dot{\varphi}_1 \\ \dot{\varphi}_2 \\ \dot{\varphi}_3 \\ \dot{\varphi}_4 \end{bmatrix} = JR(\theta) \begin{bmatrix} \dot{x}_I \\ \dot{y}_I \\ \dot{\theta} \end{bmatrix} = -\frac{1}{r_w} \begin{bmatrix} -1 & 1 & W + H \\ 1 & 1 & -(W + H) \\ -1 & 1 & -(W + H) \\ 1 & 1 & W + H \end{bmatrix} \begin{bmatrix} \cos\theta & \sin\theta & 0 \\ -\sin\theta & \cos\theta & 0 \\ 0 & 0 & 1 \end{bmatrix} \begin{bmatrix} \dot{x}_I \\ \dot{y}_I \\ \dot{\theta} \end{bmatrix} \qquad (8)$$

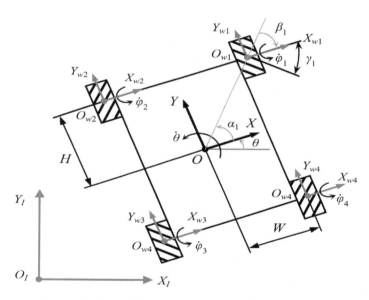

Figure 2. Four-Mecanum-wheeled mobile robot.

2.3. Kinematic Model of Multiple Mecanum-Wheeled Mobile Robot System

In Figure 3, in the global coordinate system is $X_I O_I Y_I$, when a multiple Mecanum-wheeled-robot system composed of m ($m = 1, 2, \cdots j \cdots , k$) robots co-transports an object, the poses and relative positions of these robots remain unchanged. In order to better describe the motion of the multi-robot system, a common coordinate system $X_S O_S Y_S$ is established at a specified location. $X_j O_j Y_j$ is the local coordinate system of the j-the robot of this system, which consists of n_j ($n_j = 1, 2, \cdots i \cdots , k$) Mecanum wheels. The relative position of coordinate systems $X_S O_S Y_S$ and $X_j O_j Y_j$ of each robot is determined by the geometric parameters of the multi-robot system. Let $\dot{\Phi}_{jn_j}$, J_j, and $\dot{\zeta}_j$ be the wheel rotation matrix, Jacobian matrix, and motion state in the local coordinate system of the j-th robot, respectively. The inverse kinematic equation of the j-th robot is shown in Equation (9):

$$\dot{\Phi}_{jn_j} = J_j \dot{\zeta}_j \tag{9}$$

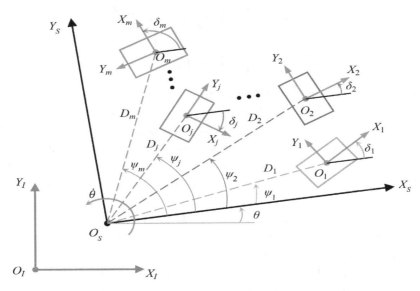

Figure 3. Multiple-robot mobile system.

After transforming the local coordinate system $X_j O_j Y_j$ of the j-th robot to the designated coordinate system $X_S O_S Y_S$, set $\dot{\zeta}_S = \begin{bmatrix} \dot{x}_S & \dot{y}_S & \dot{\theta} \end{bmatrix}^T$, and the inverse kinematic equation of the robot is as follows:

$$\dot{\Phi}_{jn_j} = J_j K_j \dot{\zeta}_S \tag{10}$$

where

$$
\begin{aligned}
K_j &= R(\delta_j) T(D_j, \psi_j) = \begin{bmatrix} \cos\delta_j & \sin\delta_j & 0 \\ -\sin\delta_j & \cos\delta_j & 0 \\ 0 & 0 & 1 \end{bmatrix} \begin{bmatrix} 1 & 0 & -D_j\sin\psi_j \\ 0 & 1 & D_j\cos\psi_j \\ 0 & 0 & 1 \end{bmatrix} \\
&= \begin{bmatrix} \cos\delta_j & \sin\delta_j & D_j\sin(\delta_j - \psi_j) \\ -\sin\delta_j & \cos\delta_j & D_j\cos(\delta_j - \psi_j) \\ 0 & 0 & 1 \end{bmatrix}
\end{aligned} \tag{11}
$$

The inverse kinematic equation of the j-th robot in global coordinate system $X_I O_I Y_I$ is

$$\dot{\Phi}_{jn_j} = J_j K_j R(\theta) \dot{\zeta}_I \tag{12}$$

In global coordinate system $X_I O_I Y_I$, the inverse kinematic equation of the multi-robot mobile system composed of m ($m = 1, 2, \cdots j \cdots, k$) robots is shown as follows:

$$
\begin{bmatrix} \dot{\Phi}_{1n_1} \\ \dot{\Phi}_{2n_2} \\ \vdots \\ \dot{\Phi}_{jn_j} \\ \vdots \\ \dot{\Phi}_{mn_m} \end{bmatrix} = \begin{bmatrix} J_1 K_1 \\ J_2 K_2 \\ \vdots \\ J_j K_j \\ \vdots \\ J_m K_m \end{bmatrix} R(\theta) \begin{bmatrix} \dot{x}_I \\ \dot{y}_I \\ \dot{\theta} \end{bmatrix} \tag{13}
$$

If the multiple-robot system only performs translational motion, its inverse kinematic equation is as follows

$$
\begin{bmatrix} \dot{\Phi}_{1n_1} \\ \dot{\Phi}_{2n_2} \\ \vdots \\ \dot{\Phi}_{jn_j} \\ \vdots \\ \dot{\Phi}_{mn_m} \end{bmatrix} = \begin{bmatrix} J_1 K_1 \\ J_2 K_2 \\ \vdots \\ J_j K_j \\ \vdots \\ J_m K_m \end{bmatrix} \begin{bmatrix} \dot{x}_I \\ \dot{y}_I \\ 0 \end{bmatrix}
\tag{14}
$$

The robot configuration composed of m same four-Mecanum-wheeled robots connected end-to-end and side-by-side is a common multi-robot system, as shown in Figure 4. The coordinate system $X_S O_S Y_S$ of the robot system is usually established on the structurally symmetric center line. $\dot{\Phi}_{j4}$, J, and K_j are the wheel velocity matrix, Jacobian matrix, and motion state transformation matrix, respectively, relative to the specified reference coordinate system $X_S O_S Y_S$ of the j-th four-Mecanum-wheeled robot. The inverse kinematic equation of the multi-robot configuration is as follows:

$$
\begin{bmatrix} \dot{\Phi}_{14} \\ \dot{\Phi}_{24} \\ \vdots \\ \dot{\Phi}_{j4} \\ \vdots \\ \dot{\Phi}_{m4} \end{bmatrix} = \begin{bmatrix} JK_1 \\ JK_2 \\ \vdots \\ JK_j \\ \vdots \\ JK_m \end{bmatrix} \begin{bmatrix} \dot{x}_S \\ \dot{y}_S \\ \dot{\theta} \end{bmatrix} = \begin{bmatrix} JT(D_1,\psi_1) \\ JT(D_2,\psi_2) \\ \vdots \\ JT(D_j,\psi_j) \\ \vdots \\ JT(D_m,\psi_m) \end{bmatrix} R(\theta) \begin{bmatrix} \dot{x}_I \\ \dot{y}_I \\ \dot{\theta} \end{bmatrix}
\tag{15}
$$

where $\dot{\Phi}_{j4} = \begin{bmatrix} \dot{\varphi}_{j1} & \dot{\varphi}_{j2} & \dot{\varphi}_{j3} & \dot{\varphi}_{j4} \end{bmatrix}^{\mathrm{T}}$, $\psi_j = \pm 90°$.

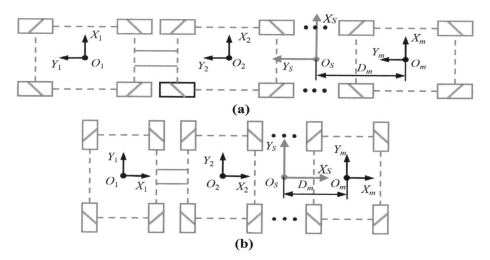

Figure 4. Combination configurations of multiple four-Mecanum-wheeled mobile platform with symmetrical structure: (**a**) tandem connected configuration; (**b**) side-by-side connected configuration.

2.4. Velocity Compensation of the Robot System

During the movement of the robot, due to manufacturing errors, wheel slippage, and other factors, there are errors between the actual velocity and the desired velocity, and the relative errors of longitudinal, lateral, and rotational angular speed are also different. Normally, the actual velocity of the robot is less than the set desired velocity. Different robots have different wheel slip rates and velocity errors due to different structural dimensions, manufacturing precision, number of wheels, deformation rate of wheel rollers, and friction coefficient between wheel and ground. For example, the precision of the roller angle of the Mecanum wheel has a great influence on the motion accuracy and velocity of the robot. In the following, taking a four-Mecanum-wheeled robot as an example, the influence of the precision of the roller angle on velocity is analyzed.

In order to analyze the influence of roller angle γ on the velocity of the robot, the partial derivative of Equation (6) with respect to γ is calculated.

$$
\begin{bmatrix} \frac{\partial \dot{\varphi}_1}{\partial \gamma} \\ \frac{\partial \dot{\varphi}_2}{\partial \gamma} \\ \frac{\partial \dot{\varphi}_3}{\partial \gamma} \\ \frac{\partial \dot{\varphi}_4}{\partial \gamma} \end{bmatrix} = -\frac{1}{r_w} \begin{bmatrix} \csc^2 \gamma & 0 & -H \csc^2 \gamma \\ -\csc^2 \gamma & 0 & H \csc^2 \gamma \\ \csc^2 \gamma & 0 & H \csc^2 \gamma \\ -\csc^2 \gamma & 0 & -H \csc^2 \gamma \end{bmatrix} \begin{bmatrix} \dot{x} \\ \dot{y} \\ \dot{\theta} \end{bmatrix}
\tag{16}
$$

If there is an error in roller angle $\Delta \gamma$ and $\gamma = 45°$, the wheel velocity will be compensated according to Equation (17):

$$
\begin{bmatrix} \Delta \dot{\varphi}_1 \\ \Delta \dot{\varphi}_2 \\ \Delta \dot{\varphi}_3 \\ \Delta \dot{\varphi}_4 \end{bmatrix} = -\frac{\Delta \gamma}{r_w} \begin{bmatrix} \csc^2 \gamma & 0 & -H \csc^2 \gamma \\ -\csc^2 \gamma & 0 & H \csc^2 \gamma \\ \csc^2 \gamma & 0 & H \csc^2 \gamma \\ -\csc^2 \gamma & 0 & -H \csc^2 \gamma \end{bmatrix} \begin{bmatrix} \dot{x} \\ \dot{y} \\ \dot{\theta} \end{bmatrix} = -\frac{\Delta \gamma}{r_w} \begin{bmatrix} 2 & 0 & -2H \\ -2 & 0 & 2H \\ 2 & 0 & 2H \\ -2 & 0 & -2H \end{bmatrix} \begin{bmatrix} \dot{x} \\ \dot{y} \\ \dot{\theta} \end{bmatrix}
\tag{17}
$$

According to Equation (17), the longitudinal motion velocity of the robot is not related to the roller angle, so it is unnecessary to compensate the wheel velocity, the lateral velocity and rotation velocity need to be compensated; and the rotation velocity compensation rate is related to the structural parameters of the robot. If the angle error of each roller of each Mecanum wheel is different, the situation will be more complicated. Because the velocity error is the result of many factors, the contribution of each factor is difficult to measure accurately. For a specific robot, its geometric size, manufacturing error, wheel deformation, and friction coefficient with the designated ground have been determined. The influence of these factors on the speed error can be determined. The velocity error may have a certain functional relationship with the motion variables, such as velocity of the robot. At different velocities, the relative errors of the robot's longitudinal, transverse, and rotational velocity can be obtained by experiments, so that the velocity compensation coefficients with different velocities can be obtained. For the multi-robot system in Figure 3, the velocity compensation equation matrix of the j-th robot is:

$$
C_j = \begin{bmatrix} f_j & 0 & 0 \\ 0 & g_j & 0 \\ 0 & 0 & u_j \end{bmatrix}
\tag{18}
$$

Then, the inverse kinematic equation of the j-th robot with velocity compensation is

$$
\dot{\Phi}'_{jn_j} = J_j C_j \dot{\zeta}_j = J_j C_j K_j \dot{\zeta}_S = J_j C_j K_j R(\theta) \dot{\zeta}_I
\tag{19}
$$

The inverse kinematic equation of this multi-robot system composed of m robots with velocity compensation is

$$
\begin{bmatrix} \dot{\Phi}'_{1n_1} \\ \dot{\Phi}'_{2n_2} \\ \vdots \\ \dot{\Phi}'_{jn_j} \\ \vdots \\ \dot{\Phi}'_{mn_m} \end{bmatrix} = \begin{bmatrix} J_1 C_1 K_1 \\ J_2 C_2 K_2 \\ \vdots \\ J_j C_j K_j \\ \vdots \\ J_m C_m K_m \end{bmatrix} R(\theta) \begin{bmatrix} \dot{x}_I \\ \dot{y}_I \\ \dot{\theta} \end{bmatrix}
\tag{20}
$$

If the multi-robot system only performs translational motion, the inverse kinematic equation with velocity compensation is

$$
\begin{bmatrix}
\dot{\Phi}'_{1n_1} \\
\dot{\Phi}'_{2n_2} \\
\vdots \\
\dot{\Phi}'_{jn_j} \\
\vdots \\
\dot{\Phi}'_{mn_m}
\end{bmatrix}
=
\begin{bmatrix}
J_1 C_1 K_1 \\
J_2 C_2 K_2 \\
\vdots \\
J_j C_j K_j \\
\vdots \\
J_m C_m K_m
\end{bmatrix}
\begin{bmatrix}
\dot{x}_I \\
\dot{y}_I \\
0
\end{bmatrix}
\tag{21}
$$

For the four-Mecanum-wheeled robot in Figure 2, according to Equations (8) and (19), the inverse kinematic equation with velocity compensation is

$$
\begin{bmatrix}
\dot{\varphi}'_1 \\
\dot{\varphi}'_2 \\
\dot{\varphi}'_3 \\
\dot{\varphi}'_4
\end{bmatrix}
= JCR(\theta)
\begin{bmatrix}
\dot{x}_I \\
\dot{y}_I \\
\dot{\theta}
\end{bmatrix}
\tag{22}
$$

3. Motion Simulation of Four-Mecanum-Wheeled Robot Virtual Prototype and Motion Test of Physical Robot

In order to further verify the correctness of the above theoretical research, the Mecanum-wheeled robot was simulated and analyzed by virtual prototyping technology. In this paper, the virtual prototype of the Mecanum-wheel mobile platform was modelled in SolidWorks, and the simulation was carried out on RecurDyn.

3.1. Creating a Four-Mecanum-Wheeled Robot Virtual Prototype for Simulation

3.1.1. Importing Robot Virtual Prototype Model and Creating Joints and Motions

Before importing into RecurDyn, the robot model should be simplified. The simplified assembly model built in SolidWorks should be imported into RecurDyn in Parasolid format. In the process of modeling, a single Mecanum wheel is defined as a subsystem in RecurDyn, which not only facilitates the hierarchical management of the model, but also enables the subsystem to be established, modified, imported, and exported independently. That can improve the reuse rate of the model, reduce the chance of model construction errors, and improve the efficiency.

According to the actual movement of the robot platform, the constraint relationship between components is created based on determining the assembly and motion relationships of components. In the robot virtual prototype simulation model, 36 revolute joints are created, as shown in Figure 5a. There are four revolute joints between the hub of the Mecanum wheel and the main body, and eight revolute joints between the rollers and the hub in a single Mecanum wheel subsystem, as shown in Figure 5b. The following are the basic dimensions of the robot model: $\gamma = 45°$, $W = 370$ mm, $H = 350$ mm, and $R = 152.4$ mm. The weight of the robot is 200 kg. By setting the angular velocity motion functions on the revolute joints between the four Mecanum wheels and the main body, the robot platform can move in simulation at desired speed. In order to prevent system anomalies caused by sudden changes in speed during the simulation, the drive function is usually applied to the revolute joints using the system's predefined STEP function. The STEP function is a third-order polynomial function built into the software and is used to define a relatively smooth load curve, which is suitable for smoothly loading drivers for mobile robot platforms.

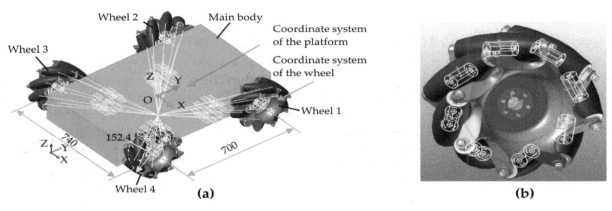

Figure 5. Simulation model of four-Mecanum-wheeled robot platform: (**a**) 3D model; (**b**) 8 revolute joints between rollers and hub in single Mecanum wheel subsystem.

3.1.2. Creating Contacts and Setting Parameters

Only by adding contacts between the rollers and the ground and setting reasonable contact parameters can the virtual prototype be simulated in accordance with the predetermined scheme. The RecurDyn contact toolkit provides rich contact models, from which Solid Contact was selected. Usually, the outer layer of the roller is vulcanized with a layer of rubber or polyurethane material. The higher the surface hardness of the roller, the stronger the bearing capacity and the higher the elastic modulus. The surface hardness of rollers with Mecanum wheels for small mobile platforms is usually 75 A Shaw hardness, and 90 A for the rollers of heavy-duty AGV. Moreover, we assume that the workplace has C30 grade concrete ground. The estimated values of the attribute parameters of the roller and the ground are shown in Table 1.

Table 1. Material property parameters of ground and roller.

Material	Elastic Modulus E (MPa)	Poisson's Ratio v
Concrete	30,000	0.20
Polyurethane	90	0.45

Based on Hertz contact theory, the stiffness coefficient between two objects can be expressed as:

$$k = \frac{4}{3}\rho^{\frac{1}{2}}E^* \qquad (23)$$

where ρ is the comprehensive radius of curvature, $\frac{1}{\rho} = \frac{1}{\rho_1} + \frac{1}{\rho_2}$; and ρ_1 and ρ_2 are the radii of curvature of the two objects at the contact; and E^* is the comprehensive elastic modulus; $\frac{1}{E^*} = \frac{1-v_1^2}{E_1} + \frac{1-v_2^2}{E_2}$; E_1, E_2 and v_1, v_2 are the elastic modulus and Poisson's ratio of the two materials, respectively.

The material property parameters of the ground and the rollers in Table 1, the radius ($\rho_1 = 10^{10}$ mm) of curvature of the horizontal ground, and the radius ($\rho_2 = 33.5$ mm) of curvature of the roller are substituted into Equation (23), and the stiffness coefficient between the two objects can be calculated. According to the theoretical calculation and feedback of simulation results, the contact parameters are set as follows: spring coefficient is 1200 N/mm, damping coefficient is 4, dynamic friction coefficient is 0.35, stiffness exponent is 2, and the rest are kept as default. On each roller of the Mecanum wheel, solid contact with the ground is set. Each Mecanum wheel is provided with 8 solid contacts on the ground, so 32 solid contacts are established for the mobile platform. The virtual prototype simulation model established through the above steps is shown in Figure 5a.

3.2. Motion Simulation and Analysis of a Single Four-Mecanum-Wheeled Robot

The motion simulations of the four-Mecanum-wheeled robot platform were carried out using 11 modes to analyze the motion characteristics. In order to better show the motion state of the robot platform during the simulation process, the trajectory tool in the software was used to display the motion track of the robot center or the specified marker point. The motion situation and trajectories of the robot in the simulation using 11 motion modes are shown in Figure 6. The simulation trajectories of the center of the robot platform in longitudinal, lateral, and oblique motion at 45° are shown in Figure 6a–c, respectively. Figure 6d–g show screenshots of the motion simulation including turning on the spot, circular motion, moving around an 8-figure curve, and sinusoidal curve motion. Figure 6h,i show the robot moving along a circle and an 8-figure curve in the oblique state. Figure 6j,k show centripetal circular motion. Figure 6l,m show centripetal motion of a 45° angle. The centripetal circular motion is a combination of translation and rotation. The curve equations of the robot moving in circular, 8-shaped, and harmonic curves are shown in Table 2.

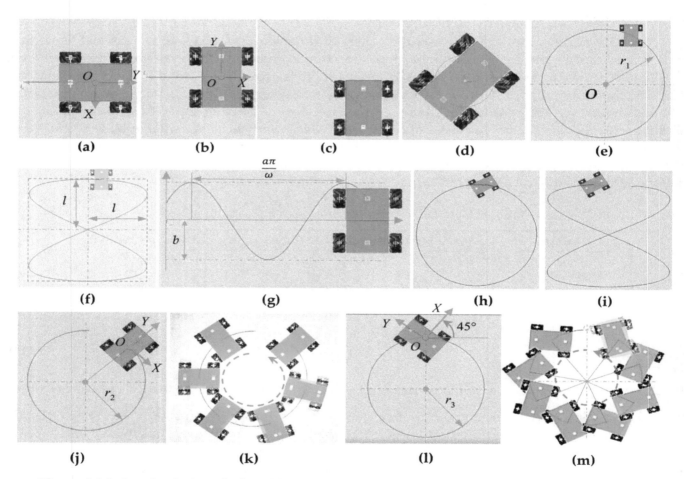

Figure 6. Motion simulation of robot: (**a**) longitudinal motion; (**b**) lateral motion; (**c**) oblique motion at 45°; (**d**) turning on the spot; (**e**) circular motion, where radius r_1 of the circle is 2500 mm; (**f**) translation motion around 8-figure curve; (**g**) translational motion along a simple harmonic motion curve; (**h**) circular translation motion of robot in oblique state; (**i**) translation motion of robot around 8-figure curve in oblique state; (**j**,**k**) centripetal circular motion, where radius r_2 of the circle is 1350 mm; (**l**,**m**) centripetal motion of 45° angle, where radius r_3 of the circle is 1350 mm.

Table 2. Trajectory equations of robot moving in circular, figure-8-shaped, and harmonic curves.

Names of Curve	Trajectory Equations in Figure 6
Circular curve	$x = r_i \sin \omega_i t$, $y = r_i \cos \omega_i t$, $\theta = 0$, where $r_1 = 2500$ mm, $r_2 = r_3 = 1350$ mm, $\omega_1 = \pi/10$, $\omega_2 = \omega_3 = \pi/5$
Figure-8-shaped curve	$x = l \sin 2\omega t$, $y = l \cos \omega t$, $\theta = 0$, where $l = 2500$ mm, $\omega = \pi/10$
Harmonic curve	$x = at$, $y = b \sin \omega t$, $\theta = 0$, where $a = 400$ mm/s, $b = 600$ mm, $\omega = 2\pi/5$

3.2.1. Analysis of Simulations of Four Simple Motions

(1) Longitudinal and Lateral Motion Simulation

Figure 7 shows the load curves of a wheel and a roller on the wheel during longitudinal movement of the mobile platform. By observing these two load curves, it can be seen that both the wheel and the roller bear periodic loads. The periodic load on the roller is derived from the roller being in contact with the ground every revolution of the wheel, and the contact duration is short and the load changes drastically. When the robot platform is in a stable motion state, the periodicity of the wheel load is mainly affected by the change of the state of contact between the roller and the ground on the body. In addition, the load on the roller is significantly greater than the load on the wheel, since the load on the entire robot is all applied to the roller in contact with the ground. The longitudinal and lateral motions of the robot are simulated with a series of different velocities, and the average velocities of the robot in stable state are obtained, as shown in Table 3. These simulation velocities are less than the corresponding set velocities. From the simulation data, the relative error of longitudinal velocity is very small; the average relative error is about −0.22%, so its correlation with velocity is not obvious. Five sets of simulation results show that the velocity loss of lateral translation is significantly greater than that of longitudinal translation when the robot runs at the same set velocity. Moreover, there is a negative correlation between the error of the robot and the set velocity in the lateral translation motion. The relative error curves of the velocities of the robot in the two motion modes are shown in Figure 8.

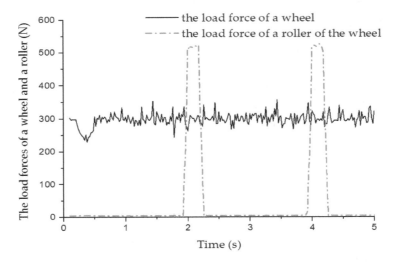

Figure 7. Wheel load and roller load in longitudinal motion simulation.

Table 3. Velocity errors and compensation coefficients in longitudinal and lateral motion simulation.

Motion Mode	Set Velocity Value (mm/s)	Average Simulation Velocity (mm/s)	Relative Error (%)	Compensation Coefficient	Average Compensation Coefficient
Longitudinal translation motion	100	99.767	−0.233	1.00234	
	200	199.564	−0.218	1.00218	
	300	299.339	−0.220	1.00221	1.002
	400	399.151	−0.212	1.00212	
	500	498.905	−0.219	1.00219	
Lateral translation motion	100	96.928	−3.072	1.03169	
	200	193.684	−3.158	1.03261	
	300	290.504	−3.165	1.03269	1.033
	400	386.930	−3.268	1.03378	
	500	483.158	−3.368	1.03486	

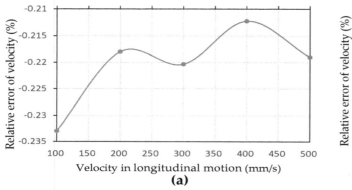
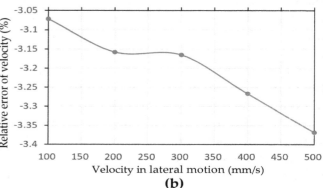

Figure 8. Relative error of velocity in (**a**) longitudinal and (**b**) lateral translation motion simulation.

Figure 9a shows the displacement curve of the robot in the x-direction in lateral motion. Figure 9c shows the x-direction velocity curve of the robot platform when it moves laterally at a set velocity of 500 mm/s. In the figure, the velocity value rises gradually from zero, which reflects the process of accelerating the robot platform from static state to stable moving state, and then the value fluctuates steadily in a range. In the simulation of lateral movement at a set speed of 500 mm/s, the maximum lateral velocity is 497.5 mm/s, the minimum velocity is 460.4 mm/s, and the average velocity is 483.2 mm/s. Figure 9b shows the longitudinal displacement and velocity curves in the y-direction during lateral motion. When the robot moves laterally in the x-direction, it fluctuates slightly in the y-direction and deviates from the x-axis direction. The velocity of the robot along the y-axis fluctuates around the zero line, as shown in Figure 9d. In 10 seconds of lateral movement, that is, when the robot moves about 5000 mm in the lateral direction, it deviates from the x-axis direction by about 4 mm. So, the longitudinal offset has little effect on the lateral motion of the robot platform.

(2) Simulation of Oblique Motion at 45°

When the robot moves in a straight line with an oblique direction of 45°, only two Mecanum wheels on one diagonal line rotate; the wheels on the other diagonal line do not rotate, but the bottom rollers on the wheels that do not turn passively rotate. Table 4 shows the simulation velocities and relative errors of the 45° oblique translation of the robot. According to the simulation data, there is loss of velocity in motion, and the velocity loss in the x- and y-directions is almost equal. The relative error of velocity in the x-direction is smaller than that of the lateral translation set at the same velocity.

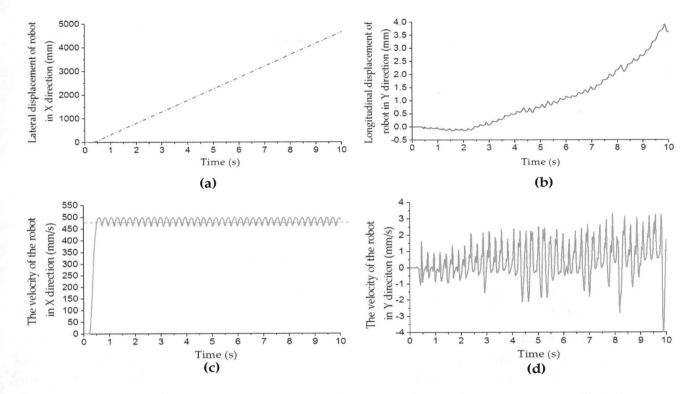

Figure 9. Displacement and velocity curves of robot in x- and y-directions during lateral motion simulation: (**a**) lateral displacement curve in x-direction; (**b**) longitudinal displacement curve in y-direction; (**c**) velocity curve in x-direction; (**d**) velocity curve in y-direction.

Table 4. Velocity error of simulation results.

Oblique 45° Motion	Set Velocity Value (mm/s)	Average Simulation Velocity in Steady State (mm/s)	Relative Error (%)
X- axis direction	200	196.566	−1.717
Y- axis direction	200	196.65	−1.675
X- axis direction	500	490.061	−1.988
Y- axis direction	500	490.164	−1.967

Figure 10a shows that the velocity of the x- and y-axis increases rapidly from 0 to a predetermined value within 0.6 s when the mobile platform moves along the oblique direction of 45°. Then the velocities in the two directions are basically the same, and fluctuate within a certain range. In the simulation, the velocity in the x- and y-axis directions is set at 500 mm/s, and the average test velocities are 490.164 mm/s and 490.061 mm/s, respectively. The curves of the two velocities are symmetrical relative to the zero line. The combination of these two directions enables the robot platform to move along the 45° direction. The trajectory curve of the robot in the oblique 45° motion is shown in Figure 10b.

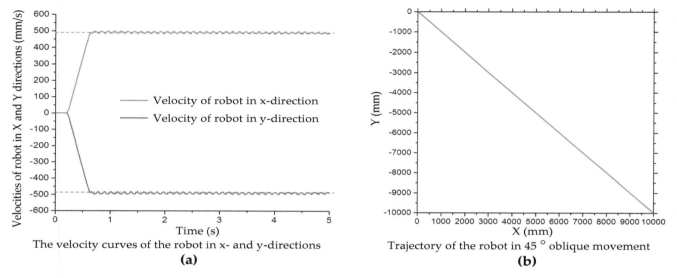

The velocity curves of the robot in x- and y-directions
(a)

Trajectory of the robot in 45 ° oblique movement
(b)

Figure 10. (**a**) Velocity curves in x- and y-directions, and (**b**) trajectory of robot platform in simulation of oblique motion at 45°.

(3) Turning on the Spot

Table 5 shows the relative errors of the angular velocity of the robot in the turning on the spot motion simulation at different angular velocities, which are positively correlated with increased velocity. The trajectory of the robot center in this simulation in five cycles is shown in Figure 11a. The trajectory has great randomness. In the first cycle, the displacement of the robot center in both x- and y-axis directions is within 1 mm. However, with the increased number of rotation cycles, the displacement tends to increase. After five cycles of rotation, the maximum displacement of the robot has exceeded 4 mm. The displacement curves of the robot center in the x- and y-axis directions in five cycles are shown in Figure 11b. It can be seen that the offset of the center coordinate of the robot increases with the increased number of rotation cycles.

Table 5. Relative error of angular velocity in simulation of turning on the spot.

Set Angular Velocity Value (rad/s)	Average Angular Velocity (rad/s)	Relative Error (%)	Compensation Coefficient
$\frac{\pi}{12}$ (=0.262)	0.257	−1.753	1.0195
$\frac{\pi}{6}$ (=0.524)	0.514	−1.920	1.0195
$\frac{\pi}{4}$ (=0.785)	0.768	−2.182	1.0221

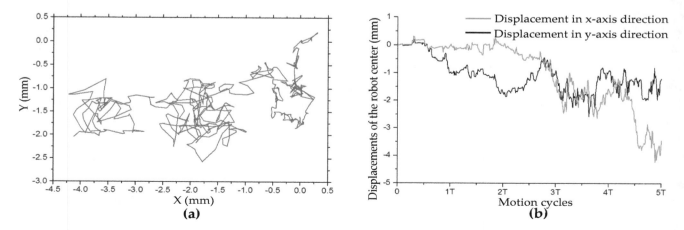

(a)

(b)

Figure 11. (**a**) Trajectory and (**b**) displacement of robot center in x- and y-axis directions in simulation of turning on the spot in five cycles.

According to the errors between simulation results and theoretical values of the velocities shown in Tables 3–5, the simulation velocities of the robot are less than the theoretical velocities. In these motion modes, the relative errors of simulation velocity vary slightly with different velocities but have a positive correlation on the whole, especially in lateral translation, oblique 45° translation, and in situ rotation. Comparing these three kinds of motion simulation, the velocity error of lateral motion is the largest, and that of longitudinal motion is the smallest. The relative error of velocity of 45° oblique motion is smaller than that of lateral motion and more than that of longitudinal motion. The relative error of angular velocity of in situ rotation is equivalent to that of 45° oblique translation motion.

Possible reasons for the errors between the simulation and theoretical set velocity are as follows: First, the wheel skidding phenomenon occurs in the process of rotation, which is common in wheeled mobile robots; slipping is most serious in the process of lateral movement. Second, the rim of the multiple rollers of the Mecanum wheel is theoretically a circle; however, in the simulation movement, deformation of the rubber layer at different positions of the rollers under load is different, and the effective radius of the wheel changes with rotation, thereby causing periodic changes even if the angular velocity is the same. Third, the roller cuts into the ground under load, which causes the radius of the wheel to be less than the theoretical radius, which is similar to the compression of the elastic roller in actual motion.

It can be seen that there are errors between the robot's velocity in motion simulation and the set velocity, especially in lateral translation motion. In order to ensure precise motion of the robot along the desired trajectory, it is necessary to compensate its velocity.

3.2.2. Velocity Compensation in Motion Simulation of Complex Trajectories

According to the simulation results of simple motion in the previous section, it can be seen that there are velocity losses in longitudinal translation, lateral translation, and rotational motion. When the robot translates along a complex trajectory, such as circular, 8-shaped, and simple harmonic motion curves, its motion can be decomposed into motion in both x- and y-directions. When the robot performs simulated motion along complex trajectories, if a theoretical velocity value is assigned to each Mecanum wheel, the robot will not be able to move to the expected position, so there is an error between the simulated and theoretical trajectory. The trajectories of the robot in the motion simulation along circular, 8-shaped, and harmonic curves are plotted with blue lines in Figure 12. These simulation curves are basically consistent with the theoretical curves, plotted with dashed red lines in Figure 12, but because of the velocity error of the robot, there are deviations between the simulation and theoretical curves. The circular and 8-shaped simulation curves are closed, so the curves deviate inward. The displacement errors of circular curves in Figure 12a in the x- and y-axis directions are 3.48% and 0.14%, respectively. The displacement errors of the 8-shaped trajectory in Figure 12b in the x- and y-axis are 2.92% and 0.28%, respectively. Relative errors of displacement of x- and y-axis directions in motion simulation along the circular and 8-shaped curves are shown in Table 6. With the increased motion period, the deviation of the simple harmonic trajectory in the x-direction increases, as shown in Figure 12c. If the robot keeps a certain tilt angle during the translational motion, there is a large displacement error in the trajectory in the lateral direction. Figure 12d,e show the trajectories of the robot in motion simulation shown in Figure 6h,i, maintaining a 60° angle during translating motion. According to Figure 12d,e, the deviation between the robot's trajectory and the theoretical curve in the oblique direction is the largest. The oblique direction is the x-direction in the local coordinate system of the robot.

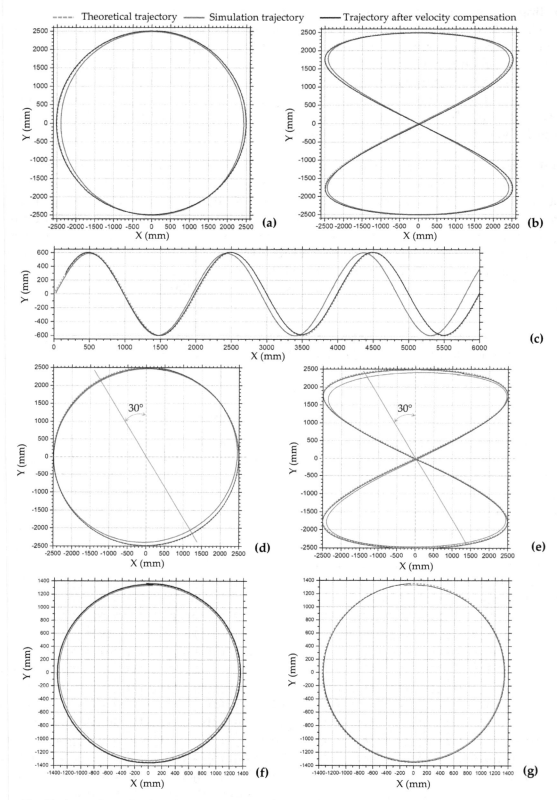

Figure 12. Trajectories of robot in motion simulation shown in Figure 6: (**a**) circular trajectory in Figure 6e; (**b**) 8-shaped trajectory in Figure 6f; (**c**) simple harmonic motion trajectory in Figure 6g; (**d**) circular trajectory in Figure 6h; (**e**) 8-shaped trajectory in Figure 6i; (**f**) centripetal circular motion in Figure 6j; (**g**) centripetal motion of 45° angle in Figure 6l.

Table 6. Relative errors of displacement in motion simulation along circular and 8-shaped curves.

Motion Mode		Maximum Displacement (mm)	Minimum Displacement (mm)	Relative Error (%)
Circular motion in	X-axis	2430.51	−2395.56	−3.48
Figure 12a	Y-axis	2494.11	−2498.98	−0.14
8-shaped curve motion in	X-axis	2423.55	−2430.37	−2.92
Figure 12b	Y-axis	2494.90	−2490.90	−0.28

Compensating the robot's velocity can reduce the trajectory error and make the trajectory closer to the theoretical trajectory. In the translational motion of the robot, only the velocity in the x- and y-axis directions needs to be compensated. According to the simulation data in Table 3, as the velocity changes, the relative error of the robot's velocity changes, but the amplitude is small. When the robot makes a curve motion, the velocities of the robot and the wheels change with the moment of change at all times. So a fixed compensation coefficient can be set for the virtual prototype model of the robot. In this simulation test, the average values of the velocity compensation coefficients of the longitudinal and lateral translation motions in Table 3 are used as the compensation coefficients in the x- and y-axis directions. The motion shown in Figure 6e–i is translation motion; according to Table 3, the velocity compensation matrix is set to C_1. The centripetal circular motion shown in Figure 6j–m is a combination of translation and rotation; therefore, it is also necessary to compensate for its rotational motion. According to Table 5, the compensation coefficient of rotational angular velocity is set to 1.0195, and the compensation coefficient of lateral velocity is set to 1.04 due to its large simulated lateral velocity. Then, the compensation matrix of velocity is set to C_2.

$$C_1 = \begin{bmatrix} f & 0 & 0 \\ 0 & g & 0 \\ 0 & 0 & u \end{bmatrix} = \begin{bmatrix} 1.033 & 0 & 0 \\ 0 & 1.002 & 0 \\ 0 & 0 & 0 \end{bmatrix}, C_2 = \begin{bmatrix} 1.04 & 0 & 0 \\ 0 & 1.002 & 0 \\ 0 & 0 & 1.0195 \end{bmatrix}$$

The trajectories of the robot in the motion simulation along the circular, 8-shaped, and simple harmonic motion curves after the velocity compensation are plotted with solid black lines in Figure 12. It can be seen from Figure 12a–e that the trajectories after velocity compensation are very consistent with the theoretical curves. The trajectories of the centripetal circular motions in Figure 6j–m are shown in Figure 12f,g, and the simulation trajectories after velocity compensation are better than those without velocity compensation. The simulation results show that this velocity compensation method is effective.

4. Motion Test of a Four-Mecanum-Wheeled Robot Physical Prototype

4.1. Motion Test System of Mecanum-Wheeled Robot Using Optitrack Optical Motion Capture System

In Section 3, RecurDyn software was used to simulate the robot's motion, and the motion simulations verified the correctness of the kinematic model of the four-Mecanum-wheeled robot and the feasibility of velocity compensation. However, they were carried out to simulate an ideal robot on ideal ground, which is quite different from real situations. The movement of a robot in a real environment is affected by many factors, including machining error of the Mecanum wheels, angle error of rollers, installation error of the four wheels, flatness of the ground, and control performance of the motor. In order to verify the correctness of the motion and motion compensation models, motion tests of the robot physical prototype were carried out. In order to measure the coordinates and trajectories of the robot during movement, the motion test system was constructed using the (high-speed motion capture cameras), shown in Figure 13a, were arranged on each side of the test area. The cameras were connected to the data and power supply by using a Gigabit Ethernet GigE/PoE OptiTrack optical motion capture system, as shown in Figure 13 [48]. Three OptiTrack Prime 13 cameras interface. All cameras were connected to a Gigabit network hub with Ethernet cables. An installed workstation with Motive optical motion capture was connect to the hub with a cable. The Motive software was used for recording, presentation, playback, and remote data services of the position data. The test used a four-Mecanum-wheeled robot prototype, shown in Figure 13b. The distance between the center lines of the front and rear wheels is 400 mm, and of the left and right wheels is 450 mm. The robot is controlled by another computer with a human–computer interaction (HCI) system. A Hand Rigid Bodies Marker Set was fixed on the robot prototype to test its space coordinates in the test space.

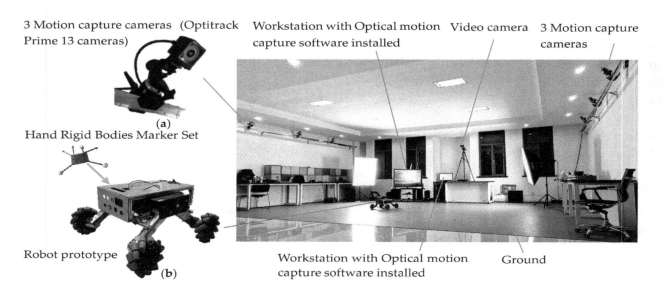

Figure 13. Test system of the Mecanum-wheeled robot using the Optitrack optical motion capture system: (**a**) the Optitrack Prime 13 cameras; (**b**) four-Mecanum-wheeled robot prototype.

4.2. Motion Test System of Mecanum-Wheeled Robot

4.2.1. Motion Shots of the Robot in Test

During the testing process, the Mecanum-wheeled robot was controlled to move in these motion modes: longitudinal translation, lateral translation, oblique 45° translation, turning on the spot, circular and 8-shaped curves, and translation motion along a simple harmonic curve. The OptiTrack optical motion capture system captured and recorded the spatial coordinates of the robot. In order to visualize the movement process, the Motion Shot app was used to create a motion picture of a series of coherent images of the robot in the same test environment. The motion picture recorded the position of the robot during movement and showed its movement track. Figure 14 shows six motion pictures of the robot in motion, including longitudinal translation (Figure 14a), lateral translation (Figure 14b), oblique 45° translation (Figure 14c), translation along a circle (Figure 14e,f), 8-shaped curve (Figure 14g,h), and translation along a simple harmonic curve (Figure 14d). The arrows in Figure 14 indicate the moving direction of the robot.

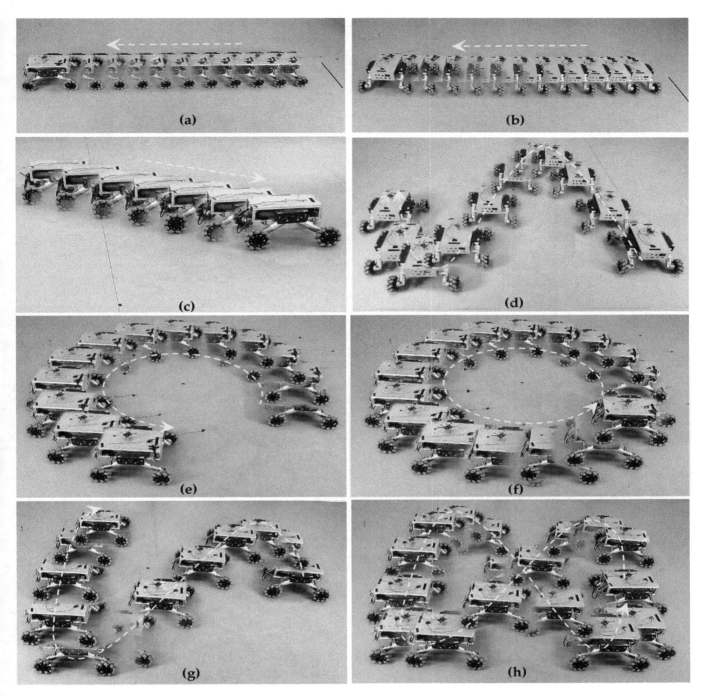

Figure 14. Motion shots of robot in motion test: (**a**) longitudinal translation; (**b**) lateral translation; (**c**) oblique 45° translation; (**d**) translation along a simple harmonic curve; (**e,f**) translation along a circle; (**g,h**) translation along an 8-shaped curve.

4.2.2. Analysis and Motion Tests of Robot in Four Simple Motions

(1) Longitudinal and Lateral Motion Tests

Regardless of the longitudinal or lateral translational motion, there is a certain velocity error of the robot, so the actual velocity is less than the set value. The test velocities and velocity errors are shown in Table 7. When the moving velocity of the robot is set to 100 mm/s, the longitudinal velocity error is 2.18% and the lateral translation velocity error is 12.38%. It can be seen that the lateral movement velocity error is much larger than the longitudinal. The lateral movement test was performed at different set velocities. It can be seen from the test results that the velocity error increased as the set velocity value increased, as shown in Table 7. The loss of velocity during the movement will have a great impact on the trajectory of the robot. Figure 15 shows the trajectory, displacement, and velocity curves of the robot in lateral translation motion test at a set velocity of 500 mm/s.

Table 7. Velocity errors and compensation coefficients in motion tests.

Motion Mode	Set Velocity Value (mm/s)	Average Test Velocity in Steady State (mm/s)	Relative Error (%)	Compensation Coefficient
Longitudinal translation motion	100	97.822	−2.178	1.022
	200	196.287	−1.857	1.019
	300	293.731	−2.090	1.021
	400	390.558	−2.361	1.024
	500	486.907	−2.619	1.027
Lateral translation motion	100	87.621	−12.379	1.141
	200	174.193	−12.904	1.148
	300	260.675	−13.108	1.151
	400	345.426	−13.644	1.158
	500	432.065	−13.587	1.157

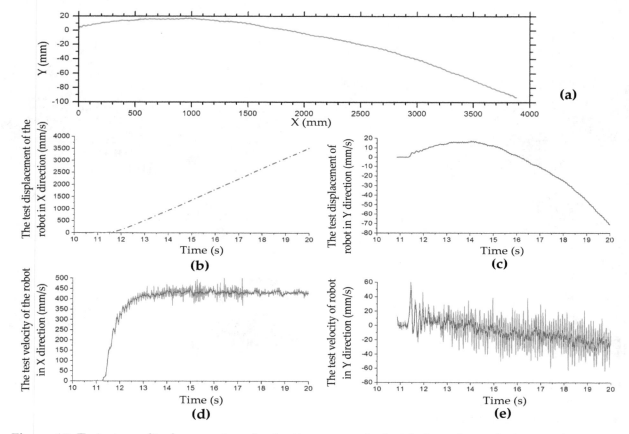

Figure 15. Trajectory, displacement, and velocity curves of robot in lateral translation motion test at a set velocity of 500 mm/s: (**a**) trajectory in lateral translation test; (**b**) lateral displacement curve in x-direction; (**c**) longitudinal displacement curve in y-direction; (**d**) test velocity curve in x-direction; (**e**) test velocity curve in y-direction.

Figure 15a shows the trajectory of the robot in the lateral translation test at a set velocity of 500 mm/s. The robot moves laterally 3800 mm in the x-axis direction and offset 100 mm in the y-axis direction. The first section of the x-direction velocity curve is an acceleration process; when the set value is reached, the velocity fluctuates at a certain value, as shown in Figure 15d. The lateral displacement curve in the x-direction is shown in Figure 15b. The test displacement and velocity curve of the robot in the y-direction in lateral translation motion is shown in Figure 15c,e, respectively. In the process of lateral translation, in the initial stage, the robot slides to one side, then changes direction, and the sliding displacement accumulates continuously. The velocity curve of the robot along the y-axis fluctuates around the zero line, resulting in a small fluctuation in the trajectory, shown in Figure 15a.

(2) Test of Oblique Motion at 45°

Figure 16a shows the test trajectory of the robot in oblique 45° translation, and Figure 16b,c show the displacement and velocity curves of the robot in the x- and y-directions in oblique 45° translation. In the first half of the 45° oblique motion, the velocity in the y-axis direction is faster than that in the x-axis direction. In the second half of the motion, the velocity in the y-axis direction decreases gradually and is less than that in the x-axis direction. The trajectory of the robot deviates from the theoretical trajectory line in the positive direction of the y-axis, and the second half of the robot gradually approaches the theoretical trajectory line. The trajectory is not a straight line, and there is a certain radian. The maximum distance of the trajectory deviating from the theoretical straight line is 207.7 mm. In this test, the set velocity of the robot is 500 mm/s and the theoretical velocity of the x- and y-directions should be 353.6 mm/s. The average test velocities of the x- and y-axes in the steady state are 319.4 mm/s and 330.7 mm/s, respectively, and the relative errors are −9.67% and −6.47%, respectively.

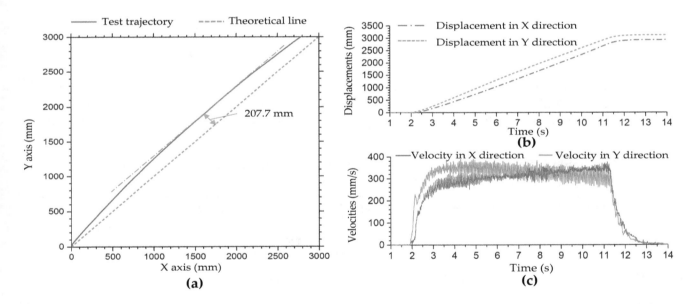

Figure 16. Oblique 45° translation motion test: (**a**) trajectories of robot in oblique 45° translation; (**b**) test displacement curves of the robot in x- and y-directions; (**c**) test velocity curves of the robot in x- and y-directions.

(3) Test of Motion of Turning on the Spot

Figure 17 shows the trajectories of points A and B on the robot in turning on the spot, where point A is very close to the center and point B is about 200 mm from the center. Figure 17a shows the trajectories of points A and B in the first cycle, and the trajectory of point B in five cycles. Figure 17b is an enlarged view of the trajectory of point A in the first cycle in Figure 17a. While the robot rotates around one circle, point B draws a perfect circular curve. There is local tortuosity in the circular curve, which indicates that the position of the geometric center of the robot slips during its rotation. The trajectories of point B of multiple rotation cycles do not coincide, and they constitute a trajectory band. By marking and drawing the trajectory of the center point of the robot, we can observe the change of the center point more directly in the course of rotation, but it is difficult to mark accurately.

Marker point A can only be located at the geometric center of the robot as far as possible. Marker point B drifts in a small range, and the trajectory of point B in one cycle is shown in Figure 17b, which indicates the displacement variation of the center point of the robot in the process of rotation, and also reflects the trend of circular trajectory caused by the small deviation from the center point (about 3 mm). Table 8 shows the loss and relative errors of the angular velocity of the robot during turning on the spot. This angular velocity error cannot be ignored in precise motion control.

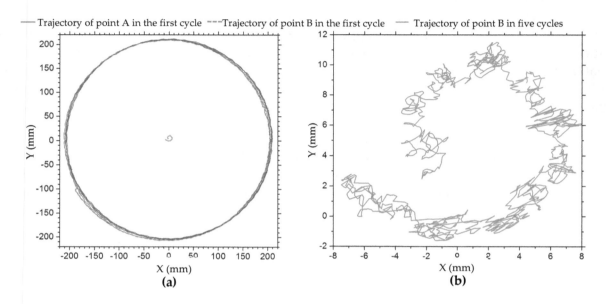

Figure 17. Trajectories of points A and B of the robot in turning on spot; point A is very close to the center, point B is about 200 mm from the center: (**a**) trajectories of point A and B in the first cycle, and the trajectory of point B in five cycles; (**b**) enlarged view of trajectory of point A in the first cycle in (a).

Table 8. Relative error of angular velocity in motion test of turning on the spot.

Set Angular Velocity Value (rad/s)	Average Angular Velocity (rad/s)	Relative Error (%)
$\frac{\pi}{6}$ (=0.524)	0.511	−2.550
$\frac{\pi}{3}$ (=1.047)	1.020	−2.467

4.2.3. Velocity Compensation in Motion Test of Complex Trajectories

Figure 18 shows trajectories of translation motion along circle, 8-shaped, and simple harmonic curves. According to the curve equations in Table 2, in the tests, the parameters of the circle curve are $r = 1000$ mm, $\omega = \pi/10$; the parameters of the 8-shaped curve are $l = 1000$ mm and $\omega = \pi/10$; and the parameters of the simple harmonic curve are $a = 300$ mm, $b = 1000$ mm and $\omega = \pi/5$. According to Table 7, the compensation coefficients of longitudinal and lateral velocity are set to 1.04 and 1.145, respectively. Then, the compensation matrix of velocity is set to C_3.

$$C_3 = \begin{bmatrix} 1.145 & 0 & 0 \\ 0 & 1.04 & 0 \\ 0 & 0 & 0 \end{bmatrix}$$

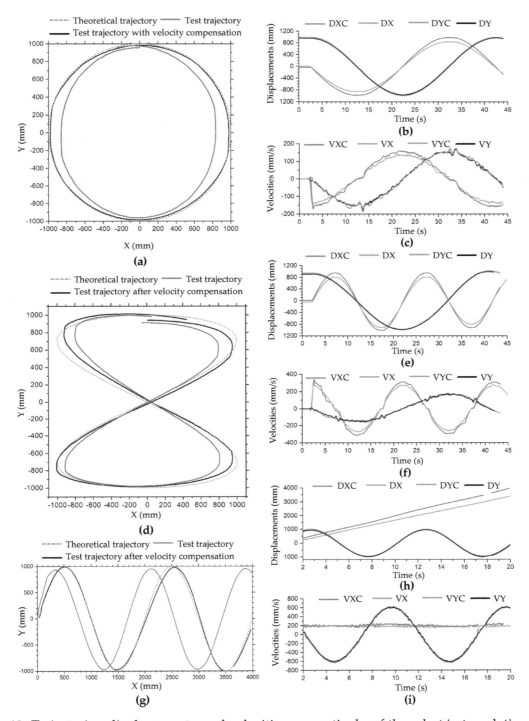

Figure 18. Trajectories, displacements, and velocities, respectively, of the robot in translation motion along (**a–c**) a circle, (**d–f**) an 8-shaped curve, and (**g–i**) a simple harmonic curve. DX or DY, displacement in x- or y-direction; DXC or DYC, displacement in x- or y-direction with velocity compensation; VX or VY, velocity in x- or y-direction; VXC or VYC, velocity in x- or y-direction with velocity compensation.

The blue curve in Figure 18a is the test trajectory of the robot in translation motion along a circle, which shifts to the inside of the theoretical circle (dashed red line) and is a vertical approximate ellipse. During the movement, the robot slips or its heading angle changes, causing the trajectory to fail to coincide. After motion compensation, the trajectory of the robot (solid black line) is very consistent with the theoretical curve, as shown in Figure 18a. From the velocity comparison curves (Figure 18c) and displacement comparison curves (Figure 18b) in the x- and y-directions, we can see that velocity and displacement in the motion test with velocity compensation at the same time are greater than the uncompensated velocity and displacement. In the test of translation motion along an 8-shaped curve, the test motion trajectory also deviates greatly from the theoretical trajectory. In the lateral direction, the displacement of the robot is obviously less than the theoretical value, as shown in Figure 18d. In the process of motion, the robot slips and the heading angle changes. Therefore, the test trajectory of the robot deflects. In the first motion cycle, the initial point and termination point of the robot cannot coincide. The displacements and velocities in translation motion along an 8-shaped curve are shown in Figure 18d,f, respectively. In the test of translational motion of a simple harmonic curve, the error of longitudinal displacement (x-axis direction) is small, while that of transverse displacement (y-axis direction) is large. With the increased motion cycles of the robot, the deviation between transverse displacement and theoretical trajectory curve increases, as shown in Figure 18g. This is consistent with the trend of dynamic simulation. The robot maintains a uniform velocity in the x-direction, and the speed in the x-direction with compensation (solid blue line) is slightly larger, as shown in Figure 18i; the displacement curve in the x-direction with motion compensation is at the top, as shown in Figure 18h. In the y-direction, the velocity and displacement of the robot change periodically, and are compensated accordingly. Table 9 shows the relative errors of displacement in motion simulation along a circle, 8-shaped curve, and simple harmonic curve in Figure 18. From Table 9, improvement of the robot's displacement accuracy using the kinematic model with velocity compensation can be observed more clearly.

Table 9. Relative errors of displacement in motion test of complex trajectories.

Motion Mode		Dmax (mm)	Dmin (mm)	DCmax (mm)	DCmin (mm)	RED (%)	REDC (%)
Circular motion in Figure 18a	X-axis	834.757	−860.636	978.495	−981.766	−15.23%	−1.99%
	Y-axis	978.196	−964.902	982.437	−987.510	−2.85%	−1.50%
8-shaped curve motion in Figure 18d	X-axis	807.439	−914.31	952.785	−1008.12	−13.91%	−1.95%
	Y-axis	995.717	−980.872	1013.374	−985.743	−1.17%	−0.04%
A simple harmonic curve in Figure 18g	Y-axis	950.934	−974.760	990.842	−1005.300	−3.715%	−0.191%

Note: Dmax, maximum displacement; Dmin, minimum displacement; DCmax, maximum displacement with velocity comparison; DCmin, minimum displacement with velocity comparison; RED, relative error of displacement; REDC, relative error of displacement with velocity comparison.

From the results of the motion compensation test, we can get the following conclusions: (1) in the real environment, the wheel slippage of a Mecanum-wheeled robot has a great impact on the robot's motion accuracy, especially lateral motion; (2) for motion with small velocity change, the constant velocity compensation matrix can have a good compensation effect, and the actual trajectories of the robot are very consistent with the theoretical curves; (3) more accurate motion compensation requires the use of a variable velocity compensation matrix, which is dynamically set according to the robot's velocity; (4) in the process of motion, the robot will deviate from its heading, which will cause its trajectory to deviate; and (5) according to the robot's parameters and environment information detected by position, attitude, and speed sensors, the motion direction and velocity can be adjusted in real time by using this kinematic model with speed compensation, which is more conducive to the robot's motion accuracy.

5. Motion Simulation of Combined Configurations of Multiple Four-Mecanum-Wheeled Robots

In Section 3, the typical motions of the four-Mecanum-wheeled robot were simulated. In Section 4, motion tests were performed using the robot physical prototype. Although the geometry of the virtual prototype used for motion simulation and the physical prototype used for testing are different, the motion test results verify the feasibility and effectiveness of motion simulation of Mecanum-wheeled robots by RecurDyn software. Therefore, the motion model of combined configurations of multiple-Mecanum-wheeled robots can also be verified by this simulation method, which can also make up for the shortcomings that we only have one physical robot sample and cannot achieve a collaborative combined multiple-robot experiment. In this section, kinematic simulations of a variety of configurations are performed to verify the kinematic model of multiple robots. According to the motion simulations in Section 3, the relative error between the simulation result and the theoretical value is small. Even without motion compensation, the simulation trajectory of the robot is very close to the theoretical curve. Therefore, in this section, in order to carry out simulations efficiently, simulations without motion compensation are first carried out for various combination configurations, and then two robot combination configurations are selected for simulation with motion compensation.

5.1. Multiple Four-Mecanum-Wheeled Robot Combination Configurations for Simulation

In this section, seven multiple robot combination configurations are used for motion simulation. The configurations and their main dimensions are shown in Figure 19. Figure 19a,b show two four-Mecanum-wheeled robots arranged laterally. The center distances of the two robots are the same; the two robots in Figure 19a are connected in the middle, called side-by-side connected configuration of two robots (SCC-2), and the two in Figure 19b are separate, called side-by-side unconnected configuration of two robots (SUC-2). Figure 19c,d show longitudinally arranged dual four-Mecanum-wheeled robots with the same center-to-center distance. The two robots in Figure 19c are connected, called tandem connected configuration of two robots (TCC-2), and the two in Figure 19d are separate, called tandem unconnected configuration of two robots (TUC-2). Figure 19e shows a tandem connected configuration of three robots (TCC-3). Figure 19f shows an arbitrary unconnected configuration of two robots arranged at an angle of 30° (AUC-2). If the two robots are connected, this is an arbitrary connected configuration (ACC-2).

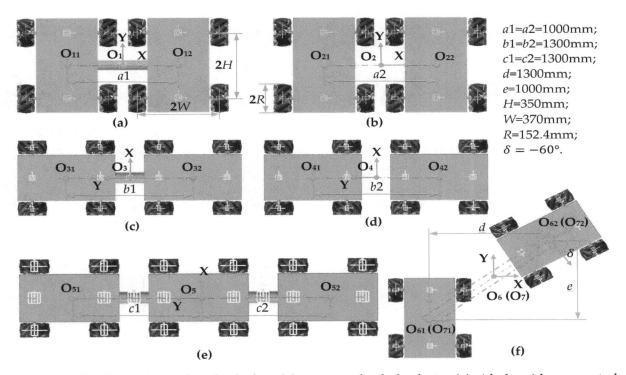

$a1=a2=1000mm$;
$b1=b2=1300mm$;
$c1=c2=1300mm$;
$d=1300mm$;
$e=1000mm$;
$H=350mm$;
$W=370mm$;
$R=152.4mm$;
$\delta = -60°$.

Figure 19. Configurations of multiple four-Mecanum-wheeled robots: (**a**) side-by-side connected configuration of two robots (SCC-2); (**b**) side-by-side unconnected configuration of two robots (SUC-2); (**c**) tandem connected configuration of two robots (TCC-2); (**d**) tandem unconnected configuration of two robots (TUC-2); (**e**) tandem connected configuration of three robots (TCC-3); (**f**) arbitrary unconnected configuration of two robots (AUC-2); if the two robots are connected, it is an arbitrary connected configuration (ACC-2).

5.2. Motion Simulation of Configurations of Multiple Four-Mecanum-Wheeled Robots

5.2.1. Simulation of Turning on Spot of Multiple Four-Mecanum-Wheeled Robots

In situ rotation motion simulation of the five configurations of robots was carried out, as shown in Figure 20. Figure 21 shows trajectories of the points on the robots in the simulation of turning on the spot after five cycles. The trajectories of O_{11} of SCC-2 in Figure 20b and of O_{31} of TCC-2 in Figure 20e are shown in Figure 21a; Figure 21b shows trajectories of O_{21} and O_{22} of SCC-2 in Figure 20c. We can see that after the robot rotates in place for many cycles, the trajectory of the marker point on the robot forms a circular trajectory range. In comparison, the trajectories of two unconnected robots (SUC-2) are relatively tortuous, as shown in Figure 21b. This is because the robots randomly move in the plane during in situ rotation. Figure 22 shows the trajectories of the center points of the combined robots and displacement curves in both x- and y-directions during the rotation. The center trajectory coverage of the unconnected robots is larger than that of the connected robots, as shown in Figure 22a,b, and their centers also have larger displacements in the x- or y-direction, as shown in Figure 22c,d. With rotation, there are certain periodic variations in displacement, and displacement has an increasing tendency, as shown in Figure 22c,d.

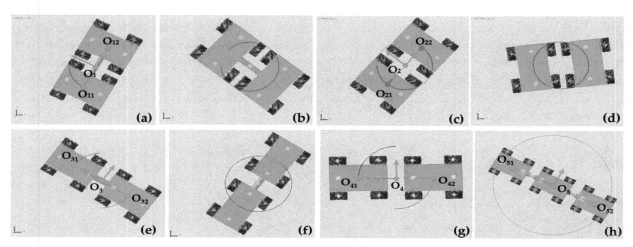

Figure 20. Simulation of turning on the spot of multiple four-Mecanum-wheeled Robots: (**a**,**b**) SCC-2; (**c**,**d**) SUC-2; (**e**,**f**) TCC-2; (**g**) TUC-2; (**h**) TCC-3.

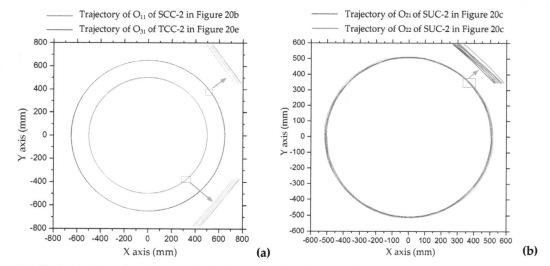

Figure 21. Trajectories of points on the robots in simulation of turning on the spot after five cycles: (**a**) O_{11} of SCC-2 in Figure 20b and O_{31} of TCC-2 in Figure 20e; (**b**) O_{21} and O_{22} of SUC-2 in Figure 20c.

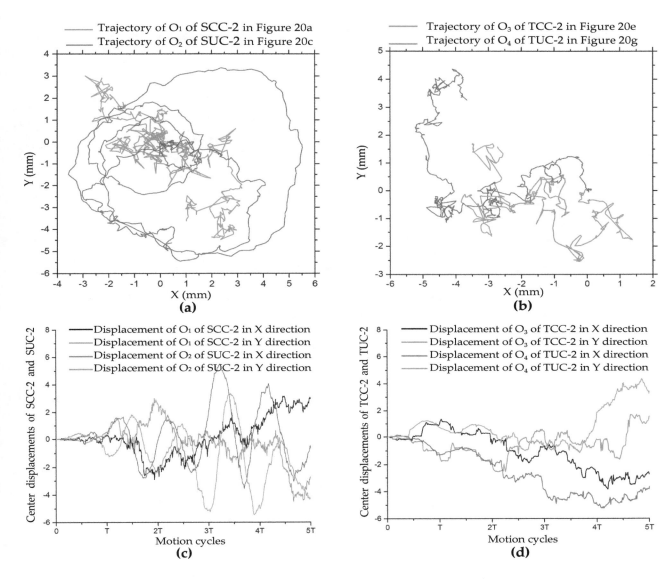

Figure 22. Trajectories of robot centers and center offset curves for in situ rotating motion: (**a**) centers O_1 of SCC-2 and O_2 of SUC-2; (**b**) centers O_3 of TCC-2 and O_4 of TUC-2; (**c**) displacement curves of SCC-2 and SUC-2 in x and y directions; (**d**) displacement curves of TCC-2 and TUC-2 in x- and y-directions.

5.2.2. Translation Simulation along a Circle of Multiple Four-Mecanum-Wheeled Robots

Three robot configurations, SUC-2, TUC-2, and AUC-2, were simulated for translational motion around a circle with a radius of 2500 mm. The simulation process, shown in Figure 23, verifies the correctness of the motion model of multiple-robot combination configurations. Figure 24a shows the trajectories of the centers O_2 of SUC-2 and O_4 of TUC-2; these two trajectories are very close, but because the simulation does not perform velocity compensation, they are smaller than the theoretical circle in the x-direction. In the process of simulation, the distance between the two robots in two configurations, SUC-2 and TUC-2, changes. The change curves of the distance between the two robots in these two configurations in the simulation process are shown in Figure 24b, which shows that the relative positions of the two robots change. From Figure 24c, it can be seen that the trajectory of center O_6 of configuration AUC-2 changes greatly during several motion cycles, and the change of distance between the two robots in this configuration is relatively large and has certain periodicity, as shown in Figure 24d.

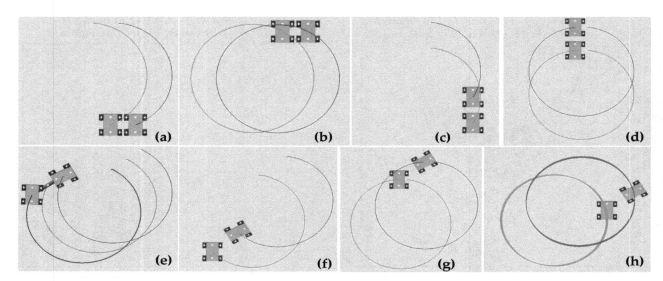

Figure 23. Translation simulation process along a circle: (**a,b**) SUC-2; (**c,d**) TUC-2; (**e**) ACC-2; (**f,g**) AUC-2; (**h**) the state of AUC-2 after multiple rotation cycles.

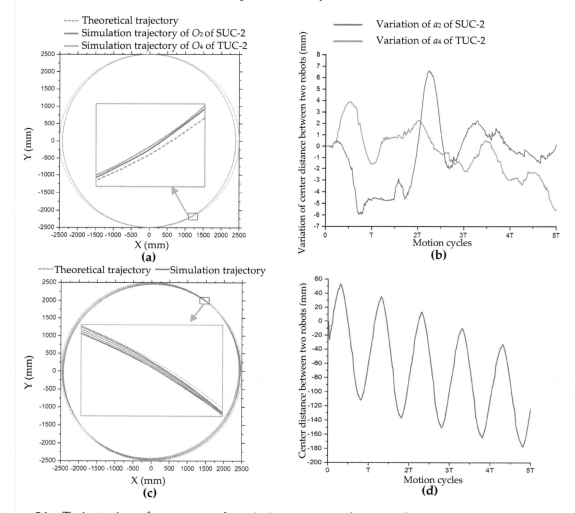

Figure 24. Trajectories of centers and variation curves of center distance between two robots in translation simulation along a circle: (**a**) trajectories of centers of SUC-2 and TUC-2; (**b**) variation curves of center distance between two robots of SUC-2 and TUC-2; (**c**) simulation trajectory of center O_6 of AUC-2; (**d**) variation curve of center distance between two robots of AUC-2.

5.2.3. Translation Simulation along an 8-Shaped Curve of Multiple Four-Mecanum-Wheeled Robots

Figure 25 shows the translation motion simulation process of the seven combination configurations in Figure 19 along an 8-shaped curve, which is defined in Table 2. Figure 26 shows the state after these robots have moved for several cycles. In Figure 27, the motion trajectories of these simulations are compared. According to Figure 27, since no velocity compensation is performed in the motion simulations, there is obvious displacement error in the x-direction for all simulation trajectories. For the combination of two robots arranged side-by-side or end-to-end, the error of their motion trajectories in the first cycle is similar whether they are connected or not, as shown in Figure 27a,b. For the combination of connected multiple robots, the error of the simulation track does not show an obvious difference, as shown in Figure 27c. After several cycles of motion, the trajectory of each moving robot will shift to a certain extent, and these trajectories form a trajectory band, as shown in Figure 26. Taking AUC-2 as an example, the trajectory band after five cycles of robot combination rotation is shown in Figure 27d.

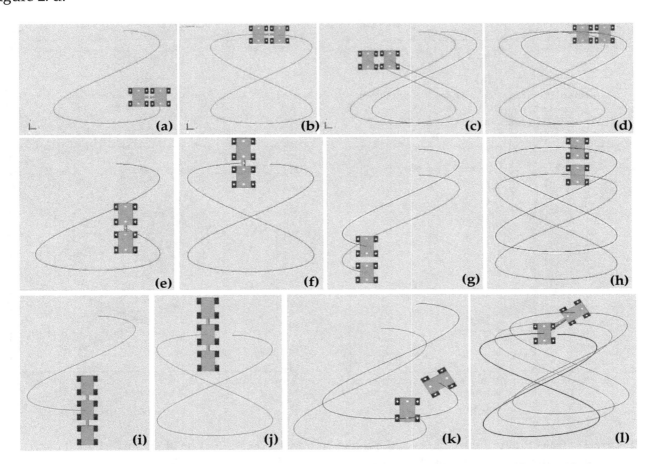

Figure 25. Translation motion simulation process of the seven configurations in Figure 19 along an 8-shaped curve: (**a,b**) SCC-2; (**c,d**) SUC-2; (**e,f**) TCC-2; (**g,h**) TUC-2; (**i,j**) TCC-3; (**k**) AUC-2; (**l**) ACC-2.

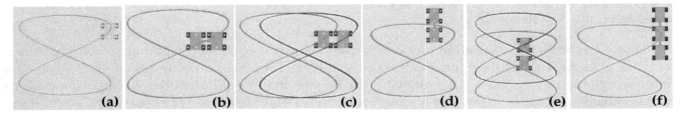

Figure 26. State of four configurations in translation motion along an 8-shaped curve after several cycles: (**a**) single four-Mecanum-wheeled robot; (**b**) SCC-2; (**c**) SUC-2; (**d**) TCC-2; (**e**) TUC-2; (**f**) TCC-3.

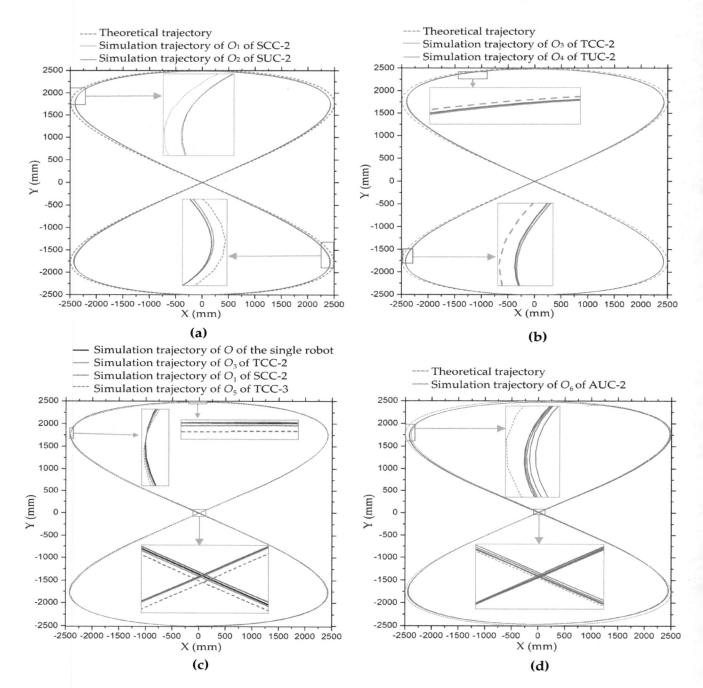

Figure 27. Trajectories of the centers of the robots in translation simulation along an 8-shaped curve: (a) centers O_1 of SCC-2 and O_2 of SUC-2; (b) centers O_3 of TCC-2 and O_4 of TUC-2; (c) multiple cycles trajectory of O_6 on AUC-2; (d) simulation trajectories of a single robot, TCC-2, SCC-2 and TCC-3.

5.2.4. Translation Simulation along Simple Harmonic Motion Curve of the Configurations of Multiple Four-Mecanum-Wheeled Robots

Figure 28 shows the translation motion simulation process of the six combination configurations in Figure 19 along a simple harmonic motion curve, which is defined in Table 2. Figure 29 shows the motion simulation trajectories of the robots. It can be seen that the simulation trajectories are close to the theoretical trajectories in the first motion cycle, but as the motion cycles increase, the error of the robot in the x-direction accumulates, causing the simulation trajectories to lag behind the theoretical trajectories. In the y-direction, trajectory errors are not accumulated, and because the longitudinal velocity errors are small, the displacements of the robots in the y-direction can almost reach the theoretical value.

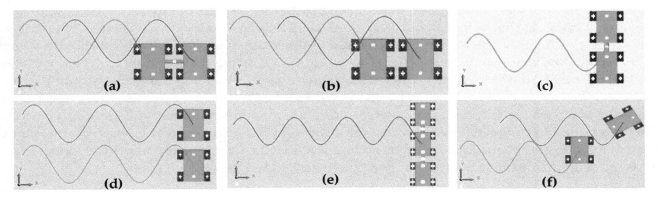

Figure 28. Translation motion simulation process of configurations along a simple harmonic motion curve: (**a**) SCC-2; (**b**) SUC-2; (**c**) TCC-2; (**d**) TUC-2; (**e**) TCC-3; (**f**) AUC-2.

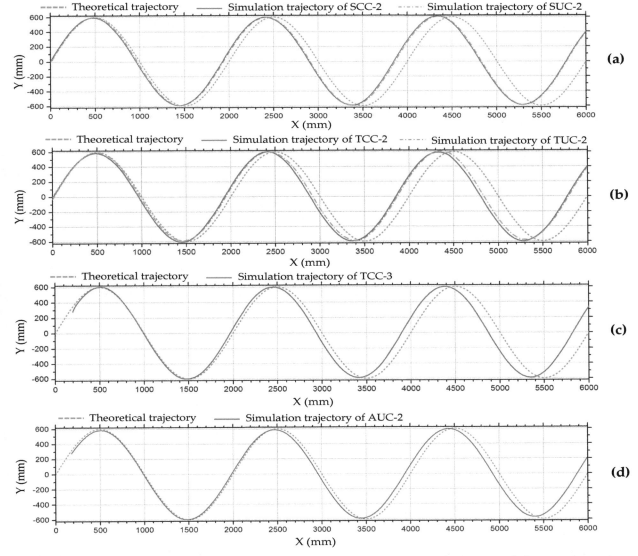

Figure 29. Simulation trajectories of the six configurations in Figure 19 along a simple harmonic motion curve: (**a**) O_1 of SCC-2 and O_2 of SUC-2; (**b**) O_3 of TCC-2 and O_4 of TUC-2; (**c**) O_5 of TCC-3; (**d**) O_6 of AUC-2.

5.3. *Motion Simulation after Velocity Compensation for TCC-3*

In Section 5.2, the motion simulations of the combination configurations of robot were carried out, and the kinematic model of the multiple-robot combination was verified. However, in the simulation experiment, there was no velocity compensation, so there were obvious errors between the robot's trajectories and the theoretical curves, especially in the x-direction of the local coordinate system. In this section, the robot configuration of TCC-3 and ACC-2 are selected and motion compensation simulation is carried out to verify the applicability of the velocity compensation model of multiple robot combination configurations. The velocity compensation coefficients used in this simulation are also obtained through the longitudinal and lateral translation motion simulation. In Figure 30, compared with the simulation curves without velocity compensation drawn by solid blue lines, the trajectories of the robot after compensation, drawn by solid black lines, are significantly improved, and are very consistent with the theoretical curves drawn by the dashed red lines.

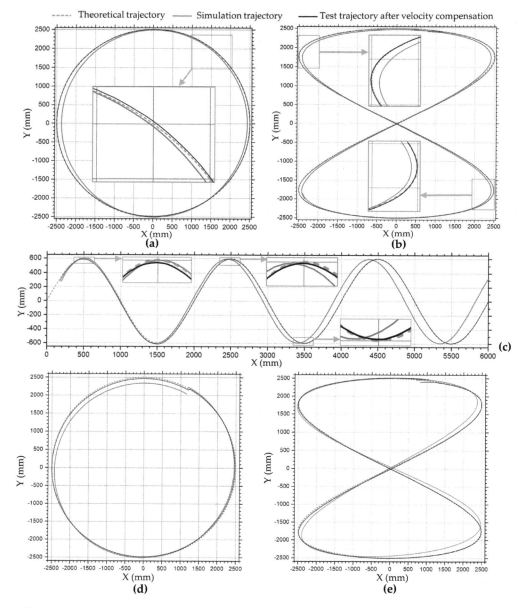

Figure 30. Trajectories of O_5 of TCC-3 and O_7 of ACC-2 in motion simulation: (**a**) circular trajectories of TCC-3; (**b**) 8-shaped trajectories of TCC-3; (**c**) simple harmonic motion trajectories of TCC-3; (**d**) circular trajectories of ACC-2; (**e**) 8-shaped trajectories of ACC-2.

6. Conclusions

In order to solve the problem of transporting large-scale goods or equipment in industry, a combination system of multiple Mecanum wheels or multiple-Mecanum-wheeled robots is usually adopted. The kinematics of this combined mobile system is the theoretical basis for the motion control, path planning, and navigation of the combined system. In a real environment, slippage of Mecanum wheels has a great impact on the robot's motion accuracy, especially lateral motion. In this work, based on studying the kinematic constraints of a single Mecanum wheel in a mobile system, a kinematic model of Mecanum-wheeled robots with multiple wheels is derived, then a kinematic model with velocity compensation of a combination mobile system composed of multiple Mecanum-wheeled robots is created. Taking four-Mecanum-wheeled robots as example, motion simulations of a virtual prototype on RecurDyn and motion tests of a physical prototype were carried out to verify the kinematic model of the robot and the motion compensation model. For motion with small velocity change, the constant velocity compensation matrix has a good compensation effect, and the actual trajectories of the robot are very consistent with the theoretical curves. Finally, the motions of a variety of combined mobile systems composed of multiple Mecanum-wheeled robots were simulated on RecurDyn. Motion simulations and tests prove that the kinematic model of single robot and multi-robot combined mobile systems is correct, and the inverse kinematic correction model with velocity compensation matrix is feasible. Through simulation or experiment, the velocity compensation coefficient of the robots can be obtained and the velocity compensation matrix can be created. This modified inverse kinematic model can effectively reduce the errors of motion caused by wheel slippage and improve the motion accuracy of the robot mobile system.

More accurate motion compensation requires the use of a variable velocity compensation matrix, which is dynamically set according to the robot's velocity. According to the robot's parameters and environment information detected by position, attitude, and speed sensors, the robot's motion direction and velocity can be adjusted in real time by using this kinematic model with speed compensation, which is more conducive to motion accuracy.

Author Contributions: Methodology, Y.L., S.G., and S.D.; validation, L.Z., Y.S., S.D., and Y.Z.; formal analysis, Y.L., S.G., and S.D.; investigation, Y.L., X.Y., and Y.Z.; writing—original draft preparation, Y.L., X.Y. and S.D.; writing—review and editing, Y.S. and L.Z.; project administration, Y.L. All authors have read and agreed to the published version of the manuscript.

Acknowledgments: We would like to thank Yong Wang for providing the OptiTrack optical motion capture system which were funded by Research Projects of General Administration of Quality Supervision, Inspection and Quarantine (2017QK002). We would like to thank Zhen Wei, Guanyu Zhou, Sihao Bi, and Shuai Li, masters of China University of Mining and Technology, for helping with the experiments.

References

1. Dang, Q.V.; Nguyen, C.T.; Rudová, H. Scheduling of mobile robots for transportation and manufacturing tasks. *J. Heuristics* **2018**, *25*, 175–213. [CrossRef]
2. Bogh, S.; Schou, C.; Rühr, T.; Kogan, Y.; Dömel, A.; Brucker, M.; Eberst, C.; Tornese, R.; Sprunk, C.; Tipaldi, G.D. Integration and assessment of multiple mobile manipulators in a real-world industrial production facility. In Proceedings of the 45th International Symposium on Robotics and the 8th German Conference on Robotics (ISR/Robotik 2014), Munich, Germany, 2–4 June 2014.
3. Chawla, V.K.; Chanda, A.K.; Angra, S. The scheduling of automatic guided vehicles for the workload balancing and travel time minimi-zation in the flexible manufacturing system by the nature-inspired algorithm. *J. Proj. Manag.* **2019**, *4*, 19–30. [CrossRef]
4. Du, L.Z.; Ke, S.F.; Wang, Z.; Tao, J.; Yu, L.Q.; Li, H.J. Research on multi-load AGV path planning of weaving workshop based on time priority. *Math. Biosci. Eng.* **2019**, *16*, 2277–2292. [CrossRef] [PubMed]
5. Chen, C.; Huy, D.T.; Tiong, L.K.; Chen, I.-M.; Cai, Y.Y. Optimal facility layout planning for AGV-based modular prefabricated manufacturing system. *Autom. Constr.* **2019**, *98*, 310–321. [CrossRef]

6.   Chen, C.; Tiong, L.K.; Chen, I.M. Using a genetic algorithm to schedule the space-constrained AGV-based prefabricated bathroom units manufacturing system. *Int. J. Prod. Res.* **2019**, *57*, 3003–3019. [CrossRef]

7.   Dehnavi-Arani, S.; Saidi-Mehrabad, M.; Ghezavati, V. An integrated model of cell formation and scheduling problem in a cellular manufacturing system considering automated guided vehicles' movements. *Int. J. Oper. Res.* **2019**, *34*, 542–561. [CrossRef]

8.   Qian, J.; Zi, B.; Wang, D.M.; Ma, Y.G.; Zhang, D. The design and development of an omni-directional mobile robot oriented to an intelligent manufacturing system. *Sensors* **2017**, *17*, 2073. [CrossRef]

9.   Sun, S.K.; Hu, J.P.; Li, J.; Liu, R.D.; Shu, M.; Yang, Y. An INS-UWB based collision avoidance system for AGV. *Algorithms* **2019**, *12*, 40. [CrossRef]

10. Xie, L.; Scheifele, C.; Xu, W.L.; Stol, K.A. Heavy-duty Omni-directional Mecanum-wheeled Robot for Autonomous Navigation: System Development and Simulation Realization. In Proceedings of the 2015 IEEE International Conference on Mechatronics (ICM), Nagoya, Japan, 6–8 March 2015; pp. 256–261.

11. Heß, D.; Künemund, F.; Röhrig, C. Linux based control framework for Mecanum based omnidirectional automated guided vehicles. In Proceedings of the World Congress on Engineering and Computer Science 2013, San Francisco, CA, USA, 23–25 October 2013.

12. Han, L.; Qian, H.H.; Chung, W.K.; Hou, K.W.; Lee, K.H.; Chen, X.; Zhang, G.H.; Xu, Y.S. System and design of a compact and heavy-payload AGV system for flexible production line. In Proceedings of the 2013 IEEE International Conference on Robotics and Biomimetics (ROBIO), Shenzhen, China, 12–14 December 2013; pp. 2482–2488.

13. Zachewicz, B. Flugzeugtransporter mecanumbasierend. Available online: https://www.youtube.com/watch?v=U1MAv7tz9wM (accessed on 25 April 2012).

14. Airbus Rolls Out Its Second A350 XWB Composite Fuselage Demonstrator. Available online: https://www.airbus.com/newsroom/news/en/2009/08/airbus-rolls-out-its-second-a350-xwb-composite-fuselage-demonstrator.html (accessed on 7 August 2009).

15. KUKA omniMove. Available online: https://www.kuka.com/en-de/products/mobility/mobile-platforms/kuka-omnimove (accessed on 10 June 2019).

16. Automation, K.-R. KUKA omniMove at Siemens plant Krefeld. Available online: https://www.youtube.com/watch?v=EvOrFgSmQoc (accessed on 2 October 2014).

17. Muir, P.F.; Neuman, C.P. Kinematic modeling of wheeled mobile robots. *J. Robot. Syst.* **1987**, *4*, 281–340. [CrossRef]

18. Muir, P.F. Modeling and Control of Wheeled Mobile Robots. Ph.D. Dissertation, Carnegie mellon University, Pittsburgh, PA, USA, 1988.

19. Muir, P.F.; Neuman, C.P. Kinematic modeling for feedback control of an omnidirectional wheeled mobile robot. In Proceedings of the 1987 IEEE International Conference on Robotics and Automation, Raleigh, NC, USA, 31 March–3 April 1987; pp. 1772–1778.

20. Campion, G.; Bastin, G.; Dandrea-Novel, B. Structural properties and classification of kinematic and dynamic models of wheeled mobile robots. *IEEE Trans. Robot. Autom.* **1996**, *12*, 47–62. [CrossRef]

21. Röhrig, C.; Heß, D.; Künemund, F. Motion controller design for a mecanum wheeled mobile manipulator. In Proceedings of the 2017 IEEE Conference on Control Technology and Applications (CCTA), Big Island, HI, USA, 27–30 August 2017; pp. 444–449.

22. Tuci, E.; Alkilabi, M.H.M.; Akanyeti, O. Cooperative Object Transport in Multi-Robot Systems: A Review of the State-of-the-Art. *Front. Robot. AI* **2018**, *5*, 59. [CrossRef]

23. Alonso-Mora, J.; Knepper, R.; Siegwart, R.; Rus, D. Local motion planning for collaborative multi-robot manipulation of deformable objects. In Proceedings of the 2015 IEEE International Conference on Robotics and Automation (ICRA), Seattle, WA, USA, 26–30 May 2015; pp. 5495–5502.

24. Alonso-Mora, J.; Baker, S.; Rus, D. Multi-robot formation control and object transport in dynamic environments via constrained optimization. *Int. J. Robot. Res.* **2017**, *36*, 1000–1021. [CrossRef]

25. Habibi, G.; Kingston, Z.; Xie, W.; Jellins, M.; McLurkin, J. Distributed centroid estimation and motion controllers for collective transport by multirobot systems. In Proceedings of the 2015 IEEE International Conference on Robotics and Automation (ICRA), Seattle, WA, USA, 26–30 May 2015; pp. 1282–1288.

26. Habibi, G.; Xie, W.; Jellins, M.; McLurkin, J. Distributed path planning for collective transport using homogeneous multi-robot systems. In *Distributed Autonomous Robotic Systems*; Springer: Tokyo, Japan, 2016; pp. 151–164.

27. Lippi, M.; Marino, A. Cooperative object transportation by multiple ground and aerial vehicles: Modeling and planning. In Proceedings of the 2018 IEEE International Conference on Robotics and Automation (ICRA), Brisbane, Australia, 21–26 May 2018; pp. 1084–1090.

28. Verginis, C.K.; Nikou, A.; Dimarogonas, D.V. Communication-based decentralized cooperative object transportation using nonlinear model predictive control. *arXiv* **2018**, arXiv:1803.07940v1.

29. Tsai, C.C.; Wu, H.L.; Tai, F.C.; Chen, Y.S. Decentralized cooperative transportation with obstacle avoidance using fuzzy wavelet neural networks for uncertain networked omnidirectional multi-robots. In Proceedings of the 2016 12th IEEE International Conference on Control and Automation (ICCA), Kathmandu, Nepal, 1–3 June 2016; pp. 978–983.

30. Wang, Z.; Yang, G.; Su, X.S.; Schwager, M. OuijaBots: Omnidirectional robots for cooperative object transport with rotation control using no communication. In *Distributed Autonomous Robotic Systems*; Springer: Cham, Switzerland, 2018; pp. 117–131.

31. Paniagua-Contro, P.; Hernandez-Martinez, E.G.; González-Medina, O.; González-Sierra, J.; Flores-Godoy, J.J.; Ferreira-Vazquez, E.D.; Fernandez-Anaya, G. Extension of Leader-Follower Behaviours for Wheeled Mobile Robots in Multirobot Coordination. *Math. Probl. Eng.* **2019**, *2019*, 1–16. [CrossRef]

32. Chu, B. Position Compensation Algorithm for Omnidirectional Mobile Robots and Its Experimental Evaluation. *Int. J. Precis. Eng. Manuf.* **2017**, *18*, 1755–1762. [CrossRef]

33. Kulkarni, S.G.; Mulay, G.N.; Patil, T.; Parkhe, C. Automated High Speed Omnidirectional Navigation Using Closed Loop Implementation of Four Wheel Holonomic Mecanum Drive. *Spvryan's Int. J. Eng. Sci. Technol. (SEST)* **2015**, *2*, 19.

34. Tian, P.; Zhang, Y.N.; Zhang, J.; Yan, N.M.; Zeng, W. Research on Simulation of Motion Compensation for 8 × 8 Omnidirectional Platform Based on Back Propagation Network. *Appl. Mech. Mater.* **2013**, *299*, 44–47. [CrossRef]

35. Udomsaksenee, J.; Wicaksono, H.; Nilkhamhang, I. Glocal Control for Mecanum-Wheeled Vehicle with Slip Compensation. In Proceedings of the 2018 15th International Conference on Electrical Engineering/Electronics, Computer, Telecommunications and Information Technology (ECTI-CON), Chiang Rai, Thailand, 18–21 July 2018; pp. 403–406.

36. Keek, J.S.; Loh, S.L.; Chong, S.H. Comprehensive Development and Control of a Path-Trackable Mecanum-Wheeled Robot. *IEEE Access* **2019**, *7*, 18368–18381. [CrossRef]

37. Adamov, B.I. A Study of the Controlled Motion of a Four-wheeled Mecanum Platform. *Russ. J. Nonlinear Dyn.* **2018**, *14*, 265–290. [CrossRef]

38. Wen, R.Y.; Tong, M.M. Mecanum wheels with Astar algorithm and fuzzy PID algorithm based on genetic algorithm. In Proceedings of the 2017 International Conference on Robotics and Automation Sciences (ICRAS), Hong Kong, China, 26–29 August 2017; pp. 114–118.

39. Kim, J.; Woo, S.; Kim, J.; Do, J.; Kim, S.; Bae, S. Inertial navigation system for an automatic guided vehicle with Mecanum wheels. *Int. J. Precis. Eng. Manuf.* **2012**, *13*, 379–386. [CrossRef]

40. Li, Y.W.; Dai, S.M.; Zhao, L.L.; Yan, X.C.; Shi, Y. Topological Design Methods for Mecanum Wheel Configurations of an Omnidirectional Mobile Robot. *Symmetry* **2019**, *11*, 1268. [CrossRef]

41. Wang, Y.Z.; Chang, D.G. Motion Performance Analysis and Layout Selection for Motion System with Four Mecanum Wheels. *J. Mech. Eng.* **2009**, *45*, 307–310. [CrossRef]

42. Wang, Y.Z.; Chang, D.G. Motion restricted condition and singular configuration for mecanum wheeled omni-directional motion system. *J. Shanghai Univ. (Nat. Sci.)* **2009**, *15*, 181–185.

43. Lin, L.-C.; Shih, H.-Y. Modeling and adaptive control of an omni-Mecanum-wheeled robot. *Intell. Control Autom.* **2013**, *4*, 166–179. [CrossRef]

44. Taheri, H.; Qiao, B.; Ghaeminezhad, N. Kinematic model of a four mecanum wheeled mobile robot. *Int. J. Comput. Appl.* **2015**, *113*, 6–9. [CrossRef]

45. Becker, F.; Bondarev, O.; Zeidis, I.; Zimmermann, K.; Abdelrahman, M.; Adamov, B. An approach to the kinematics and dynamics of a four-wheel Mecanum vehicle. *Sci. J. IFToMM Probl. Mech.* **2014**, *2*, 27–37.

46. Abdelrahman, M.; Zeidis, I.; Bondarev, O.; Adamov, B.; Becker, F.; Zimmermann, K. A description of the dynamics of a four wheel mecanum mobile system as a basis for a platform concept for special purpose vehicles for disabled persons. In Proceedings of the 58th Ilmenau Scientific Colloquium, Ilmenau, Germany, 8–12 September 2014.

47.   Gao, X.; Wang, Y.; Zhou, D.; Kikuchi, K. Floor-cleaning robot using omni-directional wheels. *Ind. Robot Int. J.* **2009**, *36*, 157–164. [CrossRef]

48.   Li, Y.; Dai, S.; Shi, Y.; Zhao, L.; Ding, M. Navigation simulation of a Mecanum wheel mobile robot based on an improved A* algorithm in Unity3D. *Sensors* **2019**, *19*, 2976. [CrossRef]

3

Non-Linear Lumped-Parameter Modeling of Planar Multi-Link Manipulators with Highly Flexible Arms

Ivan Giorgio [1,2,*,†] **and Dionisio Del Vescovo** [1,2,†]

[1] Department of Mechanical and Aerospace Engineering, SAPIENZA Università di Roma, via Eudossiana 18, 00184 Rome, Italy; dionisio.delvescovo@uniroma1.it
[2] Research center on Mathematics and Mechanics of Complex Systems, Università degli studi dell'Aquila, 67100 L'Aquila, Italy
* Correspondence: ivan.giorgio@uniroma1.it
† These authors contributed equally to this work.

Abstract: The problem of the trajectory-tracking and vibration control of highly flexible planar multi-links robot arms is investigated. We discretize the links according to the Hencky bar-chain model, which is an application of the lumped parameters techniques. In this approach, each link is considered as a kinematic chain of rigid bodies, and suitable springs are added in order to model bending resistance. The control strategy employed is based on an optimal input pre-shaping and a feedback of the joint angles to treat the effects of undesired disturbances. Some numerical examples are given to show the potentialities of the proposed control, and a comparison with a standard collocated Proportional-Derivative (PD) control strategy is performed. In particular, we study the cases of a linear and a parabolic trajectory with a polynomial time law chosen to minimize the onset of possible vibrations.

Keywords: nonlinear flexible beams; discrete modeling; underactuated robots; optimal preshaping input

---

1. Introduction

Although the literature on flexible robotic manipulators is very varied and covers different aspects of the dynamic analysis and control of these mechanical systems [1–6], in industrial practice, the robotic arms are still treated as rigid multi-body systems. Only in aerospace, for evident weight issues, this strong assumption has long since been removed. In this context, however, moderately flexible links are considered to simplify the analysis and synthesis of rest-to-rest motion, trajectory-tracking and vibration control, resulting in a degradation of performance.

In this paper, we want to address the problem of considering planar multi-link highly flexible robotic arms whose transversal deflections are very large, and therefore, no linearization procedure is adoptable (refer to [7,8] for some relevant real applications). For these systems, it is possible to model the links with the continuous non-linear beam model of Euler–Bernoulli, i.e., the elastica theory [9–15]. However, as this continuous one-dimensional model is characterized by infinite degrees of freedom, it is rather difficult to analyze the manipulator's motion and to design the related control. For this reason, usually a spatial discretization of the continuous model is resorted to. Typically, three methods of discretization are used: the assumed modes technique; the Finite Element Method (FEM); and the lumped parameters method. The assumed modes approach is an intrinsically linear method, and therefore, it is applicable only to a one-link system with small flexibility or multi-link systems in which the task time is much greater than the characteristic period of oscillation of the first modes. The latter assumption is strongly requested because a progressive linearization must be carried out in the neighborhood of the generic current configuration, which indeed varies with the motion [16,17]. The finite element method can be seen as a special case of the assumed modes technique, but it can be generalized to the nonlinear case and can therefore cover also the cases taken into account for the present study. As an example, we mention Du et al. [18], who address the problem of a non-linear 3D flexible manipulator with FE analysis. The discretization by lumped parameters [19,20] is one of the earliest methods and is characterized by an extreme simplicity of modeling and computational advantages. Moreover, being intrinsically nonlinear, it does not introduce any approximation due to some kind of linearization [21,22]. Although the finite element method converges faster than that of lumped parameters and requires a lower number of degrees of freedom to obtain the same accuracy, to deal with nonlinear cases, it is necessary to make use of a specific, rather complex formulation, which could be familiar to experts of computational mechanics, but less accessible to a larger audience (see, e.g., [23]). It should be noted, indeed, that the commercial FEM codes are rather lacking in dealing with the problem of large deflections of beams, to the best of the authors' knowledge. There are many papers that analyze the accuracy of the lumped parameters approach from different points of view, and in all of them, it is possible to find that the error in the approximation with such a method is inversely proportional at least to the first power of the number of elements used [24,25]. In the case of a cantilever beam subject to flexural oscillations, the inverse square law for the error can be verified without difficulty [26]. This means that with a rather simple modeling method, it is possible to reach the desired degree of accuracy simply by appropriately selecting the number of elements; a thumb rule is to use 13 elements per wave-length to be analyzed [26]. In the past, often the lumped-parameter approach was underestimated because assigning spring lumped constants was considered not straightforward [3]. Unfortunately, this is the result of a certain compartmentalization of expertise that sometimes occurs [27]. Indeed, it is widely known to those dealing with beam homogenization how to properly assign these discretized constants of stiffness [28–33].

In this paper, we will use the lumped parameters approach because of its simplicity and versatility. In this formulation, an elastic rod is discretized into a set of rigid segments, which are free to rotate relative to their adjacent neighbors. Springs located in these joints give the system the ability to resist bending. The case study can be classified as an under-actuated system. Indeed, the considered multi-link arm is subject to a lower number of actuators than degrees of freedom. The particular formulation adopted, among the various advantages, could make use of widely established results in classical robotics for this type of problem since in fact the model adopted is a kinematic chain of rigid bodies (see, e.g., [34–36] and the references therein).

System flexibility leads to vibration and, in turn, to an imprecise positioning due mainly to a non-minimum phase character of the system. As an illustrative numerical example, we consider a control strategy for the problem of the trajectory-tracking in the framework of the input shaping control [37–39] with a feedback for the stabilization of the response. In particular, to obtain the command torques, instead of using proper filtering as is usually done, we formulate an optimal control problem with the aim of minimizing the positioning error of the tip manipulator (see, e.g., [40]). This variational approach has been adopted for trajectory planning both for flexible [41,42] and for rigid arms [6]. Herein, instead of obtaining a trajectory with the desired mechanical characteristics, the best possible input command is found to follow a given trajectory.

The paper is organized as follows: Section 2 is devoted to describing the adopted discrete method, which is applied to a planar two-link flexible arm. Section 3 reports on numerical simulations for some trajectory-tracking cases and a comparison with a standard control approach. The paper ends with the conclusions and some future perspectives.

2. Dynamic Modeling of the Flexible Robot Manipulator

To address the complex problem of trajectory control of the tip for flexible robot manipulators, we propose to study an elemental prototype case whose behavior is rich enough to easily extend the obtained results to more generalized situations, namely more links or manipulators subject to a 3D motion (see, e.g., [43] for the analogous 3D formulation). Specifically, a planar horizontal robot manipulator constituted of two highly flexible links is considered. Each link of the manipulator is characterized by a length ℓ_i (with $i = \{1,2\}$), a uniform distribution of stiffness and mass density and is driven by an actuator, with mass m_{hi}, inertia J_{hi} and supplying a torque τ_i. The manipulator eventually may carry a tip payload of mass m_p and inertia J_p.

To model this system, we consider a lumped-parameter discretization. Since the manipulator is made up of two links, which can be modeled in the range of large deflections by the elastica theory, we adopt the well-known Hencky technique [28,29,43–45] to discretize the system and, then, using a more comfortable Lagrangian mechanical system, to study the motion of the robot and to design the trajectory tracing. Therefore, the considered discrete system consists of two articulated chains of n_i rigid rods of length η connected to each other by means of zero-torque hinges, also known as 'pseudo-joints'. At each joint of the same link, a rotational spring is placed in order to model the resistance to being bent of the arm (see Figure 1). In other words, the torques provided by these springs represent the spatial discretization of the internal actions of the links, namely the bending moment. The Lagrangian coordinates, which describe the configurations of the manipulator are $\Phi_j(t)$ (with $j = \{1,2,\ldots n_1 + n_2\}$). In particular, they represent the orientation of the rigid rods with respect to the x-axis. Moreover, the following definitions are useful to specify the angles of the actuation:

$$\begin{cases} \vartheta_1 = \Phi_1 \\ \vartheta_2 = \Phi_{n_1+1} - \Phi_{n_1} \end{cases} \tag{1}$$

for the two links, and the relative angles:

$$\begin{cases} \varphi_{h-1}^{(1)} = \Phi_h - \Phi_{h-1} & \text{with } h = 2,\ldots n_1 \\ \varphi_{k-1-n_1}^{(2)} = \Phi_k - \Phi_{k-1} & \text{with } k = n_1 + 2,\ldots n_1 + n_2 \end{cases} \tag{2}$$

which are relevant to define the deformation energy of the links whose label is reported between parentheses as a superscript. Indeed, the relative angles can be used to describe a discrete point-wise curvature. The two links prior to deformation are straight. Their mass is discretized with lumped masses m_i at the boundaries of each rigid segment by dividing the mass of each segment at its ends. The position of each point mass can be written as:

$$\begin{cases} x_j(t) = \sum_{k=1}^{j} \eta \cos(\Phi_k) \\ y_j(t) = \sum_{k=1}^{j} \eta \sin(\Phi_k) \end{cases} \tag{3}$$

and by a differentiation with respect to time, the velocities of the point masses are given by:

$$\begin{cases} \dot{x}_j(t) = -\sum_{k=1}^{j} \eta \dot{\Phi}_k \sin(\Phi_k) \\ \dot{y}_j(t) = \sum_{k=1}^{j} \eta \dot{\Phi}_k \cos(\Phi_k) \end{cases} \tag{4}$$

The equations of the motion can be derived from the Lagrangian:

$$\mathscr{L} = \mathfrak{K} - \Psi \tag{5}$$

where $\mathfrak{K}$ and Ψ are the kinetic and potential energies of the system, respectively. Particularly, the kinetic energy can be obtained by the sum of three contributions, i.e., a term due to the links:

$$\mathfrak{K}_\ell = \sum_{j=1}^{n_1+n_2} \frac{1}{2} m_j \left\{ \left[\sum_{k=1}^{j} \eta \dot{\Phi}_k \sin(\Phi_k) \right]^2 + \left[\sum_{k=1}^{j} \eta \dot{\Phi}_k \cos(\Phi_k) \right]^2 \right\} \tag{6}$$

a term due to the two actuators:

$$\mathfrak{K}_h = \frac{1}{2}J_{h1}\dot{\Phi}_1^2 + \frac{1}{2}m_{h2}\left(\dot{x}_{n_1}^2 + \dot{y}_{n_1}^2\right) + \frac{1}{2}J_{h2}\dot{\Phi}_{n_1+1}^2 \tag{7}$$

and the last term for the payload:

$$\mathfrak{K}_p = \frac{1}{2}m_p\left(\dot{x}_{n_1+n_2}^2 + \dot{y}_{n_1+n_2}^2\right) + \frac{1}{2}J_p\dot{\Phi}_{n_1+n_2}^2 \tag{8}$$

The elastic potential energy Ψ is assumed to be:

$$\Psi = \sum_{i=1}^{2}\sum_{j=1}^{n_i-1}\kappa_{bj}^{(i)}\left[1 - \cos(\varphi_j^{(i)})\right] \tag{9}$$

where a lumped bending stiffness $\kappa_{bj}^{(i)} = YJ_{\ell_i}/\eta$ associated with the rotational springs is introduced using the elastic modulus of the beam's material, Y, and the second moment of the area of the beam's cross-section, J_{ℓ_i} [46,47]. Note that the potential energy in Equation (9) is positive definite, and in the continuum limit, i.e., for η tending to zero, the expression (9) becomes an energy density, which is quadratic in the curvature of the beam axis [46], in accord with the elastica theory. As a first approximation, we also consider a viscous dissipation (see, e.g., [48,49]), introducing the Rayleigh dissipation function:

$$\mathfrak{R} = \sum_{i=1}^{2}\sum_{j=1}^{n_i-1}\frac{1}{2}c_{bj}^{(i)}\left(\dot{\varphi}_j^{(i)}\right)^2 \tag{10}$$

$c_{bj}^{(i)}$ being a lumped viscous coefficient. This type of dissipation can be associated with the rate of the bending deformation; thus with similar reasoning used for the elastic bending mode deformation, it is possible to evaluate the lumped dissipation with the expression: $c_{bj}^{(i)} = C_vJ_{\ell_i}/\eta$, where the material parameter C_v is the viscous coefficient of the beam material, which can be experimentally identified (see, e.g., [50]).

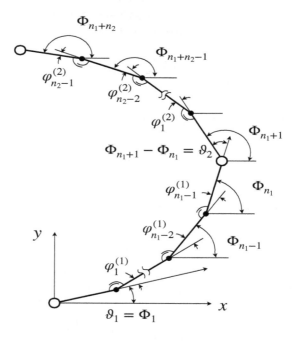

Figure 1. Discrete system for the planar two-link flexible robot.

The Euler–Lagrange equations of motion, thus, are:

$$\frac{\partial}{\partial t}\left(\frac{\partial \mathfrak{K}}{\partial \dot{\Phi}_i}\right) - \frac{\partial \mathfrak{K}}{\partial \Phi_i} + \frac{\partial \Psi}{\partial \Phi_i} + \frac{\partial \mathfrak{R}}{\partial \dot{\Phi}_i} = Q_i \quad \text{for } i = 1 \ldots n_1 + n_2 \tag{11}$$

The virtual work of the torques applied by actuators at the basis of each link is:

$$\delta W = \tau_1 \delta \vartheta_1 + \tau_2 \delta \vartheta_2 = \tau_1 \delta \Phi_1 + \tau_2 \left(\delta \Phi_{n_1+1} - \delta \Phi_{n_1}\right) \tag{12}$$

hence, the only generalized forces different from zero are:

$$Q_1 = \tau_1, \quad Q_{n_1+1} = \tau_2, \quad Q_{n_1} = -\tau_2 \tag{13}$$

One of the performance specifications in a control system should be a satisfactory regulation against disturbances. Here, to test the effectiveness of the proposed control, we take into consideration two kinds of disturbances, due to the actual realization of the actuators, that add up to the torques introduced in Equation (13); specifically, friction torques, which arise at the actuated joints, and cogging torques, typical for electrical motors. The former disturbance is described by a Lund–Grenoble model [51], which is able to capture most of the major nonlinear effects involved in the considered case such as pre-sliding displacement, stick-slip motion, the Stribeck effect, and so forth; the latter disturbance, due to the interaction between the permanent magnets of the rotor and the slots of the stator, is modeled by means of a constitutive relationship, experimentally identified, i.e., a periodic function of the relative position between the stator and the rotor of the motor [52]. An evolution rule for the friction torques τ_{fi} is assumed as follows:

$$\frac{d\tau_{fi}}{dt} = k_i \dot{\vartheta}_i \left(1 - \frac{\tau_{fi}}{\tau_L(\dot{\vartheta}_i)} \text{sign}(\dot{\vartheta}_i)\right) \tag{14}$$

where $\tau_L(\dot{\vartheta}_i) = \tau_C + (\tau_S - \tau_C)\exp\left[-\left(\dot{\vartheta}_i/\nu_s\right)^2\right]$ is the limit torque introduced to take into account the Stribeck effect. In detail, $\tau_S = 0.2$ Nm is the static friction; $\tau_C = 0.1$ Nm is the Coulomb friction torque; and $\nu_s = 0.1$ s^{-1} represents the Stribeck velocity. The cogging torques are evaluated as a function of the angles ϑ_i as follows:

$$\tau_{cgi} = T_{cg} \sum_{k=1}^{6} B_k \sin(n_p\, k\, \vartheta_i) \tag{15}$$

where $T_{cg} = 0.1$ Nm is the amplitude of the torque, $n_p = 2$ is the number of poles of the motors and the coefficients of the trigonometric polynomial B_k are assumed to be $\{-0.7937, -0.3586, -0.0341, 0.0039, -0.0016, -0.0064\}$ in the performed simulations.

3. Numerical Examples for Some Trajectory-Tracking Cases

In this section, a control scheme for trajectory-tracking and vibration control of flexible arms is used to show the potentialities of the proposed formulation. This approach is based on an optimal design of the command torques applied to the actuated joints that aim at following the desired trajectory and reducing vibrations. The control strategy, thus, includes a feedforward control based on such a command input and a feedback control that stabilizes the $2R$ flexible robot along the desired trajectory by a joint-based collocated Proportional-Derivative (PD) controller [39]. Precisely, the optimal control technique is used to produce input profiles for the torques acting on the flexible system as described below. Since the command shaping technique does not require additional sensors or actuators, this technique is particular attractive in order to have a hardware apparatus for the control characterized by minimal equipment.

As a first step, we plan a desired trajectory $x_d(t)$ connecting the ends of each link from arbitrary initial points to desired final points in a given time interval $\mathcal{I} = [0, t_f]$, i.e., we set

$x_d(t) = (x_{n_1,d}, y_{n_1,d}, x_{n_1+n_2,d}, y_{n_1+n_2,d})$. Then, denoting the actual trajectory of the ends of each link, evaluated on the solution of Equation (11) in $\mathcal{I}$ without disturbances, with $\tilde{x}(t) = (\tilde{x}_{n_1}, \tilde{y}_{n_1}, \tilde{x}_{n_1+n_2}, \tilde{y}_{n_1+n_2})$, the optimal control problem for the design of the input torques can be formulated as follows:

Find the torques $\boldsymbol{\tau}(t) = (\tau_1, \tau_2)$ as real-valued smooth functions defined on $\mathcal{I}$, which minimize the continuous-time cost functional:

$$J(\tilde{\boldsymbol{\Phi}}, \boldsymbol{\tau}, t) = \frac{1}{2} \int_0^{t_f} (\tilde{x} - x_d)^\top R\, (\tilde{x} - x_d)\, \mathrm{d}t \tag{16}$$

subject to the dynamic constraints that $\tilde{x}(t)$ is computed on the solution of Equation (11) with given initial conditions. R is a constant diagonal positive definite weight matrix.

In detail, we directly minimize the functional J varying the torques $\tau_1(t)$ and $\tau_2(t)$ instead of solving the Euler–Lagrange equations, which may laboriously be obtainable by means of calculus of variations. Therefore, representing the torques τ_i in a discrete way as follows:

$$\bar{\tau}_i(t) = \tau_{0i}(t) + w(t) \sum_{h=0}^{n_d} a_h^{(i)} p_h(t) \tag{17}$$

where $\tau_{0i}(t)$ is a reference torque, $p_h(t)$ are interpolation functions defined on $\mathcal{I}$ and $w(t)$ is a proper window function, the problem (16) results in finding the coefficients $a_h^{(i)}$ that minimize the functional J evaluated with the approximated shapes $\bar{\tau}_i(t)$. The particular form chosen for Equation (17) is based on the idea of finding an approximate solution for the considered problem, by starting from the exact solution of a related, simpler problem and then adding a correction. The first term $\tau_{0i}(t)$, indeed, is the required torque evaluated for the 2R rigid robot, while the other term represents the correction needed to solve the primary problem. In particular, the function $w(t)$ is conceived of in order to account for a correction of the torques, which tends to zero at the beginning and at the end of the interval $\mathcal{I}$. In this way, a jump in the torques, responsible for the onset of possible vibrations, can be avoided. A possible choice of this function $w(t)$ is the Welch window, defined as:

$$w(t) = \begin{cases} 4\frac{t}{t_f}\left(1 - \frac{t}{t_f}\right) & \text{for } t \in \mathcal{I} \\ 0 & \text{outside } \mathcal{I} \end{cases} \tag{18}$$

Indeed, this window is designed, as many others, to moderate the sudden changes of a rectangular window and, thus, to improve dynamic range. Regarding the p_h functions, thinking of a Taylor expansion properly truncated to express the torque corrections, we consider $p_h(t) = t^h$. Of course, the corrections can be expressed in alternative ways, for example a truncated Fourier series can also be a valid representation. In both cases, a convergence analysis is needed to determine how many terms should be taken into account to minimize the truncation error.

Once the optimal torques $\tau_{id}(t)$ are obtained by solving the problem (16), the complete control strategy can be expressed in the following way:

$$\tau_i = \tau_{di} + K_{Pi}(\vartheta_{di} - \vartheta_i) + K_{Di}(\dot{\vartheta}_{di} - \dot{\vartheta}_i) \tag{19}$$

where $(\vartheta_i, \dot{\vartheta}_i)$ are related to the actual trajectory for the joint angles, directly measured by the motor encoders, while $(\vartheta_{di}, \dot{\vartheta}_{di})$ correspond to the desired trajectory, which is computed off-line by a numerical simulation of the manipulator performed using the optimal torques $\tau_{di}(t)$ as input. To choose the PD gains, a standard technique can be employed based on setting the natural frequencies, which govern the speed of response, as well as taking into account the saturation of each actuator (for a detailed description see, e.g., [53]).

To illustrate the potentialities of the proposed control strategy, we examine some representative examples in which the manipulator tip is constrained to move along two different paths, namely a straight line segment and a piece of parabolic curve.

In the first case, the desired coordinates for the end effector are assumed to be:

$$\begin{cases} x_{n_1+n_2,d}(t) = \ell_1 + \ell_2 - s(t) \cos\left(\arctan\left(\frac{\ell_1}{\ell_1+\ell_2}\right)\right) \\ y_{n_1+n_2,d}(t) = s(t) \sin\left(\arctan\left(\frac{\ell_1}{\ell_1+\ell_2}\right)\right) \end{cases} \tag{20}$$

in which the time function $s(t)$ is a Peisekah polydyne, which is expressed as follows:

$$s(t) = A_d \left[126\left(\frac{t}{t_f}\right)^5 - 420\left(\frac{t}{t_f}\right)^6 + 540\left(\frac{t}{t_f}\right)^7 - 315\left(\frac{t}{t_f}\right)^8 + 70\left(\frac{t}{t_f}\right)^9\right] \tag{21}$$

where $A_d = \sqrt{\ell_1^2 + (\ell_1 + \ell_2)^2}$ and the task time is set to $t_f = 1$ s. The well-known time law (21) is chosen because the values of its derivatives with respect to time up to fourth order at the initial and final times are all zero. This feature is particularly desired to avoid exciting vibrations. Here, the initial configuration provides the robotic arm arranged along the x axis completely unfolded, while the final arrangement of the arm is characterized by having the end effector in the position of coordinates $(0, \ell_1)$.

In the second case, the desired coordinates for the tip manipulator follow the parabola:

$$\begin{cases} x_{n_1+n_2,d}(t) = (\ell_1 + \ell_2) - s(t)(\ell_1 + \ell_2) \\ y_{n_1+n_2,d}(t) = \ell_1 s(t)^2 \end{cases} \tag{22}$$

where the function $s(t)$ is given by Equation (21) with $A_d = 1$ and $t_f = 1.25$ s. The initial and final configurations are set up as in the previous case.

In all the cases, we extend the desired time interval for a while in order to stabilize, in the optimization stage, the solution at the final configuration. The desired coordinates of the intermediate joint, $(x_{n_1,d}, y_{n_1,d})$, have been calculated considering the robotic arm as a $2R$ rigid robot for both cases addressed.

Equations (11) are numerically solved by means of the computing system Simulink considering a $2R$ flexible arm with links of length $\ell_1 = \ell_2 = 0.5$ m and having a rectangular cross-section of size 50×2 mm with a second moment of area $J_{\ell_1} = J_{\ell_2} = 3.33 \times 10^{-11}$ m^4. The links are discretized using 99 rigid segments. The Young modulus of the links is $Y = 200$ GPa, and the mass of each of them is 0.3925 kg. The viscous coefficients are assumed to be $c_b^{(1)} = c_b^{(2)} = 0.15$ Nms. The actuators are characterized by masses $m_{h1} = m_{h2} = 1$ kg and moments of inertia $J_{h1} = J_{h2} = 0.1$ kg m^2, while the payload has $m_p = 0.1$ kg and $J_p = 0.005$ kg m^2. For the implementation of the optimal problem, we assume the non-vanishing elements of R to be $R_{11} = R_{22} = 10$ and $R_{33} = R_{44} = 100$ to give more importance to the tip error.

Figures 2a and 3a show the torques τ_{0i} evaluated for the $2R$ rigid robot, while Figures 2b and 3b display the second term of Equation (17) obtained as a result of the optimization problem.

The used coefficients of the feedback loops are $K_{P1} = 25$ Nm, $K_{P2} = 22$ Nm, $K_{D1} = 0.42$ Nms and $K_{D2} = 0.22$ Nms.

In Figures 4–7 are compiled the results obtained for the two examined cases, respectively for the linear and the parabolic case.

In particular, Figures 4 and 6 show the reference path (dashed black line) for the tip of the arm in which the start and end points are highlighted with a circle and a star, respectively. These figures also exhibit the actual trajectories of the two ends of the links, as well as some intermediate deformed configurations for the links. The deformations of the links are clearly in the range of large deflections, and therefore, any linearization procedure is not allowed in the investigated cases.

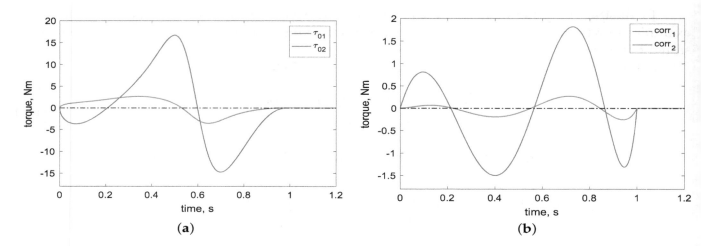

Figure 2. Linear trajectory case: (**a**) reference rigid torques; (**b**) correction torques.

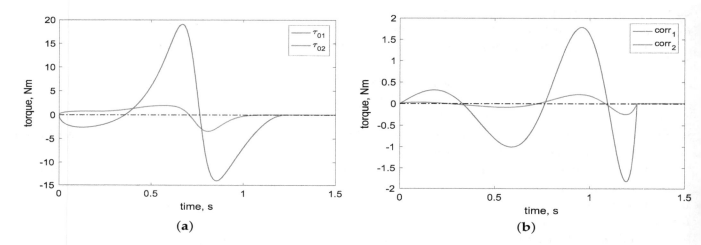

Figure 3. Parabolic trajectory case: (**a**) reference rigid torques; (**b**) correction torques.

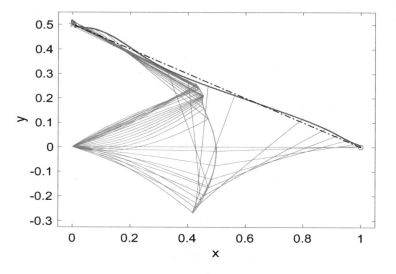

Figure 4. Stroboscopic motion for linear trajectory case.

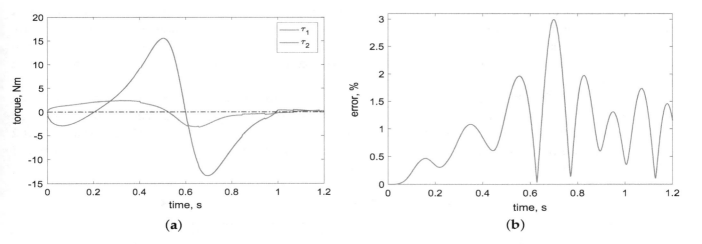

Figure 5. Linear trajectory case: (**a**) actual torques; (**b**) relative tracking error for the tip.

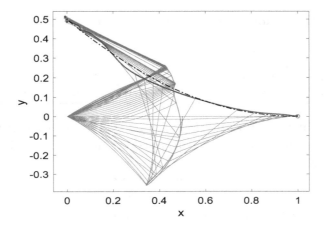

Figure 6. Stroboscopic motion for the parabolic trajectory case.

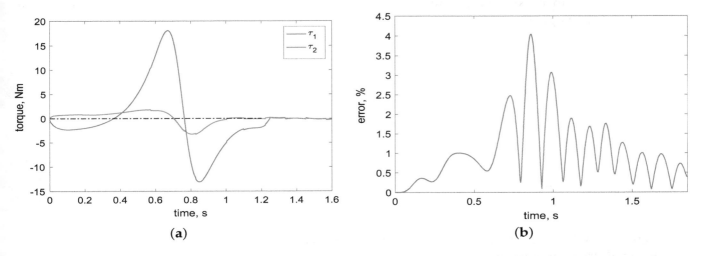

Figure 7. Parabolic trajectory case: (**a**) actual torques; (**b**) relative tracking error for the tip.

Figures 5a and 7a display the actual joint torques resulting from the control strategy of a feedforward with an optimal input command and a feedback of the signals, angles and angular velocities, acquired from the joints. These plots evince the difference between the two joint torques, as well as the power required by the motors and therefore their size. Of course, once the time law (21) has been set, to limit the maximum torque that can be provided, it is possible to change the task time. Thus, we set the task times for the considered cases in order to limit the torques at reliable values.

Figures 5b and 7b exhibit the tip error in following the desired trajectory for the three studied cases. We can see from the graphs that the errors normalized to the full length of the robotic arm were less than 4.5% despite the great deformability of the links and the shortness of the task time. Tables 1 and 2 summarize the coefficients, obtained minimizing the functional J, to represent the torques in terms of the interpolation polynomial functions considered for the analyzed cases.

Table 1. Optimal torque coefficients for the linear trajectory case.

	$a_0^{(i)}$	$a_1^{(i)}$	$a_2^{(i)}$	$a_3^{(i)}$	$a_4^{(i)}$	$a_5^{(i)}$	$a_6^{(i)}$
Link 1	113.65	−590.40	102.40	210.97	3612.9	−3800.3	−0.00178
Link 2	−0.1364	105.65	−780.47	1570.4	−848.80	−108.65	−0.01214

Table 2. Optimal torque coefficients for the parabolic trajectory case.

	$a_0^{(i)}$	$a_1^{(i)}$	$a_2^{(i)}$	$a_3^{(i)}$	$a_4^{(i)}$	$a_5^{(i)}$	$a_6^{(i)}$	$a_7^{(i)}$	$a_8^{(i)}$
Link 1	0.8090	1.114	−11.239	−1.681	1.297	19.485	14.592	−6.116	−15.732
Link 2	0.2139	−1.434	2.659	0.0668	−9.700	10.966	3.453	−5.661	−0.2856

The order of the interpolating polynomial was fixed by increasing it subsequently until the error obtained by the optimization process stabilized. To perform the minimization, we employ a MATLAB code that makes use of the function *fminsearch*.

To estimate the efficiency of the proposed approach, we compare the used control strategy with a standard PD control with a feedback of the angular joints $(\vartheta_i, \dot{\vartheta}_i)$ using a desired trajectory evaluated for the linear and parabolic rigid cases, respectively, Equations (20) and (22) and the time law (21). Figure 8 shows the relative tracking error of the manipulator tip in both examined cases with the PD feedback gains: $K_{P1} = 25$ Nm, $K_{P2} = 22$ Nm, $K_{D1} = 4.2$ Nms and $K_{D2} = 2.2$ Nms. The tracking performances obtained exhibit a maximum relative error of about 30%, much greater than the optimal pre-shaping input approach.

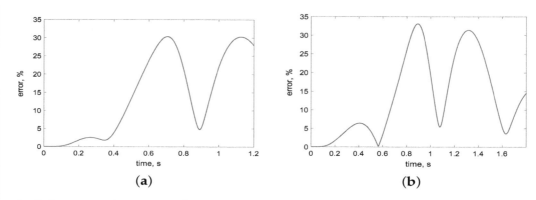

(a)   (b)

Figure 8. Relative tracking error for the tip with collocated PD control: linear trajectory case (**a**); parabolic trajectory case (**b**).

4. Conclusions

In this article, we propose a lumped parameters modeling approach for flexible robotic manipulators in the nonlinear regime. In this context, the method of the assumption of the modes, typically used, cannot be utilized because the hypotheses on which it is based are inadequate for the given problem. We propose to use this formulation for its great simplicity of modeling and for its intrinsic nonlinear nature, instead of the finite element method, which requires a greater modeling effort even if it has a faster convergence speed. Furthermore, an isogeometric formulation, still not fully affirmed as an alternative to the standard FEM, can also improve the results because at equal accuracy, it requires a lesser number of degrees of freedom and seems to be promising for further advances (see, e.g., [54–56] for recent developments).

The number of degrees of freedom that should be treated in the nonlinear case of large link deflections does not allow us to opt for a well-established control strategy such as online computed-torque; therefore, a different control approach must be developed. We present a control approach based on a pre-shaping input that, instead of using appropriately designed signal filters, produces a feedforward command signal for the motors using an optimal problem in which a functional, properly defined, is minimized on the basis of the positioning error of the end effector of the manipulator. In order to stabilize the response close to the desired one, a feedback signal is used together to make the system less sensitive to external disturbances. The employed control strategy is therefore more suitable to treat a greater number of degrees of freedom and can be implemented with minimal hardware equipment. The system considered is under-actuated, and therefore, it is not possible to obtain a perfect positioning of the manipulator tip. However, the simulated numerical cases show that under very strict operating conditions, it is possible to obtain a trajectory tracking with an error of less than about 4.5%. A comparison with a standard feedback PD control strategy shows that, with the optimal pre-shaping input, it is possible to achieve a better tip positioning.

The optimal problem in this paper has been solved numerically to explore the possibilities offered by the proposed method. The preliminary results achieved are quite encouraging, and therefore, as a future research direction, it would be interesting to address the optimal problem rigorously with the calculus of the variations and, in this way, characterize the minimum error obtainable accurately in the various operating conditions.

Author Contributions: I.G. and D.D.V. conceived and designed the experiments; I.G. and D.D.V. performed the experiments; I.G. and D.D.V. analyzed the data; I.G. and D.D.V. contributed analysis tools; I.G. and D.D.V. wrote the paper.

References

1. Cannon, R.H., Jr.; Schmitz, E. Initial experiments on the end-point control of a flexible one-link robot. *Int. J. Robot. Res.* **1984**, *3*, 62–75.
2. Tosunoglu, S.; Lin, S.H.; Tesar, D. Accessibility and controllability of flexible robotic manipulators. *J. Dyn. Syst. Meas. Control* **1992**, *114*, 50–58.
3. Sayahkarajy, M.; Mohamed, Z.; Mohd Faudzi, A.A. Review of modeling and control of flexible-link manipulators. *Proc. Inst. Mech. Eng. Part I J. Syst. Control Eng.* **2016**, *230*, 861–873.
4. Önsay, T.; Akay, A. Vibration reduction of a flexible arm by time-optimal open-loop control. *J. Sound Vib.* **1991**, *147*, 283–300.
5. Meckl, P.; Seering, W. Experimental evaluation of shaped inputs to reduce vibration for a cartesian robot. *J. Dyn. Syst. Meas. Control* **1990**, *112*, 159–165.
6. Gregory, J.; Olivares, A.; Staffetti, E. Energy-optimal trajectory planning for robot manipulators with holonomic constraints. *Syst. Control Lett.* **2012**, *61*, 279–291.
7. Book, W.J. Controlled motion in an elastic world. *J. Dyn. Syst. Meas. Control* **1993**, *115*, 252–261.
8. Katzschmann, R.K.; Marchese, A.D.; Rus, D. Autonomous object manipulation using a soft planar grasping manipulator. *Soft Robot.* **2015**, *2*, 155–164.
9. Antman, S.S. Kirchhoff's problem for nonlinearly elastic rods. *Q. Appl. Math.* **1974**, *32*, 221–240.
10. Steigmann, D.J.; Faulkner, M.G. Variational theory for spatial rods. *J. Elast.* **1993**, *33*, 1–26.

11. Greco, L.; Cuomo, M. Consistent tangent operator for an exact Kirchhoff rod model. *Contin. Mech. Thermodyn.* **2015**, *27*, 861–877.

12. Altenbach, H.; Bîrsan, M.; Eremeyev, V.A. Cosserat-type rods. In *Generalized Continua from the Theory to Engineering Applications*; Springer: Vienna, Austria, 2013; pp. 179–248.

13. Eugster, S.R.; Hesch, C.; Betsch, P.; Glocker, C. Director-based beam finite elements relying on the geometrically exact beam theory formulated in skew coordinates. *Int. J. Numer. Methods Eng.* **2014**, *97*, 111–129.

14. Della Corte, A.; dell'Isola, F.; Esposito, R.; Pulvirenti, M. Equilibria of a clamped Euler beam (Elastica) with distributed load: Large deformations. *Math. Models Methods Appl. Sci.* **2017**, *27*, 1391–1421.

15. Spagnuolo, M.; Andreaus, U. A targeted review on large deformations of planar elastic beams: Extensibility, distributed loads, buckling and post-buckling. *Math. Mech. Solids* **2018**. [CrossRef]

16. Giorgio, I.; Della Corte, A.; Del Vescovo, D. Modelling flexible multi-link robots for vibration control: Numerical simulations and real-time experiments. *Math. Mech. Solids* **2017**. [CrossRef]

17. Low, K.H. Solution schemes for the system equations of flexible robots. *J. Field Robot.* **1989**, *6*, 383–405.

18. Du, H.; Lim, M.; Liew, K. A nonlinear finite element model for dynamics of flexible manipulators. *Mech. Mach. Theory* **1996**, *31*, 1109–1119.

19. Wang, Y.; Huston, R.L. A lumped parameter method in the nonlinear analysis of flexible multibody systems. *Comput. Struct.* **1994**, *50*, 421–432.

20. Šalinić, S. An improved variant of Hencky bar-chain model for buckling and bending vibration of beams with end masses and springs. *Mech. Syst. Signal Process.* **2017**, *90*, 30–43.

21. Yoshikawa, T.; Hosoda, K. Modeling of flexible manipulators using virtual rigid links and passive joints. *Int. J. Robot. Res.* **1996**, *15*, 290–299.

22. Konno, A.; Uchiyama, M. Vibration suppression control of spatial flexible manipulators. *Control Eng. Pract.* **1995**, *3*, 1315–1321.

23. Garcea, G.; Trunfio, G.A.; Casciaro, R. Mixed formulation and locking in path-following nonlinear analysis. *Comput. Methods Appl. Mech. Eng.* **1998**, *165*, 247–272.

24. Livesley, R. The equivalence of continuous and discrete mass distributions in certain vibration problems. *Q. J. Mech. Appl. Math.* **1955**, *8*, 353–360.

25. Leckie, F.A.; Lindberg, G.M. The effect of lumped parameters on beam frequencies. *Aeronaut. Q.* **1963**, *14*, 224–240.

26. Duncan, W.J. A critical examination of the representation of massive and elastic bodies by systems of rigid masses elastically connected. *Q. J. Mech. Appl. Math.* **1952**, *5*, 97–108.

27. Dell'Isola, F.; Bucci, S.; Battista, A. Against the fragmentation of knowledge: The power of multidisciplinary research for the design of metamaterials. In *Advanced Methods of Continuum Mechanics for Materials and Structures*; Springer: Singapore, 2016; pp. 523–545.

28. Jawed, M.K.; Novelia, A.; O'Reilly, O.M. *A Primer on the Kinematics of Discrete Elastic Rods*; Springer: Cham, Switzerland, 2018.

29. Zhang, H.; Wang, C.M.; Challamel, N. Buckling and vibration of Hencky bar-chain with internal elastic springs. *Int. J. Mech. Sci.* **2016**, *119*, 383–395.

30. Alibert, J.J.; Della Corte, A.; Seppecher, P. Convergence of Hencky-type discrete beam model to Euler inextensible elastica in large deformation: Rigorous proof. In *Mathematical Modelling in Solid Mechanics*; Springer: Singapore, 2017; pp. 1–12.

31. Alibert, J.J.; Della Corte, A.; Giorgio, I.; Battista, A. Extensional Elastica in large deformation as Γ-limit of a discrete 1D mechanical system. *Z. Angew. Math. Phys.* **2017**, *68*. [CrossRef]

32. Battista, A.; Della Corte, A.; dell'Isola, F.; Seppecher, P. Large deformations of 1D microstructured systems modeled as generalized Timoshenko beams. *Z. Angew. Math. Phys.* **2018**, *69*. [CrossRef]

33. Khakalo, S.; Balobanov, V.; Niiranen, J. Modelling size-dependent bending, buckling and vibrations of 2D triangular lattices by strain gradient elasticity models: Applications to sandwich beams and auxetics. *Int. J. Eng. Sci.* **2018**, *127*, 33–52.

34. De Luca, A.; Mattone, R.; Oriolo, G. Control of underactuated mechanical systems: Application to the planar 2R robot. In Proceedings of the 35th IEEE Conference on Decision and Control, Kobe, Japan, 13 December 1996; IEEE: Piscataway, NJ, USA, 1996; Volume 2, pp. 1455–1460.

35. De Luca, A.; Oriolo, G. Trajectory planning and control for planar robots with passive last joint. *Int. J. Robot. Res.* **2002**, *21*, 575–590.

36. De Luca, A.; Mattone, R.; Oriolo, G. Stabilization of an underactuated planar 2R manipulator. *Int. J. Robust Nonlinear Control IFAC-Affil. J.* **2000**, *10*, 181–198.

37. Mohamed, Z.; Martins, J.; Tokhi, M.; Da Costa, J.S.; Botto, M. Vibration control of a very flexible manipulator system. *Control Eng. Pract.* **2005**, *13*, 267–277.

38. Singhose, W.; Singer, N.; Seering, W. Comparison of command shaping methods for reducing residual vibration. In Proceedings of the European Control Conference, Rome, Italy, 5–8 September 1995; pp. 1126–1131.

39. De Luca, A.; Caiano, V.; Del Vescovo, D. Experiments on Rest-to-rest Motion of a Flexible Arm. In *Experimental Robotics VIII*; Springer: Berlin, Germany, 2003; pp. 338–349.

40. Pepe, G.; Carcaterra, A.; Giorgio, I.; Del Vescovo, D. Variational Feedback Control for a nonlinear beam under an earthquake excitation. *Math. Mech. Solids* **2016**, *21*, 1234–1246.

41. Boscariol, P.; Gasparetto, A. Model-based trajectory planning for flexible-link mechanisms with bounded jerk. *Robot. Comput.-Integr. Manuf.* **2013**, *29*, 90–99.

42. Boscariol, P.; Gasparetto, A. Optimal trajectory planning for nonlinear systems: Robust and constrained solution. *Robotica* **2016**, *34*, 1243–1259.

43. Turco, E. Discrete is it enough? The revival of Piola–Hencky keynotes to analyze three-dimensional Elastica. *Contin. Mech. Thermodyn.* **2018**. [CrossRef]

44. Rubinstein, D. Dynamics of a flexible beam and a system of rigid rods, with fully inverse (one-sided) boundary conditions. *Comput. Methods Appl. Mech. Eng.* **1999**, *175*, 87–97.

45. Wang, C.M.; Zhang, H.; Gao, R.P.; Duan, W.H.; Challamel, N. Hencky bar-chain model for buckling and vibration of beams with elastic end restraints. *Int. J. Struct. Stab. Dyn.* **2015**, *15*. [CrossRef]

46. Dell'Isola, F.; Giorgio, I.; Pawlikowski, M.; Rizzi, N.L. Large deformations of planar extensible beams and pantographic lattices: Heuristic homogenization, experimental and numerical examples of equilibrium. *Proc. R. Soc. A* **2016**, *472*. [CrossRef]

47. Turco, E.; dell'Isola, F.; Cazzani, A.; Rizzi, N.L. Hencky-type discrete model for pantographic structures: Numerical comparison with second gradient continuum models. *Z. Angew. Math. Phys.* **2016**, *67*. [CrossRef]

48. Altenbach, H.; Eremeyev, V.A. On the constitutive equations of viscoelastic micropolar plates and shells of differential type. *Math. Mech. Complex Syst.* **2015**, *3*, 273–283.

49. Cuomo, M. Forms of the dissipation function for a class of viscoplastic models. *Math. Mech. Complex Syst.* **2017**, *5*, 217–237.

50. Dietrich, L.; Lekszycki, T.; Turski, K. Problems of identification of mechanical characteristics of viscoelastic composites. *Acta Mech.* **1998**, *126*, 153–167.

51. De Wit, C.C.; Olsson, H.; Astrom, K.J.; Lischinsky, P. A new model for control of systems with friction. *IEEE Trans. Autom. Control* **1995**, *40*, 419–425.

52. Buechner, S.; Schreiber, V.; Amthor, A.; Ament, C.; Eichhorn, M. Nonlinear modeling and identification of a dc-motor with friction and cogging. In Proceedings of the Industrial Electronics Society, IECON 2013-39th Annual Conference of the IEEE, Vienna, Austria, 10–13 November 2013; IEEE: Piscataway, NJ, USA, 2013; pp. 3621–3627.

53. Dawson, D.M.; Abdallah, C.T.; Lewis, F.L. *Robot Manipulator Control: Theory and Practice*; CRC Press: Boca Raton, FL, USA, 2003.

54. Greco, L.; Cuomo, M.; Contrafatto, L.; Gazzo, S. An efficient blended mixed B-spline formulation for removing membrane locking in plane curved Kirchhoff rods. *Comput. Methods Appl. Mech. Eng.* **2017**, *324*, 476–511.

55. Cazzani, A.; Malagù, M.; Turco, E. Isogeometric analysis of plane-curved beams. *Math. Mech. Solids* **2016**, *21*, 562–577.

56. Niiranen, J.; Balobanov, V.; Kiendl, J.; Hosseini, S. Variational formulations, model comparisons and numerical methods for Euler–Bernoulli micro-and nano-beam models. *Math. Mech. Solids* **2017**. [CrossRef]

4

Performance-Based Design of the C̲RS-R̲RC Schoenflies-Motion Generator

Raffaele Di Gregorio [1,*], Mattia Cattai [1] and Henrique Simas [2]

[1]   Department of Engineering, University of Ferrara, 44100 Ferrara, Italy; mattia.cattai@gmail.com

[2]   Raul Guenther laboratory of Applied Robotics, Department of Mechanical Engineering, Federal University of Santa Catarina, Florianópolis, SC 88040-900, Brazil; henrique.simas@ufsc.br

*   Correspondence: raffaele.digregorio@unife.it.

Abstract: Rigid-body displacements obtained by combining spatial translations and rotations around axes whose direction is fixed in the space are named Shoenflies' motions. They constitute a 4-dimensional (4-D) subgroup, named Shoenflies' subgroup, of the 6-D displacement group. Since the set of rotation-axis' directions is a bi-dimensional space, the set of Shoenflies' subgroups is a bi-dimensional space, too. Many industrial manipulations (e.g., pick-and-place on a conveyor belt) require displacements that belong to only one Schoenflies' subgroup and can be accomplished by particular 4-degrees-of-freedom (4-DOF) manipulators (Shoenflies-motion generators (SMGs)). The first author has recently proposed a novel parallel SMG of type C̲RS-R̲RC [1]. Such SMG features a single-loop architecture with actuators on the base and a simple decoupled kinematics. Here, firstly, an organic review of the previous results on this SMG is presented; then, its design is addressed by considering its kinetostatic performances. The adopted design procedure optimizes two objective functions, one (global conditioning index (GCI)) that measures the global performance and the other (CI_{min}) that evaluates the worst local performance in the useful workspace. The results of this optimization procedure are the geometric parameters' values that make the studied SMG have performances comparable with those of commercial SMGs. In addition, a realistic 3D model that solves all the manufacturing doubts with simple and cheap solutions is presented.

Keywords: parallel robot; Shoenflies-motion generator; dimensional synthesis; kinetostatic performances; conditioning index

1. Introduction

Shoenflies' motions are rigid-body displacements obtained by combining spatial translations and rotations around axes with a fixed direction. They constitute a 4-dimensional (4-D) subgroup, named Shoenflies' subgroup, of the 6-D displacement group [1,2]. Since the set of rotation-axis' directions is a bi-dimensional space, the set of Shoenflies' subgroups is a bi-dimensional space, too. Many industrial manipulations (e.g., pick-and-place on a conveyor belt) require displacements that belong to only one Schoenflies' subgroup and can be accomplished by particular 4-degrees-of-freedom (4-DOF) manipulators, named Shoenflies-motion generators (SMGs) [3].

The serial robot SCARA, presented in 1981, is the most known SMG, but many alternative serial architectures can be conceived [3]. The main drawback of serial architectures is the need of actuating

---

[1]   Hereafter, R, P, U, S, and C stand for revolute pair, prismatic pair, universal joint, spherical pair, and cylindrical pair, respectively. With reference to a parallel architecture, which features the frame (base) and the end effector (platform) connected to each other by a number of kinematic chains (limbs), a string of capital letters denotes the sequence of joint types that are encountered by moving from the base to the platform on a limb. The hyphen separates the strings of the limbs and the underlining indicates the actuated joints. A serial architecture has only one limb and is denoted by only one string.

joints that connect mobile links, which brings to increase the mobile masses and, simultaneously, to reduce the dynamic performances. Parallel architectures can solve this issue and parallel SMGs have been presented (see, for instance [4–12]), too. Most of the proposed parallel SMGs feature four limbs (e.g., [5–7,9,12]) and one actuator per limb located on the base. Other parallel SMGs are simply obtained by adding a double Cardan shaft (i.e., a limb of RUPUR type), which connects the base to the platform, in a translational parallel manipulator.

The main drawback of parallel SMGs is their complex multi-loop structure that drastically reduces their workspace, usually brings cumbersome kinematics and control algorithms, and, often, does not allow a full rotation of the end effector (platform). Nevertheless, adopting two-limbed (i.e., single-loop) architectures with serial [4,13] or hybrid [8,11] limbs with two actuators per limb reduces the structure complexity while keeping the actuators on the base. Some design tricks [14,15] can yield end-effector's full rotations, and some architectures [12,13] can give the possibility of decoupling position and orientation (decoupled kinematics), which allows simpler and more intuitive control strategies. Moreover, not-overconstrained architectures [4] make it possible to avoid jamming without using small tolerances during manufacturing.

The first author has recently proposed a novel not-overconstrained parallel SMG of type CRS-RRC [13]. Such SMG features a single-loop architecture with actuators on the base and a simple decoupled kinematics. Here, firstly, an organic review of the previous results on this SMG is presented; then, its design is addressed by considering its kinetostatic performances. This design procedure will yield a realistic 3D model that solves all the manufacturing doubts with simple and cheap solutions.

The paper is organized as follows. Sections 2 and 3 state the notations and review the previous results presented in [13,16]. Section 4 addresses the analysis of the kinetostatic performances and the design based on them. Section 5 discusses the results and Section 6 draws the conclusions.

2. Notations and Background

Figure 1 shows the SMG of type CRS-RRC presented in [13]. The axes of the R and C pairs are all parallel. The platform is connected to the base through two limbs: one of type CRS and the other of type RRC. The actuated C pair of the CRS limb is obtained by mean of a PR chain with the sliding direction of the P pair parallel to the axis of the R pair. Such a kinematic chain can be actuated by keeping the motors on the base and using commercial components as shown in Figure 2. Also, the second actuated R pair of the RRC limb can be moved from the base by simply using a toothed-belt transmission.

With reference to Figure 1, $O_b x_b y_b z_b$ is a Cartesian reference fixed to the base; the direction of the z_b coordinate axis is the same as the R and C pair axes. In the platform, a_p is the constant distance of the center of the S pair from the axis of the passive C pair and is equal to the length of the segment $A_p A_2$. O_p is the reference point of the platform, and h is the length of the segment $A_p O_p$. The coordinates, $(x_p, y_p, z_p)^T$, of O_p, measured in $O_b x_b y_b z_b$, locate the position of the platform; whereas, the angle, φ, between the segment $A_p A_2$ and a line parallel to x_b and passing through A_p uniquely determines the platform orientation. In the CRS limb, point B_2 lies on the axis of the actuated C pair and is fixed to the output link of the C pair. The linear variable of the actuated C pair is the signed distance, d, of B_2 from O_b. The plane parallel to the $x_b y_b$ plane and passing through B_2 intersects the axis of the passive R pair at D_2, the axis of the passive C pair at A_p, and the axis parallel to the z_b axis and passing through the center of the S pair at A_2. a_3 and a_4 are the lengths of the segments $B_2 D_2$ and $D_2 A_2$, respectively; whereas, θ_3 and θ_4 are the angular variable of the actuated C pair and the joint variable of the passive R pair, respectively. The point A_p is fixed to the platform. In the RRC limb, the actuated-joint variables are θ_1 and θ_2. The $x_b y_b$ plane intersects the axis of the R pair adjacent to the base at B_1, the axis of the other R pair at D_1, and the axis of the passive C pair at A_1. a_0, a_1 and a_2 are the lengths of the segments $O_b B_1$, $B_1 D_1$ and $D_1 A_1$, respectively. The RRC limb constrains the platform to perform Schoenflies motions with rotation axis parallel to the z_b axis; whereas, the CRS limb controls the coordinate z_p of point O_p through its linear variable, d, and, independently, the platform orientation through its angular variable, θ_3.

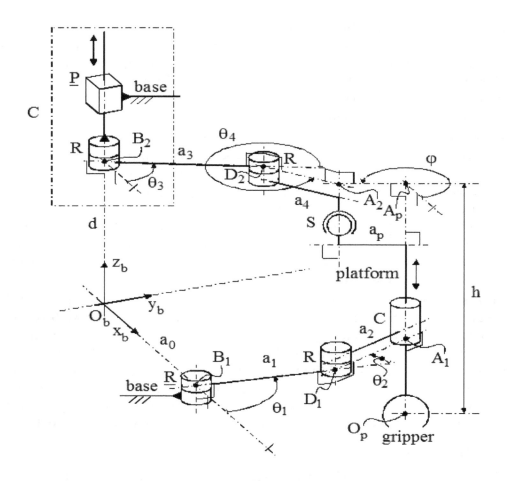

Figure 1. The Shoenflies-motion generators (SMG) of type CRS-RRC.

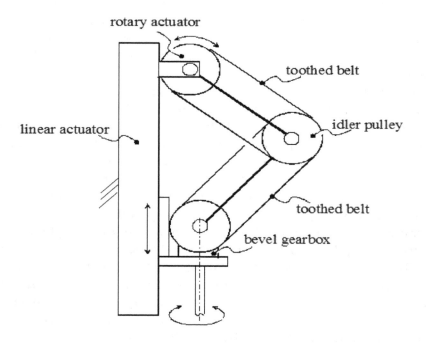

Figure 2. Constructive scheme of an actuated C pair with motors on the frame.

According to these notations, the 4-tuples $\mathbf{q} = (\theta_1, \theta_2, \theta_3, d)^T$ and $\kappa = (x_p, y_p, z_p, \varphi)^T$ collect the actuated-joint (input) variables and the platform-pose (output) variables, respectively. The inspection of Figure 1 yields the following closure equations

$$x_p = a_0 + a_1 \cos \theta_1 + a_2 \cos (\theta_1 + \theta_2), \tag{1a}$$

$$y_p = a_1 \sin \theta_1 + a_2 \sin (\theta_1 + \theta_2), \tag{1b}$$

$$z_p = d - h \tag{1c}$$

$$(x_p + a_p \cos \varphi - a_3 \cos \theta_3)^2 + (y_p + a_p \sin \varphi - a_3 \sin \theta_3)^2 = a_4{}^2 \tag{1d}$$

If $\mathbf{q}$ is assigned (direct position analysis (DPA)), Equation (1a–c) yield a unique platform position; whereas, Equation (1d) yields two platform orientations (i.e., the DPA has two solutions that share the same platform position). Vice versa, if κ is assigned (inverse position analysis (IPA)), Equation (1c) yields only one value for d; whereas, Equation (1a,b,d) yield at most four values for $(\theta_1, \theta_2, \theta_3)$ which can be computed through explicit formulas (see [13] for the proof and the formulas), that is, the IPA has at most four solutions which share the same value of d.

The time derivatives of Equation (1a–d) yield

$$\dot{x}_p = -[a_1 \sin \theta_1 + a_2 \sin(\theta_1 + \theta_2)]\dot{\theta}_1 - a_2 \sin(\theta_1 + \theta_2) \dot{\theta}_2 \tag{2a}$$

$$\dot{y}_p = [a_1 \cos \theta_1 + a_2 \cos(\theta_1 + \theta_2)]\dot{\theta}_1 + a_2 \cos(\theta_1 + \theta_2) \dot{\theta}_2 \tag{2b}$$

$$\dot{z}_p = \dot{d} \tag{2c}$$

$$m_x \dot{x}_p + m_y \dot{y}_p + \dot{\varphi} a_p (m_y \cos \varphi - m_x \sin \varphi) = \dot{\theta}_3 a_3 (m_y \cos \theta_3 - m_x \sin \theta_3) \tag{2d}$$

where $m_x = x_p + a_p \cos \varphi - a_3 \cos \theta_3$ and $m_y = y_p + a_p \sin \varphi - a_3 \sin \theta_3$.

Since the RRC limb forbids the platform to perform displacements that do not belong to the above-mentioned Schoenflies' subgroup, the studied SMG has no constraint singularities[2] and $\dot{\kappa} = (\dot{x}_p, \dot{y}_p, \dot{z}_p, \dot{\varphi})^T$ uniquely identifies the platform twist. As a consequence, Equation (2a–d) are sufficient to relate the platform twist to the actuated-joint rates, $\dot{\mathbf{q}}$, that is, they are the instantaneous input-output relationship (InI/O).

The InI/O of a manipulator is a linear mapping between platform twists and actuated-joint rates whose coefficient matrices (Jacobians) only depend on the configuration of the manipulator. The configurations (singularities) where these Jacobians have not full rank make the linear mapping not bijective and have relevant kinetostatic implications [17–21].

According to the above-deduced InI/O, if $\dot{\mathbf{q}}$ is assigned, $\dot{\mathbf{O}}_p = (\dot{x}_p, \dot{y}_p, \dot{z}_p)^T$ is always uniquely determined, but $\dot{\varphi}$ is not determined when the segments A_2D_2 and A_2A_p are aligned (i.e., a parallel singularity[3] occurs). In addition, if $\dot{\kappa}$ is assigned, only two geometric conditions make one or more actuated-joint rates undetermined (i.e., a serial singularity[4] occurs): (i) the 2-tuple $(\dot{\theta}_1, \dot{\theta}_2)$ is not

[2] Parallel manipulators with mobility lower than six that are designed to make the platform move inside a given displacement subgroup may have configurations (constraint singularities) where the platform can perform instantaneous displacements that do not belong to that subgroup [17].
[3] Parallel singularities, also named type-II singularities [18], usually occur inside the workspace. They are configurations where the platform can perform instantaneous motions with locked actuators. At a type-II singularity, a load (even infinitesimal) applied to the platform needs infinitely-high generalized torques, in at least one actuator, to be balanced.
[4] Serial singularities, also named type-I singularities [18], lie on (and identify) the workspace boundary. They are configurations where the platform stands still while the actuated joints perform instantaneous motions. At a type-I singularity, the platform can carry loads without needing that the actuators provide generalized torques to balance them.

determined when the segments A_1D_1 and D_1B_1 are aligned, and (ii) $\dot{\theta}_3$ is not determined when the segments A_2D_2 and D_2B_2 are aligned.

3. Workspace Analysis

According to the adopted notations, the geometric constants of the C̲RS-R̲RC SMG, which affect the SMG behavior, are a_0, a_1, a_2, a_3, a_4, a_p, and h (Figure 1). Two of the authors, in [16], analyzed some workspace characteristics of this SMG. Such analysis brought to size some of those constants. This section summarizes and reviews the results deduced in [16].

If the two actuated R pairs of the R̲RC limb and the linear variable, d, of the actuated C pair are locked (i.e., by keeping z_p and A_1 fixed), the segments B_2A_p $(=O_bA_1 = \sqrt{x_p^2 + y_p^2})$, B_2D_2 $(=a_3)$, D_2A_2 $(=a_4)$, and A_pA_2 $(=a_p)$ behave like frame, input link, coupler, and follower, respectively, of a four-bar linkage. θ_3 and φ are input and output variables, respectively, of this four-bar linkage, and the singularities of this four-bar correspond to the above-identified parallel singularity and serial singularity (ii). If this four-bar satisfies Grashof's law [22] and A_pA_2 is the shortest bar, the platform can perform a complete rotation (i.e., the angle φ has no limitation). Nevertheless, only a double-crank four-bar can guarantee a full control of the platform rotation.

Let $\boldsymbol{p}$ and p denote the position vector (A_1-O_b) and its magnitude $(=\sqrt{x_p^2 + y_p^2})$, respectively. From an analytic point of view, the four-bar is a double-crank (i.e., the platform can perform a complete rotation fully controlled by θ_3) if and only if

$$(a_p + p) \leq (a_3 + a_4), \tag{3a}$$

$$|a_p - p| \geq |a_3 - a_4|, \tag{3b}$$

$$p = \min\{p, a_3, a_4, a_p\}. \tag{3c}$$

Inequalities (3a,b) impose that the platform can perform a full rotation and condition (3c) make the four-bar linkage double-crank.

Since there is no reason to have a_3 different from a_4 and a possible scaling factor does not affect the analysis, the choice $a_3 = a_4 = 1$ length unit (l.u.) is adopted. With this assumption, inequality (3b) become $|a_p - p| \geq 0$ and is identically satisfied; whereas, inequality (3a) and condition (3c) become

$$p \leq a_p \leq 2 - p \tag{4}$$

The blue area of Figure 3 indicates the values of the 2-tuple (p, a_p) that satisfy inequalities (4). Figure 3 shows that the maximum value, p_{Gr}, that p can assume depends only on a_p: when the values of a_p increase, p_{Gr} increases for $a_p \leq 1$ and, then, decreases for $a_p > 1$. The maximum value of p_{Gr} is 1 l.u. and is obtained with $a_p = 1$ l.u. Since p_{Gr} is the radius (see Figure 1) of the circle centered at O_b that is the region (double-crank region) where A_1 must be located to have a double-crank linkage, the choice $a_p = 1$ l.u. is adopted to maximize the double-crank region.

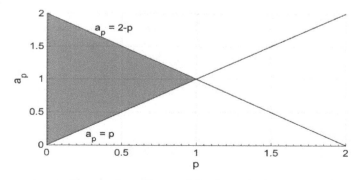

Figure 3. The blue region indicates the values of the 2-tuple (p, a_p) that satisfy inequalities (4).

Let Δd (=$d_{max} - d_{min}$) be the maximum linear stroke of the actuated C pair. The reachable workspace [23] referred to O_p is a right cylinder with height Δd, whose cross section is the intersection between two circles: one centered at O_b with radius $a_3 + a_4 + a_p$ and the other centered at B_1 with radius $a_1 + a_2$. The cross section of the reachable workspace is maximized if the smaller circle is located inside the other, that is, if

$$a_0 + a_1 + a_2 \leq a_3 + a_4 + a_p \tag{5a}$$

or

$$a_3 + a_4 + a_p \leq a_1 + a_2 - a_0. \tag{5b}$$

In addition, if such section contains the double-crank region, the dexterous workspace [23] referred to O_p is maximum, and it is equal to a right circular cylinder obtained by translating the double-crank region along the z_b axis of Δd. This condition is satisfied if

$$p_{Gr} + a_0 \leq a_1 + a_2, \tag{6a}$$

$$|a_1 - a_2| \leq a_0 - p_{Gr}, \tag{6b}$$

$$p_{Gr} \leq a_3 + a_4 + a_p. \tag{6c}$$

At the border of the double-crank region (i.e., for $p = p_{Gr}$), the four-bar linkage can still make the platform perform a complete rotation, but the linkage encounters two times the parallel singularity condition during the platform rotation. Since the integrity of parallel manipulators can be preserved only by keeping them work out and far from parallel singularities, a safe free-from-singularity dexterous workspace requires $p << p_{Gr}$.

Outside the double-crank region (i.e., for $p > p_{Gr}$), the four-bar linkage does not satisfy Grashof's law any longer; as a consequence, it can only be a rocker-rocker four-bar linkage and the rotation range, $\Delta\varphi$, of the platform is limited. Figure 4 shows the above-defined not-Grashof four-bar linkage at the configurations corresponding to the extreme values of φ. In a rocker-rocker four-bar, the whole $\Delta\varphi$ is swept by moving from a serial singularity (ii) to a parallel singularity (Figure 4a) or vice versa (Figure 4b). In both the cases, $\Delta\varphi$ does not change, but the configurations the mechanism passes through are different. The analysis of Figure 4 gives the following simple analytic formula for $\Delta\varphi$

$$\Delta\varphi = \pi + \cos^{-1}\left[\frac{(a_4 + a_p)^2 + p^2 - a_3^2}{2p(a_4 + a_p)}\right] - \cos^{-1}\left[\frac{(a_3 + a_4)^2 - p^2 - a_p^2}{2p\,a_p}\right] \tag{7}$$

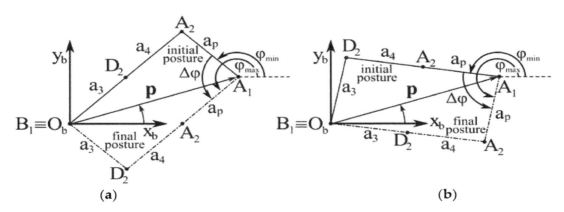

(a)   (b)

Figure 4. The above-defined four-bar linkage, with a p value that makes it a not-Grashof four-bar, at the configurations corresponding to the extreme values of φ: **(a)** $\Delta\varphi$ is swept by moving from a serial singularity (ii) to a parallel singularity, and **(b)** vice versa.

Formula (7) highlights that $\Delta\varphi$ depends only on p (i.e., the direction of the position vector *p* does not affect $\Delta\varphi$) and decreases when p increases. This symmetry with respect to the z_b axis and the fact that the decrease of $\Delta\varphi$ with the increase of p is not sharp [16] make wide regions of the reachable workspace that are out of the double-crank region usable for manipulation tasks that need not a complete platform rotation.

4. Kinetostatic-Performance Analysis and Dimensional Synthesis

The previous sections brought to choose $a_3 = a_4 = a_p = 1$ l.u. All the other geometric constants (i.e., a_0, a_1, a_2, and h) can be used to match the adoption of a useful workspace adequate to industrial tasks with satisfactory kinetostatic properties. Even though the CRS-RRC SMG can perform some tasks outside the above-defined dexterous workspace, the choice of the useful workspace has to take into account the generality of the industrial tasks of an SMG. Consequently, in this case, the useful workspace is chosen as a right circular cylinder with axis passing through O_b (see Figure 1) and radius, p_{uw}, that satisfies the condition $p_{uw} << p_{Gr}$, which makes it coincide with a safe free-from-singularity dexterous workspace.

The determination of p_{uw} can be done by imposing that the transmission angle [22,24], μ, of the above-defined four-bar linkage has an acceptable value during the platform rotation. In general, tasks that require the application of relevant forces to the gripper during motion (e.g., machining tasks like drilling) need values of $|\mu - 90|$ (°) lower than 50° and low friction in the kinematic pairs. Other manipulation tasks with reduced force interaction can accept $|\mu - 90|$ values lower than 70°; whereas, when force interaction is not present (e.g., pick-and-place tasks) $|\mu - 90|$ could be even larger than 70°. The formulas (see [24]) that give the minimum, μ_{min}, and the maximum, μ_{max}, transmission angles, when particularized to the studied case, become ($\mu \in [0°, 180°]$)

$$\mu_{min} = \cos^{-1}\left[\frac{a_4^2 + a_p^2 - (p - a_3)^2}{2\,a_4\,a_p}\right] \tag{8a}$$

$$\mu_{max} = \cos^{-1}\left[\frac{a_4^2 + a_p^2 - (p + a_3)^2}{2\,a_4\,a_p}\right] \tag{8b}$$

which, for $a_3 = a_4 = a_p = 1$ l.u., give the diagrams of Figure 5.

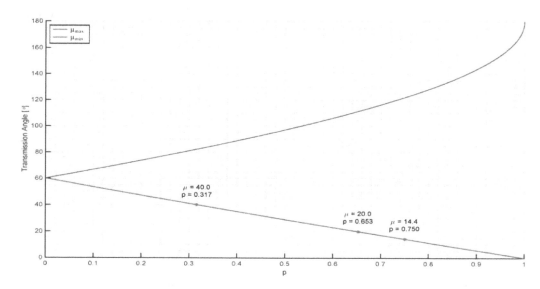

Figure 5. The minimum, μ_{min} (Equation (8a)), and the maximum, μ_{max} (Equation (8b)), transmission angles as a function of p for $a_3 = a_4 = a_p = 1$ (l.u.).

Figure 5 shows that $|\mu - 90| \leq 50°$ corresponds to $p = 0.317$ and $|\mu - 90| \leq 70°$ corresponds to $p = 0.653$; whereas, for $p = 0.75$, $|\mu - 90|$ exceeds $70°$ only in a small neighborhood of μ_{min} which is equal to $14.4°$.

Since SMGs are mainly employed in pick-and-place tasks, the choice $p_{uw} = 0.75$ is adopted by confining the force interaction tasks in region with $p \leq 0.653$ or with $p \leq 0.316$ according to the type of task.

A commercial SMG is the pickstar YS02N of Kawasaki [25]. Its useful workspace is a right circular cylinder with a base radius of 300 mm and a height of 200 mm. Hereafter, this right circular cylinder is chosen as useful workspace for the dimensional synthesis of the CRS-RRC SMG. This choice together with the previous choice $p_{uw} = 0.75$ yield 1 (l.u.) = 400 mm, that is, $a_3 = a_4 = a_p = 400$ mm.

Figure 6 shows the region of the $x_b y_b$ plane swept by the above-defined four-bar linkage with $a_3 = a_4 = a_p = 400$ mm when point A_1 is moved along the x_b axis from O_b (i.e., $A_1 = (0, 0)$) to the border of the useful workspace (i.e., $A_1 = (300, 0)$ mm). The whole region swept by the four-bar when the direction of the motion of A_1 changes can be obtained by making the region highlighted in Figure 6 rotate around O_b in the $x_b y_b$ plane. Such rotation yields a circle with a radius of 700 mm. Consequently, by taking into account the physical sizes of the links, the choice $a_0 = 800$ mm is adopted in order to avoid interferences between the frame and the four-bar during motion.

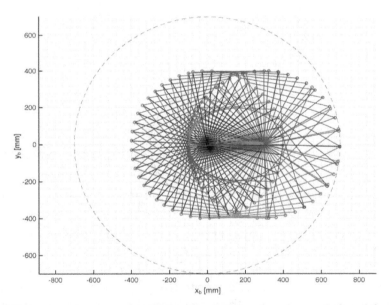

Figure 6. Region of the $x_b y_b$ plane swept during motion by the above-defined four-bar linkage with $a_3 = a_4 = a_p = 400$ mm and a useful workspace with a base radius of 300 mm.

4.1. Measure of the Kinetostatic Performances

The matrix form of system (2) (i.e., of the InI/O) is

$$\mathbf{J}_k \dot{\boldsymbol{\kappa}} = \mathbf{J}_q \dot{\mathbf{q}} \tag{9}$$

with

$$\mathbf{J}_k = \begin{bmatrix} 1 & 0 & 0 & 0 \\ 0 & 1 & 0 & 0 \\ 0 & 0 & 1 & 0 \\ m_x & m_y & 0 & a_p(m_y \cos \varphi - m_x \sin \varphi) \end{bmatrix} \tag{10a}$$

$$\mathbf{J_q} = \begin{bmatrix} -[a_1\sin\theta_1 + a_2\sin(\theta_1+\theta_2)] & -a_2\sin(\theta_1+\theta_2) & 0 & 0 \\ [a_1\cos\theta_1 + a_2\cos(\theta_1+\theta_2)] & a_2\cos(\theta_1+\theta_2) & 0 & 0 \\ 0 & 0 & 0 & 1 \\ 0 & 0 & a_3(m_y\cos\theta_3 - m_x\sin\theta_3) & 0 \end{bmatrix} \tag{10b}$$

Kinetostatic performances of manipulators can be evaluated by using indices built by using the Jacobians $\mathbf{J_k}$ and $\mathbf{J_q}$ of Equation (9). Two of these indices are the "conditioning index" (CI) [26] of the matrix $\mathbf{J} = (\mathbf{J_q})^{-1}\mathbf{J_k}$, which is defined as the inverse of the condition number of $\mathbf{J}$, and its average value on the useful workspace, named "global conditioning index" (GCI). The computation of the CI and GCI needs that the entries of the Jacobians $\mathbf{J_k}$ and $\mathbf{J_q}$ be homogeneous [9,27–29]. Jacobian homogenization can be obtained with the introduction of a characteristic length, λ, through a change of variables [27], even though which λ should be used is an open problem.

In the studied case, the introduction of the new homogeneous variables $\dot{\boldsymbol{\kappa}}_h = \left(\frac{\dot{x}_p}{\lambda}, \frac{\dot{y}_p}{\lambda}, \frac{\dot{z}_p}{\lambda}, \dot{\varphi}\right)^T$ and $\dot{\mathbf{q}}_h = \left(\dot{\theta}_1, \dot{\theta}_2, \dot{\theta}_3, \frac{\dot{d}}{\lambda}\right)^T$ into Equation (9) yields

$$\mathbf{J_h}\dot{\boldsymbol{\kappa}}_h = \dot{\mathbf{q}}_h \tag{11}$$

where $\mathbf{J_h} = (\mathbf{J_{qh}})^{-1}\mathbf{J_{kh}}$ with

$$\mathbf{J_{kh}} = \begin{bmatrix} 1 & 0 & 0 & 0 \\ 0 & 1 & 0 & 0 \\ 0 & 0 & 1 & 0 \\ \frac{m_x}{\lambda} & \frac{m_y}{\lambda} & 0 & \frac{a_p}{\lambda^2}(m_y\cos\varphi - m_x\sin\varphi) \end{bmatrix} \tag{12a}$$

$$\mathbf{J_{qh}} = \begin{bmatrix} \frac{-[a_1\sin\theta_1 + a_2\sin(\theta_1+\theta_2)]}{\lambda} & -\frac{a_2}{\lambda}\sin(\theta_1+\theta_2) & 0 & 0 \\ \frac{[a_1\cos\theta_1 + a_2\cos(\theta_1+\theta_2)]}{\lambda} & \frac{a_2}{\lambda}\cos(\theta_1+\theta_2) & 0 & 0 \\ 0 & 0 & 0 & 1 \\ 0 & 0 & \frac{a_3}{\lambda^2}(m_y\cos\theta_3 - m_x\sin\theta_3) & 0 \end{bmatrix} \tag{12b}$$

$\mathbf{J_h}$ is a homogeneous Jacobian with dimensionless entries. The condition number, χ, of $\mathbf{J_h}$ is, by definition, $\chi = \|\mathbf{J_h}\|\,\|\mathbf{J_h}^{-1}\|$ where $\|(\cdot)\|$ stands for any matrix norm of the argument, if the Frobenius norm [21] is adopted $\|\mathbf{J_h}\| = \sqrt{\mathrm{trace}(\mathbf{J_h}\mathbf{J_h^T})}$. Consequently, the conditioning index, CI $= 1/\chi$, depends both on the SMG configuration and on the geometric constants a_1 and a_2 that have not been determined, yet. The CI ranges from 0, at parallel singularities to 1 at isotropic configurations (i.e., SMG configurations where $\mathbf{J_h}$ is proportionate to the identity matrix), which are the farthest from parallel singularities. The CI is a local index that measures the kinetostatic performance of the SMG at a configuration; whereas, the GCI (i.e., its average value on the useful workspace) gives a score to the kinetostatic performance of the SMG and could be used to compare different manipulators.

Optimizing the kinetostatic performances of a manipulator by using the CI and the GCI means determining the available geometric constants (in this case, a_1 and a_2) so that the minimum value, CI_{min}, of the CI and the GCI are as high as possible for that architecture. Such optimization is presented in the following subsection.

4.2. Dimensional Synthesis

The admissible values of a_1 and a_2 must satisfy inequalities (6a,b) with $a_0 = 800$ mm and $p_{Gr} = a_p = 400$ mm. Moreover, a reasonable choice for the RRC limb would be $a_2 \leq a_1$. All these inequalities are satisfied in the region highlighted in blue of Figure 7.

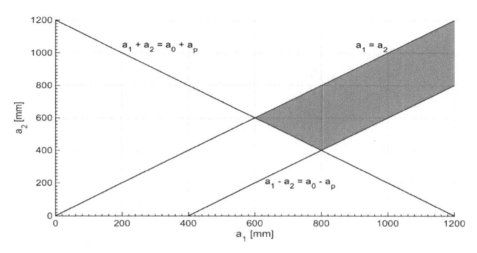

Figure 7. The values of (a_1, a_2) that satisfy inequalities (6a), (6b) and $a_2 \leq a_1$ with $a_0 = 800$ mm and $p_{Gr} = a_p = 400$ mm are those belonging to the blue region.

By setting $\lambda = 400$ mm (i.e., equal to the length of a_p, a_3 and a_4 and to the arithmetic mean of the diameter and the height of the useful workspace), a numerical algorithm has been used to compute the GCI and the CI_{min} referred to the useful workspace for each admissible values of (a_1, a_2). The results of these computations are shown in Figure 8.

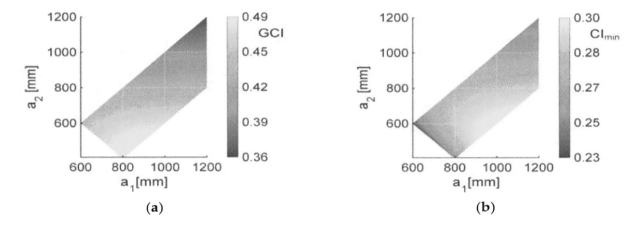

(a)   (b)

Figure 8. Kinetostatic performance as a function of a_1 and a_2: (a) global conditioning index (GCI) evaluated on the useful workspace, and (b) CI_{min} in the useful workspace.

The analysis of the GCI values displayed in Figure 8a reveals that the maximum GCI is equal to 0.48795 and is obtained with $a_1 = 869$ mm and $a_2 = 469$ mm, which correspond to $CI_{min} = 0.28682$ (Figure 8b). Also, Figure 8a shows that the maximum GCI falls in a smooth region (more or less flat) which allows large variation of (a_1, a_2) with small reductions of the GCI.

On the other side, the analysis of the CI_{min} values displayed in Figure 8b reveals that the maximum CI_{min} is equal to 0.29806 and is obtained with $a_1 = 993$ mm and $a_2 = 593$ mm, which correspond to GCI = 0.4763 (Figure 8a). Moreover, Figure 8b shows that also the maximum CI_{min} falls in a smooth region (more or less flat). These results bring to choose $a_1 = 950$ mm and $a_2 = 600$ mm which correspond to a good compromise with GCI = 0.482 and $CI_{min} = 0.29743$ that are values near enough to their maxima. The chosen values of a_1 and a_2 yield the minimum CI values at each A_1 position inside the useful workspace shown in Figure 9. Figure 9 highlights that most of the useful workspace has $CI \geq 0.45$, which makes the CI distribution acceptable.

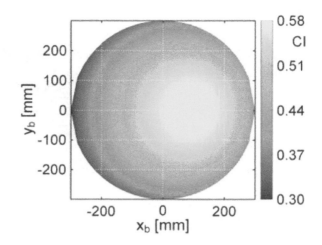

Figure 9. Minimum conditioning index (CI) values at each A_1 position inside the useful workspace for $a_1 = 950$ mm and $a_2 = 600$ mm.

4.3. The 3D Model

The dimensional synthesis, with useful workspace assigned as a right circular cylinder with a radius of 300 mm and a height of 200 mm, brought to choose the following values of the geometric constants: $a_0 = 800$ mm, $a_1 = 950$ mm, $a_2 = 600$ mm, $a_3 = 400$ mm, $a_4 = 400$ mm, and $a_p = 400$ mm. Figure 10 shows the region of the $x_b y_b$ plane swept by the <u>RRC</u> limb, with $a_0 = 800$ mm, $a_1 = 950$ mm, and $a_2 = 600$ mm, when point A_1 is moved on the whole circular boundary of the useful workspace and the limb is assembled in either of the two assembly modes the IPA identifies [13]. The analysis of Figure 10 highlights that no link interference occurs with the chosen geometric data. By combining Figures 6 and 10 the overall region of the $x_b y_b$ plane that is swept by both the limbs is obtained. Figure 11 shows such region.

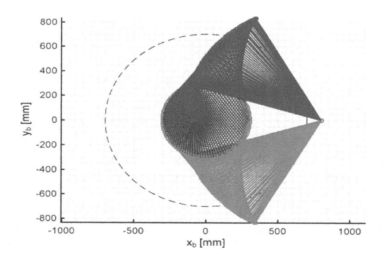

Figure 10. Region of the $x_b y_b$ plane swept by the RRC limb, with $a_0 = 800$ mm, $a_1 = 950$ mm, and $a_2 = 600$ mm, when point A_1 is moved on the whole circular border of the useful workspace and the limb is assembled in either of the two assembly modes the inverse position analysis (IPA) identifies.

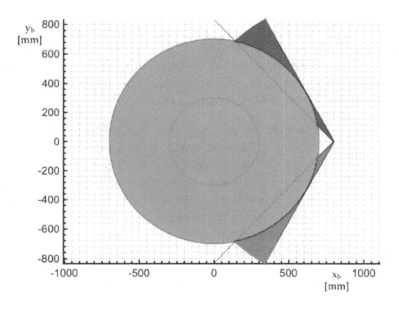

Figure 11. Overall region of the $x_b y_b$ plane that is swept by both the limbs.

In order to solve all the doubts about the actual feasibility of the machine, a CAD model of the CRS-RRC SMG has been built with the chosen geometric data. Figure 12 shows the 3D view (Figure 12a) and the lateral view (Figure 12b) of this model. Such model has an L-shaped base, only commercial actuators that are all mounted on the base, rolling bearings in all the R pairs to avoid jamming of the above-defined four-bar linkage at low transmission angles, a spherical roller bearings that implements the S-pair constraint[5], and a linear ball bearing that implements the constraint of the passive C-pair. The bevel gearbox and the toothed belts (Figures 2 and 12) are commercial products, too.

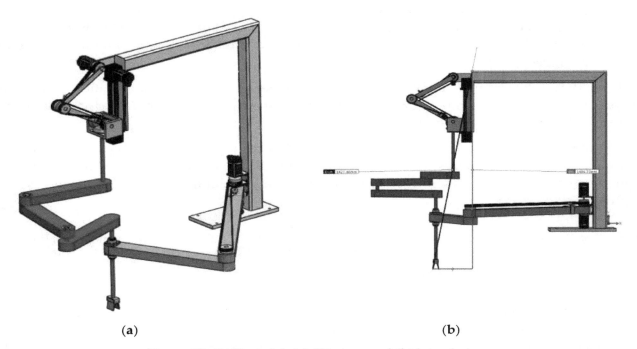

(a)   (b)

Figure 12. CAD model: (a) 3D view, and (b) lateral view.

---

[5]   In the studied SMG, the S pair has the only role of compensating errors of parallelism among the R-pair axes.

5. Summary of the Results and Discussion

The kinematic analysis [13] of the C̲RS-R̲RC SMG showed that both the DPA and IPA are solvable with simple and explicit formulas, and that platform's translation and rotation are decoupled. Therefore, the control algorithms are easy to implement and require very low computation time, which makes it possible to build a fast online control system.

The above-reviewed workspace analysis (see also Ref. [16]) highlighted that the reachable workspace has axially-symmetric properties. In particular, a safe free-from-singularity dexterous workspace, which is a right circular cylinder, is easy to identify in a wider region of the reachable workspace where the platform can perform complete rotations. Moreover, parallel singularities occur only for two known platform orientations and, in the reachable workspace, they are located on a circle that is the boundary of the dexterous workspace. Thus, the platform can pass through the parallel-singularity locus simply by changing its orientation and, if necessary, can accomplish particular tasks in region where the platform rotation is bounded.

The kinetostatic analysis based on the transmission angle and the conditioning index revealed that the machine can be so dimensioned that the GCI and the CI_{min} are acceptable in a useful workspace, equal to those of commercial SMGs, with transmission angles adequate to the generality of industrial tasks.

The CAD model of the C̲RS-R̲RC SMG showed that its manufacture needs only commercial components. This result proves the actual feasibility of the machine with cheap production processes. Moreover, it highlights that the L-shaped base when fixed on a rotating frame allows an easy way for setting up the machine with respect to different machining planes (e.g., belt conveyors).

6. Conclusions

Kinetostatic indices have been used to complete the dimensional synthesis of the SMG of type C̲RS-R̲RC previously presented by the first author. In particular, a safe free-from-singularity useful workspace equal to that of a commercial SMG (the pickstar YS02N of Kawasaki [25]) has been chosen and all the geometric constants of this machine have been determined so that the minimum CI and the GCI are maximized.

Then, the so-determined geometric constants have been used to build a realistic 3D model that involves only commercial components, and implements an original architecture for actuated C pairs that brings all the actuators on the frame.

The obtained results positively close the validation of the machine concept and open to the structural and dynamic analyses/checks necessary to match static and dynamic requirements for an assigned payload. These analyses together with stiffness and accuracy analyses are the next steps of this project.

Author Contributions: Data curation, M.C. and H.S.; Formal analysis, R.D.G.; Funding acquisition, R.D.G. and H.S.; Investigation, R.D.G., M.C. and H.S.; Methodology, R.D.G.; Project administration, R.D.G.; Software, M.C.

References

1. Hervé, J.M. The Lie group of rigid body displacements, a fundamental tool for mechanism design. *Mech. Mach. Theory* **1999**, *34*, 719–730. [CrossRef]

2. Bottema, O.; Roth, B. *Theoretical Kinematics*; Courier Dover Publications: Amsterdam, The Netherlands, 1979; ISBN 0-444-85124-0.

3. Lee, C.-C.; Hervé, J.M. Type Synthesis of Primitive Schoenflies-Motion Generators. *Mech. Mach. Theory* **2009**, *44*, 1980–1997. [CrossRef]

4.   Lee, C.-C.; Hervé, J.M. Isoconstrained Parallel Generators of Schoenflies Motion. *ASME J. Mech. Robot.* **2011**, *3*, 021006. [CrossRef]

5.   Pierrot, F.; Company, O. H4: A New Family of 4-dof Parallel Robots. In Proceedings of the IEEE/ASME International Conference on Advanced Intelligent Mechatronics (AIM'99), Atlanta, GA, USA, 19–23 September 1999; pp. 508–513.

6.   Krut, S.; Company, O.; Benoit, M.; Ota, H.; Pierrot, F. I4: A new parallel mechanism for SCARA motions. In Proceedings of the 2003 International Conference on Robotics and Automation, Taipei, Taiwan, 14–19 September 2003; pp. 1875–1880.

7.   Nabat, V.; de la O. Rodriguez, M.; Company, O.; Krut, S.; Pierrot, F. Par4:very high speed parallel robot for pick-and-place. In Proceedings of the 2005 IEEE/RSJ International Conference on Intelligent Robots and Systems, Edmonton, AB, Canada, 2–6 August 2005; pp. 553–558.

8.   Angeles, J.; Caro, S.; Khan, W.; Morozov, A. The Design and Prototyping of an Innovative Schoenflies Motion Generator. *Proc. IMechE-Part C J. Mech. Eng.* **2006**, *220*, 935–944. [CrossRef]

9.   Li, Z.; Lou, Y.; Zhang, Y.; Liao, B.; Li, Z. Type Synthesis, Kinematic Analysis, and Optimal Design of a Novel Class of Schoenflies-Motion Parallel Manipulators. *IEEE Trans. Autom. Sci. Eng.* **2013**, *10*, 674–686.

10.   Kong, X.; Gosselin, C.M. Type Synthesis of 3T1R 4-dof Parallel Manipulators Based on Screw Theory. *IEEE Trans. Robot. Autom.* **2004**, *20*, 181–190. [CrossRef]

11.   Company, O.; Pierrot, F.; Nabat, V.; Rodriguez, M. Schoenflies Motion Generator: A New Non Redundant Parallel Manipulator with Unlimited Rotation Capability. In Proceedings of the 2005 IEEE International Conference on Robotics and Automation, Barcelona, Spain, 18–22 April 2005; pp. 3250–3255.

12.   Richard, P.-L.; Gosselin, C.M.; Kong, X. Kinematic Analysis and Prototyping of a Partially Decoupled 4-DOF 3T1R Parallel Manipulator. *ASME J. Mech. Des.* **2007**, *129*, 611–616. [CrossRef]

13.   Di Gregorio, R. A Novel Single-Loop Decoupled Schoenflies-Motion Generator: Concept and Kinematics Analysis. In *Advances in Service and Industrial Robotics*; Series: Mechanisms and Machine Science; Ferraresi, C., Quaglia, G., Eds.; Springer: New York, NY, USA, 2018; Volume 49, pp. 11–18. ISBN 978-3-319-61275-1. [CrossRef]

14.   Pierrot, F.; Company, O.; Krut, S.; Nabat, V. Four-Dof PKM with Articulated Travelling-Plate. In *Parallel Kinematic Machines in Research and Practice, Proceedings of the the 5th Chemnitz Parallel Kinematics Seminar, Chemnitz, Germany, 25–26 April 2006*; Neugebauer, R., Ed.; Verlag Wissenschaftliche Scripten: Auerbach, Germany, 2006; Volume 33, pp. 677–694. ISBN 9783937524405. Available online: https://hal-lirmm.ccsd.cnrs. fr/lirmm-00105558 (accessed on 14 September 2019).

15.   Simas, H.; Di Gregorio, R. Kinematics of a Particular 3T1R Parallel Manipulator of Type 2PRPU. In Proceedings of the ASME 2017 International Design Engineering Technical Conferences and Computers and Information in Engineering Conference, Cleveland, OH, USA, 6–9 August 2017; Paper No. DETC2017-67174. Volume 5A, ISBN 978-0-7918-5817-2. [CrossRef]

16.   Simas, H.; Di Gregorio, R. Workspace Analysis and Dimensional Synthesis of the PRRS-RRC Shoenflies-Motion Generator. In *Mechanism Design for Robotics, Proceedings of the 4th IFToMM Symposium on Mechanism Design for Robotics, Udine, Italy, 11–13 September 2018*; Gasparetto, A., Ceccarelli, M., Eds.; Springer: New York, NY, USA, 2018; pp. 2011–2018.

17.   Zlatanov, D.; Bonev, I.A.; Gosselin, C.M. Constraint singularities of parallel mechanisms. In Proceedings of the 2002 IEEE International Conference on Robotics and Automation (ICRA 2002), Washington, DC, USA, 11–15 May 2002; pp. 496–502.

18.   Gosselin, C.M.; Angeles, J. Singularity analysis of closed-loop kinematic chains. *IEEE Trans. Robot. Autom.* **1990**, *6*, 281–290. [CrossRef]

19.   Ma, O.; Angeles, J. Architecture singularities of platform manipulators. In Proceedings of the 1991 IEEE International Conference on Robotics and Automation (ICRA 1991), Sacramento, CA, USA, 9–11 April 1991; pp. 1542–1547.

20.   Zlatanov, D.; Fenton, R.G.; Benhabib, B. A unifying framework for classification and interpretation of mechanism singularities. *ASME J. Mech. Des.* **1995**, *117*, 566–572. [CrossRef]

21.   Meyer, C.D. *Matrix Analysis and Applied Linear Algebra*; Society for Industrial & Applied Mathematics (SIAM): Philadelphia, PA, USA, 2000; ISBN 978-0898714548.

22.   Shigley, J.E.; Uicker, J.J. *Theory of Machines and Mechanisms*, 2nd ed.; McGraw-Hill: New York, NY, USA, 1995; pp. 33–36. ISBN 9780071137475.

23. Siciliano, B.; Sciavicco, L.; Villani, L.; Oriolo, G. *Robotics: Modelling, Planning and Control*; Springer: London, UK, 2009; p. 85. ISBN 978-1-84628-641-4.

24. Pennestrì, E.; Valentini, P.P. A review of simple analytical methods for the kinematic synthesis of four-bar and slider-crank function generators for two and three prescribed finite positions. In *Buletin Stiintific Seria Mecanica Aplicata*; University of Pitesti: Piteşti, Romania, 2009; pp. 128–143. [CrossRef]

25. Kawasaki. Y-Series: Ultra-High Speed Delta Robot. 2012. Available online: http://larraioz.com/_lib/pdf/Kawasaki/DS_Robot_YS002N.pdf (accessed on 16 July 2018).

26. Gosselin, C.; Angeles, J. A global performance index for the kinematic optimization of robotic manipulators. *ASME J. Mech. Des.* **1991**, *113*, 220–226. [CrossRef]

27. Ma, O.; Angeles, J. Optimum architecture design of platform manipulators. In Proceedings of the 5th International Conference Advanced Robotics (ICAR1991), Pisa, Italy, 19–22 June 1991; pp. 1130–1135.

28. Kim, S.; Ryu, J. New dimensionally homogeneous Jacobian matrix formulation by three end-effector points for optimal design of parallel manipulators. *IEEE Trans. Robot. Autom.* **2003**, *19*, 731–736.

29. Kim, S.M.; Kim, W.; Yi, B.-J. Kinematic Analysis and Design of a New 3T1R 4-DOF Parallel Mechanism with Rotational Pitch Motion. In Proceedings of the 2009 IEEE/RSJ International Conference on Intelligent Robots and Systems, St. Louis, MO, USA, 10–15 October 2009; pp. 5167–5172.

5

Motion Investigation of a Snake Robot with Different Scale Geometry and Coefficient of Friction

Naim Md Lutful Huq, Md Raisuddin Khan *, Amir Akramin Shafie, Md Masum Billah and Syed Masrur Ahmmad

Department of Mechatronics Engineering, International Islamic University Malaysia, Selangor 53100, Malaysia; lutful@gmail.com (N.M.L.H.); aashafie@iium.edu.my (A.A.S.); masum.uia@gmail.com (M.M.B.); bdmasrur91@gmail.com (S.M.A.)
* Correspondence: raisuddin@iium.edu.my.

Abstract: Most snakes in nature have scales at their ventral sides. The anisotropic frictional coefficient of the ventral side of the snakes, as well as snake robots, is considered to be responsible for their serpentine kind of locomotion. However, little work has been done on snake scales so far to make any guidelines for designing snake robots. This paper presents an experimental investigation on the effects of artificial scale geometry on the motion of snake robots that move in a serpentine manner. The motion of a snake robot equipped with artificial scales with different geometries was recorded using a Kinect camera under different speeds of the actuating motors attached to the links of the robot. The results of the investigation showed that the portion of the scales along the central line of the robot did not contributed to the locomotion of the robot, rather, it is the parts of the scales along the lateral edges of the robot that contributed to the motion. It was also found that the lower frictional ratio at low slithering speeds made the snake robot motion unpredictable. The scales with ridges along the direction of the snake body gave better and more stable motion. However, to get the peg effect, the scales needed to have a very high lateral to forward friction ratio, otherwise, significant side slipping occurred, resulting in unpredictable motion.

Keywords: snake robot; snake scale; scale geometry; friction ratio; serpentine motion

---

1. Introduction

Snakes possess a unique feature of locomotion that no other creature has. Although legs and wheels give very effective and efficient motion, this creature has the unique ability to move through terrains that are almost impossible to travel through by limbs or wheels. Thus, the ability to move through almost all kinds of surfaces puts snakes in a superior position in terms of locomotion. Snakes have four basic types of locomotion: concertina, serpentine, sidewinding, and rectilinear. Among these, the serpentine locomotion is a faster and a very efficient one for normal conditions. Thus, this motion can be proven exceptionally beneficial in snake robot applications such as search and rescue [1,2], firefighting [3,4], assisting in surgery [5,6], surveillance [7,8] exploring unknown territory, and so forth.

The mechanics of serpentine locomotion were first discovered and explained by J. Gray [9] in 1946. According to Gray, the serpentine locomotion of snakes is due to the lateral friction of the scales of a snake that work as supporting pegs. After the ground breaking discovery of J. Gray, Hu et al. [10] carried out several investigations on real snakes and established the theory experimentally. Some more works on biological snakes that were not focused on the analysis or investigation of the forces applied by the snakes took place. Jayne [11,12] was more focused on the muscular activities during the different types of locomotion with extensive work on serpentine and sidewinding locomotion. Jayne provided extraordinary information regarding the optimization of the actuation method and the energy for the future developments of robots. Meanwhile Miller [13] simulated the motion dynamics of snake locomotion. Later on, Jayne and Davis [14] developed a relationship between the tunnel width and the speed, or, in other words, the movable space available and the speed of the snake using the kinematics of the concertina locomotion of snake. Marvi and his group studied the friction augmentation by the snakes in concertina locomotion [15].

The archive of information on snake robots dates back to the year 1993 when S. Hirose developed the idea of the locomotion of a snake robot [16], encouraging researchers in working consistently to improve and optimize their artificial snakes. For instance, Wang, Osborne, and Alben [17] and Jing and Alben [18] optimized locomotion on an inclined surface. Instead of real snakes, Kyriakopoulos, Migadis, and Sarrigeorgidis [19] worked on the kinematics of snake robots and provided a design and motion planning in 1999. Prautsch and Mita developed a theory for the dynamic position control [20] for snake robots. Khan et al. [21] first developed a scale based snake trying to mimic the serpentine motion. Later on, they put forward their previous work by investigating the snake robot moving with the serpentine method of locomotion [22]. Apart from that, Marvi et al. [23] presented a snake inspired robot segment based on concertina locomotion in 2011. It was capable of changing the angle of attack of its scales with the change in the slope angle while climbing uphill. They also worked on sidewinding with minimal slips for both the snake and the snake robot on sandy slopes [24]. Varesis, Diamantopoulos, and Tzes [25] conducted an experiment on robots with serpentine locomotion moving through an inclined surface [23]. However, none of the works on artificial snake scales considered the geometry of the scales to investigate the motion of the snake robots. As a continuation of the above efforts, this paper presents the effects of the scale geometry and the friction factor on the locomotion of a snake robot.

This paper is organized as follows. The experimental setup and data acquisition system are described in Section 2, while the effect of the surface properties on motion and the effect of the surface on the directional stability are presented in Sections 3 and 4, respectively. Section 5 draws the conclusions of the paper.

2. The Experimental SETUP and Data Acquisition

A snake robot made of nine links, eight servo-motors, and a microcontroller, as shown in Figure 1, was used in this research. The microcontroller was programmed for controlling the motion of the servo-motors and propelling the snake in a serpentine motion. The snake's scales were attached at the ventral side of the snake robot, while markers were attached on the dorsal side to keep track of the motion of the robot.

Figure 1. The snake robot with markers on the dorsal side.

Four different types of scales were used to investigate the effect of scale geometry and the coefficient of friction on the motion of the snake robot. Photographs and the geometry of the scales are shown in Figure 2a–d. On an average, the span of the scales was 20 mm, except for the scale in Figure 2d. The scales shown in Figure 2a were fabricated from Acrylonitrile Butadiene Styrene (ABS) using a 3D printer while the scales shown in Figure 2b,c were made of ordinary plastic shells and externally threaded cylinder (Cylinder_1), splitting the cylinders into two halves. In Figure 2d, a scale made from an externally threaded cylinder with a lower friction factor than Cylinder_1 (we call it Cylinder_2) extending between the left and right edge of a link is shown.

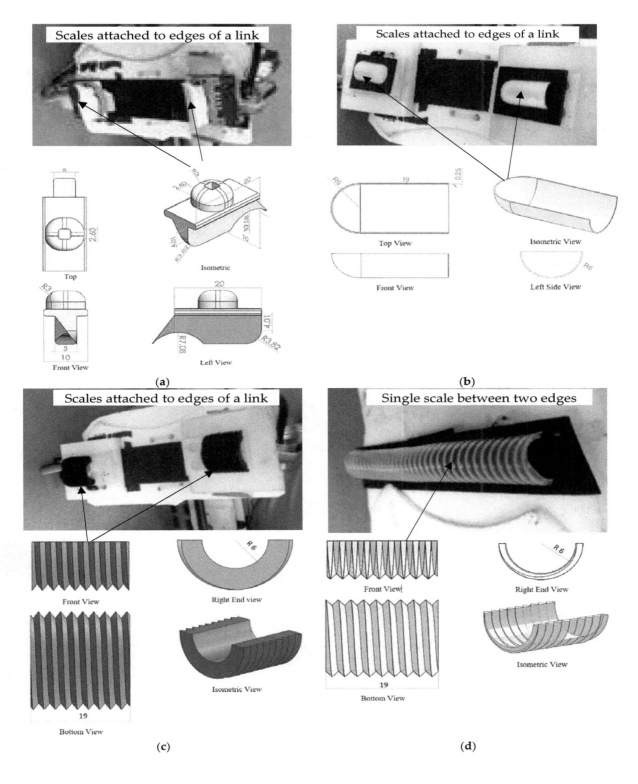

Figure 2. The photographs of the snake scales attached to the links of the snake robot and orthographic views of the scales (**a**) Designed scale; (**b**) Unthreaded scale (half-shell); (**c**) Threaded scale (Cylinder_1); (**d**) Threaded scale (Cylinder_2).

As the goal of this research is to investigate the effect of scale geometry on the motion of snake robots, the robot was equipped with one type of scale at a time. One pair of scales were attached at the lateral edges of each link of the robot. The snake robot was run on three different floor surfaces, namely, Surface 1 (Cloth), Surface 2 (artificial lather 1), and Surface 3 (artificial leather 2). The static friction coefficients between the snake scales and the surfaces are presented in Table 1. One of the main features of the snake scales was the anisotropic friction coefficients along the tangential and normal directions. Frictional anisotropy were achieved through the design of the snake scales of the snake robot.

Table 1. The static friction coefficients between the scales and different surfaces.

Scale/Surface	Fiction Coefficients								
	Surface 1			Surface 2			Surface 3		
	Tangential	Normal	Friction Ratio	Tangential	Normal	Friction Ratio	Tangential	Normal	Friction Ratio
Designed Scale	-	-	-	0.244	0.303	**1.243**	0.273	0.318	1.166
Half Shell	0.265	0.293	1.103	0.273	0.318	1.166	0.303	0.382	**1.261**
Threaded cylinder_1	0.310	0.399	1.286	0.187	0.273	**1.460**	0.201	0.288	1.433
Threaded cylinder_2	0.270	0.310	1.150	0.201	0.288	**1.433**	0.303	0.366	1.207

Figure 3 illustrates the robot orientation and the positions of the scales L1 through R4. Where 'L' denotes the left side and 'R' denotes the right side looking from the top of the robot. A Microsoft Kinect XBOX was attached on a frame above the plane of the motion to capture the motion and the shape of the robot during its motion. A photograph of the experimental setup is shown in Figure 4.

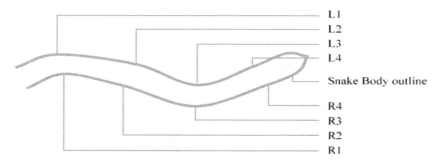

Figure 3. The snake's shape and scale placement (L and R stands for left and Right respectively).

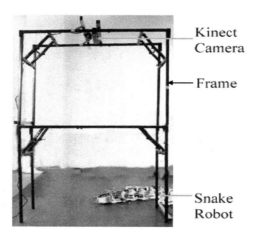

Figure 4. The experimental setup for the snake's motion capture.

3. The Effect of the Surface Properties on Motion

The otion data of the snake robot are presented in Table 2.

Table 2. The motion characteristics of the snake robot with different scales on different surfaces at different motor speeds.

Scale and Surface	High				Medium				Low			
	Cycle/s	Dis/Cy (cm)	V (cm/s)	θ (°)	Cycle/s	Dis/Cy (cm)	V (cm/s)	θ (°)	Cycle/s	Dis/Cy (cm)	V (cm/s)	θ (°)
Designed Scale on Surface 1	X	X	X	X	X	X	X	X	X	X	X	X
Designed Scale on Surface 2	0.483	3.667	1.770	2.471	0.437	3.066	1.338	0.620	0.348	2.521	0.878	1.986
Designed Scale on Surface 3	0.403	6.127	2.470	5.730	0.424	4.000	1.697	4.158	0.428	3.880	1.662	1.210
Half Shell on Surface 1	0.455	1.368	0.622	2.174	0.429	1.986	0.853	5.265	0.306	1.997	0.611	6.457
Half Shell on Surface 2	0.455	2.678	1.218	2.381	0.429	2.433	1.043	1.673	0.307	2.163	0.663	2.710
Half Shell on Surface 3	0.455	2.710	1.234	0.762	0.427	2.184	0.932	5.508	0.306	1.087	0.333	1.035
Thread Cylinder_1 on Surface 1	0.457	6.381	2.918	5.638	0.429	5.607	2.403	8.607	0.333	4.455	1.483	0.533
Thread Cylinder_1on Surface 2	0.479	3.561	1.707	0.634	0.429	3.041	1.306	1.719	0.333	3.300	1.099	5.330
Thread Cylinder_1on Surface 3	0.484	5.298	2.567	6.437	0.429	4.245	1.819	5.989	0.332	3.366	1.116	2.178
Thread Cylinder_2 on Surface 1	0.483	5.046	2.435	2.411	0.429	3.614	1.549	7.666	0.305	3.394	1.036	4.181
Thread Cylinder_2 on Surface 2	0.483	3.656	1.765	0.816	0.428	3.112	1.333	4.098	0.349	3.079	1.076	5.207
Thread Cylinder_2 on Surface 3	0.482	4.139	1.996	5.600	0.428	2.358	1.008	0.818	0.349	2.400	0.837	3.757
Thread Cylinder_2 at Center on Surface 2	0.484	0.622	0.301	5.398	-	-	-	-	-	-	-	-
Thread Cylinder_2 at Center on Surface 3	-	-	-	-	-	-	-	-	-	-	-	-

where, Cycle/s = the fraction of a complete cycle completed by the robot; Dis/Cy = the distance traveled by the robot in one cycle; V = the average velocity of the robot; θ = the angle between the initial and final orientations of the robot; High, medium, slow = the cycle speed of the robot.

Table 2 compares four motion elements like the number of cycles per second (seven cycles were taken for each observation), the distance traveled by the robot per cycle, the average velocity of the robot, and the deviation from the initial axis or orientation of the robot after each cycle corresponding to three different motor speeds, namely, high speed, medium speed, and low speed. The three types of motions (high, medium, and low) of the robot were the cycle speed of the sine wave those were created by controlling the speed of the 8 actuating motors. The symbol 'X' in the first row of this table means no data, that is, the robot was unable to move without damaging itself on surface 1 with the designed scale as the scale got stuck in the surface. The velocity of the robot should depend on the frequency of the sine wave of the snake robot. This is evident when the velocities of the robot under three types of motion are compared under the three types of motion. The highest velocity (2.918 cm/s) was achieved in the case of the cylinder_1 on surface 1 corresponding to the high speed of the motors. This surface combination also produces the highest velocity corresponding to the medium and low speed of the motors. The angular deviation of the robot θ from its initial orientation, in the case of the high and medium speeds, are 5.638° and 8.607°, respectively, which are quite high compared to the 0.533° that happened in the case of low speed. The friction ratio for this surface combination is lower than that of the cylinder_1 on surface 2, however, the coefficient of friction along the normal direction for cylinder_1 on surface 1 is the highest (μ = 0.399). Thus, it may be concluded that the higher value of the coefficient of friction along the normal direction plays a significant role in increasing the speed of the robot.

Among the four types of scales and the three types of surfaces, the friction coefficient of surface 2 gives the maximum possible friction ratios between the tangential and normal friction coefficient (Table 1), while the threaded cylinder_1 on surface 2 gives the highest friction ratio among them all. The comparison of the robot velocities of the designed scale on surface 2 and that of cylinder_1 on surface 2 showed very similar results. The robot velocities achieved in these cases are almost half that of cylinder_1 on surface 1 for all types of motor speed. However, the angular deviations are significantly low except for the low speed of the motors. The half shell on surface 3 has the highest movement with better stability (lowest angular deviation) than others, while the smooth half shell gives the minimum distance per cycle. It is noticed that the half shell had the highest friction ratio on surface 3. The last piece of data is a very important one, this data is for the threaded cylinder_2 on surface 2 (the one with the highest friction ratio) where the scales are placed at the mid surface (centerline) of the links instead of at the edges of the links. It shows that it does not move even a centimeter, yet, the same scale on the same surface gave the best output among all other combinations. Thus, it can be concluded that snake scales need to be placed at the edges of the links for the better motion of the robot, and it is unnecessary to place the snake scales along the centerline of the links.

4. The Effect of the Surface on Directional Stability

The poses of the robot after each cycle of motion are presented in Figure 5. Each line in these figures is the connecting line between the head and tail of the robot after each cycle of actuation. Thus, these lines are presenting the change in position as well as change in the orientation with respect to the x-y axis. The initial orientation of the robot was parallel to the x axis.

It is observed in Figure 5a–l that the surfaces and the scales have profound effects on the directional stability and motion for the movement of the robot.

Due to practical reason it was not possible to run the robot on surface 1 using the designed scale, thus, the data is absent. However, some important phenomena can be observed from the rest of the figures. All the scales that have threads on it (other than Figure 5c–e,l) are superior in terms of stability in the motion direction, as well as the distance covered. On top of that, when they move on the surface 1 (cloth) (Figure 5f,i) which has an anisotropic friction property within itself, cover higher distance. On the other hand, the half shell (Figure 5c–e) did not have much of a promising outcome in terms of the directional stability and velocity. On surface 2, the motion direction is better maintained than that on surface 3.

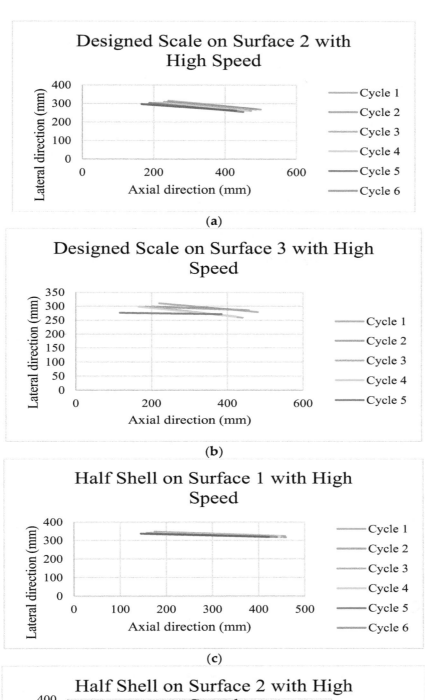

(a)

(b)

(c)

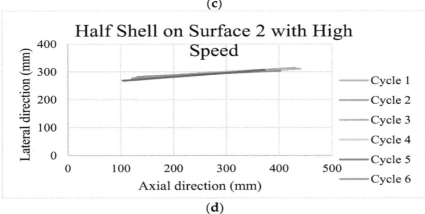

(d)

Figure 5. *Cont.*

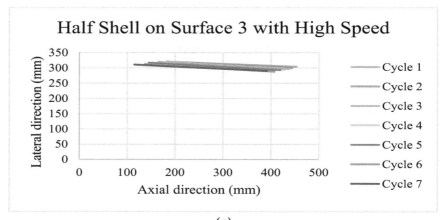

(e)

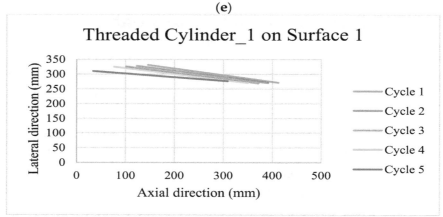

(f)

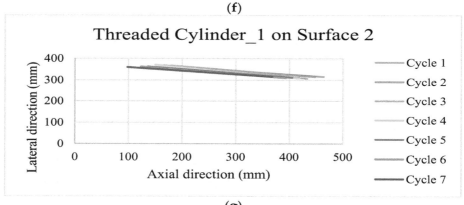

(g)

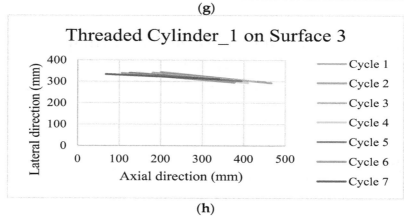

(h)

Figure 5. *Cont.*

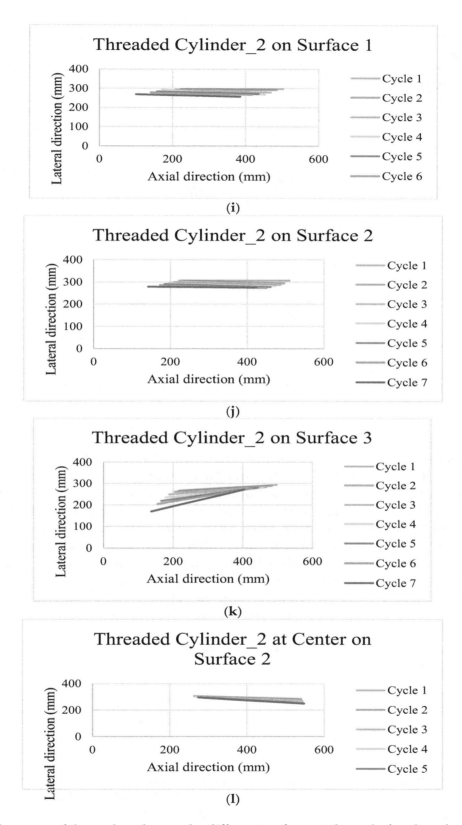

Figure 5. The poses of the snake robot on the different surfaces at the end of each cycle of actuation.

All the orientations of the snake robot after the last cycle of actuation found in Figure 5a–l are presented in Figure 6. A very important finding that is observed in the last set of data in Figure 5l is shown circled in red in Figure 6. This is the situation when the scales were placed in the middle of the snake body leaving the edges of the link free and floating. There is no significant movement as well as no control of direction in this case. In fact, the robot moved in the backward direction instead of moving forward, however, the same threaded cylinder 2 with the scale at the outer edges of the links gave better robot motion.

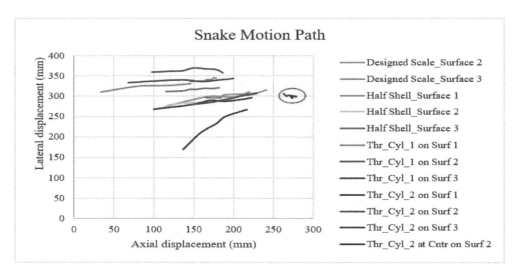

Figure 6. The comparison of the motion paths of the snake head on different surfaces.

5. Conclusions

This work makes an effort to discover the relationship between the different variables of the snake scales, like friction ratio, scale geometry, and the span of the scales beneath a snake robot that undergoes serpentine motion. The comparison of the motion shows that scales with a higher friction ratio set at the outer edge of the snake robot helps to generate better and more stable motion. On the other hand, scales only at the center of the belly, irrespective of the friction ratio, does not help the robot to generate any good motion or any forward motion at all. Scales have a ridge like texture on it along the direction of the snake body, which, when placed at the outer edge of the belly, helps generate stable and faster motion. It is predictable from the above experiments that the higher friction ratio will help achieve higher and better motion. Motions with the actuating motors at lower cycle speeds with a scale-surface pair having a lower friction ratio are very unpredictable in nature.

Author Contributions: Writing –Original Draft Preparation, Naim Md. Lutful Huq; Supervision, Funding Acquisition & Project administration, Md. Raisuddin Khan and Amir Akramin Shafie; Software, Md. Masum Billah; Methodology, Syed Masrur Ahmmad.

Acknowledgments: The authors profoundly acknowledge the Ministry of Higher Education (MOE), Malaysia for funding this research through the Fundamental Research Grant Scheme (FRGS).

References

1. Casper, J.; Murphy, R.R. Human-robot interactions during the robot-assisted urban search and rescue response at the world trade center. *IEEE Trans. Syst. Man Cybern. Part B Cybern.* **2003**, *33*, 367–385. [CrossRef] [PubMed]
2. Wolf, A.; Brown, H.B.; Casciola, R.; Costa, A.; Schwerin, M.; Shamas, E.; Choset, H. A mobile hyper redundant mechanism for search and rescue tasks. In Proceedings of the 2003 IEEE/RSJ International Conference on Intelligent Robots and Systems, Las Vegas, NV, USA, 27–31 October 2003; Volume 3, pp. 2889–2895.
3. Liljeback, P.; Stavdahl, O.; Beitnes, A. SnakeFighter-development of a water hydraulic firefighting snake robot. In Proceedings of the 9th International Conference on Control, Automation, Robotics and Vision, ICARCV'06, Singapore, 5–8 December 2006; pp. 1–6.

4. Cai, L.; Zhang, R. Design and Research of Intelligent Fire-Fighting Robot. *Adv. Mater. Res.* **2013**, *823*, 358–362. [CrossRef]

5. Ho, C.; Chen, M.; Lien, C. Machine Vision-Based Intelligent Fire Fighting Robot. *Key Eng. Mater.* **2010**, *450*, 312–315. [CrossRef]

6. Degani, A.; Choset, H.; Wolf, A.; Zenati, M.A. Highly articulated robotic probe for minimally invasive surgery. In Proceedings of the 2006 IEEE International Conference on Robotics and Automation, ICRA 2006, Orlando, FL, USA, 15–19 May 2006; pp. 4167–4172.

7. Tully, S.; Kantor, G.; Zenati, M.A.; Choset, H. Shape estimation for image-guided surgery with a highly articulated snake robot. In Proceedings of the 2011 IEEE/RSJ International Conference on Intelligent Robots and Systems (IROS), San Francisco, CA, USA, 25–30 September 2011; pp. 1353–1358.

8. Granosik, G.; Hansen, M.G.; Borenstein, J. The OmniTread serpentine robot for industrial inspection and surveillance. *Ind. Robot* **2005**, *32*, 139–148. [CrossRef]

9. Granosik, G.; Borenstein, J.; Hansen, M.G. Serpentine robots for industrial inspection and surveillance. In *Industrial Robotics: Programming, Simulation and Applications*; I-Tech Publ.: London, UK, 2007; pp. 633–662.

10. Gray, J. The mechanism of locomotion in snakes. *J. Exp. Biol.* **1946**, *23*, 101–120. [PubMed]

11. Hu, D.L.; Nirody, J.; Scott, T.; Shelley, M.J. The mechanics of slithering locomotion. *Proc. Natl. Acad. Sci. USA* **2009**, *106*, 10081–10085. [CrossRef] [PubMed]

12. Jayne, B.C. Muscular mechanisms of snake locomotion: An electromyographic study of the sidewinding and concertina modes of *Crotaluscerastes, Nerodiafasciata* and *Elapheobsoleta*. *J. Exp. Biol.* **1988**, *140*, 1–33. [PubMed]

13. Jayne, B.C. Muscular mechanisms of snake locomotion: An electromyographic study of lateral undulation of the Florida banded water snake (*Nerodiafasciata*) and the yellow rat snake (*Elapheobsoleta*). *J. Morphol.* **1988**, *197*, 159–181. [CrossRef] [PubMed]

14. Miller, G.S. The motion dynamics of snakes and worms. In Proceedings of the ACM Siggraph Computer Graphics and Interactive Techniques, Atlanta, GA, USA, 1–5 August 1988; Volume 22, pp. 169–173.

15. Jayne, B.C.; Davis, J.D. Kinematics and performance capacity for the concertina locomotion of a snake (*Coluber constrictor*). *J. Exp. Biol.* **1991**, *156*, 539–556.

16. Marvi, H.; Hu, D.L. Friction enhancement in concertina locomotion of snakes. *J. R. Soc. Interface* **2012**, *9*, 3067–3080. [CrossRef] [PubMed]

17. Hirose, S. *Biologically Inspired Robots*; Oxford University Press: Oxford, UK, 1993.

18. Wang, X.; Osborne, M.T.; Alben, S. Optimizing snake locomotion on an inclined plane. *Phys. Rev. E* **2014**, *89*. [CrossRef] [PubMed]

19. Jing, F.; Alben, S. Optimization of two-and three-link snakelike locomotion. *Phys. Rev. E* **2013**, *87*. [CrossRef] [PubMed]

20. Kyriakopoulos, K.J.; Migadis, G.; Sarrigeorgidis, K. The NTUA snake: Design, planar kinematics, and motion planning. *J. Robot. Syst.* **1999**, *16*, 37–72. [CrossRef]

21. Prautsch, P.; Mita, T. Control and analysis of the gait of snake robots. In Proceedings of the 1999 IEEE International Conference on Control Applications, Kohala Coast, HI, USA, 22–27 August 1999; Volume 1, pp. 502–507.

22. Khan, R.; Watanabe, M.; Shafie, A.A. Kinematics model of snake robot considering snake scale. *Am. J. Appl. Sci.* **2010**, *7*, 669. [CrossRef]

23. Marvi, H.; Gong, C.; Gravish, N.; Astley, H.; Travers, M.; Hatton, R.L.; Mendelson, J.R., III; Choset, H.; Hu, D.L.; Goldman, D.I. Sidewinding with minimal slip: Snake and robot ascent of sandy slopes. *Science* **2014**, *346*, 224–229. [CrossRef] [PubMed]

24. Khan, R.; Billah, M.; Huq, N.; Helmi, M.; Ahmmed, S. Investigation on Data Extraction Trends for Snake Robot. *Procedia Comput. Sci.* **2014**, *42*, 145–152. [CrossRef]

25. Varesis, O.; Diamantopoulos, C.; Tzes, A. Experimental studies of serpentine motion control of snake robots at inclined planes. In Proceedings of the 2016 24th Mediterranean Conference on Control and Automation (MED), Athens, Greece, 21–24 June 2016; pp. 737–742.

6

Spacecraft Robot Kinematics using Dual Quaternions

Alfredo Valverde *,† **and Panagiotis Tsiotras** †

School of Aerospace Engineering, Georgia Institute of Technology, Atlanta, GA 30313, USA; tsiotras@gatech.edu
* Correspondence: avalverde3@gatech.edu
† These authors contributed equally to this work.

Abstract: In recent years, there has been a growing interest in servicing orbiting satellites. In most cases, in-orbit servicing relies on the use of spacecraft-mounted robotic manipulators to carry out complicated mission objectives. Dual quaternions, a mathematical tool to conveniently represent pose, has recently been adopted within the space industry to tackle complex control problems during the stages of proximity operations and rendezvous, as well as for the dynamic modeling of robotic arms mounted on a spacecraft. The objective of this paper is to bridge the gap in the use of dual quaternions that exists between the fields of spacecraft control and fixed-base robotic manipulation. In particular, we will cast commonly used tools in the field of robotics as dual quaternion expressions, such as the Denavit-Hartenberg parameterization, or the product of exponentials formula. Additionally, we provide, via examples, a study of the kinematics of different serial manipulator configurations, building up to the case of a completely free-floating robotic system. We provide expressions for the dual velocities of the different types of joints that commonly arise in industrial robots, and we end by providing a collection of results that cast convex constraints commonly encountered by space robots during proximity operations in terms of dual quaternions.

Keywords: spacecraft; robotics; dual quaternions; kinematics

1. Introduction

Robots are increasingly present in our daily lives, with their many uses ranging from simple vacuuming devices to complex manufacturing robotic arms. This growth is sustained by the continuous development of faster and better software and hardware, as well as strong theoretical advances in the areas of kinematics, dynamics, computer vision, sensing, etc. The space industry, owing to obvious reasons having to do with the unfriendliness of the space environment to humans, relies heavily on the use of robotics systems. In fact, interplanetary robotic exploration is at the core of NASA's Jet Propulsion Laboratory, and a wide variety of companies and governmental agencies are currently developing space-rated robotic manipulation systems for on-orbit satellite servicing [1,2].

The field of robotics is a well-established one. Lately, in the field of fixed-base robotics, progress in the area of kinematics and dynamics has mainly focused on ease of use, and speed and performance improvements [3,4]. The combination between the study of robots and their use in space, i.e., space robotics, must find common ground between the techniques used in both. For example, quaternions are the representation of choice when it comes to attitude parameterization for spacecraft control and estimation, while SE(2)/SE(3) and the Spatial Vector Algebra [5] are the dominant tools of choice in the fixed-base robotic community. Therefore, with the recent advent of dual quaternions, it is only natural to explore the use of a pose (i.e., position and attitude) representation tool for spacecraft control and estimation in order to study robotic systems mounted on a spacecraft.

Dual quaternion algebra is an extension of the well-known quaternion algebra. The former is used to study rigid body pose while the latter is used extensively to study just the attitude of a rigid body. Dual quaternions have recently seen a proliferation in their use for spacecraft control [6–9]. Several factors have contributed to the recent interest in dual quaternions in spacecraft control. First, the similarities between quaternion and dual quaternion-based spacecraft controllers and estimators [10] make dual quaternions an appealing tool for the practitioner who is familiar with the (standard) quaternion algebra. Next, dual quaternions naturally encode position information, thus avoiding the artificial separation of rotational and translational motion during control, which becomes essential during proximity operations or robotic servicing missions. More recently, dual quaternions have been used extensively for the dynamic modeling of ground-based robotic manipulators, providing an even stronger argument towards their use for dynamic modeling of spacecraft-mounted robotic manipulators [11].

While dual quaternions have been used for serial robot kinematic design [12–14], and for kinematic manipulation of points, vectors, lines, screws and planes [15], few references incorporate velocity information into their study of kinematics. Leclercq et al. [16] studied robot kinematics in the context of human motion using dual quaternions, yielding one of the most complete references to study kinematic chains with dual quaternions. More recently, Quiroz-Omaña and Adorno [17] have made use of dual quaternions in the context of robotic manipulation on a non-holonomic base. The methodologies exhibited in [16,17], however, do not take advantage of the well-known and convenient dual quaternion expression for kinematics, which could avoid manually taking time derivatives of pose expressions. A possible reason for this is the lack of a systematic manner to represent the combined linear and angular velocities of joints in dual algebra, and instead relying on the explicit derivative of pose-like expressions.

Works in the fields of dynamics and spacecraft control have settled on an understanding of the construction of dual velocities [18,19], which can be extended to provide generic expressions for the dual velocities of rigid bodies, or even of the different types of joints that may appear in a serial kinematic chain. In fact, Özgür and Mezouar [20] make use of said representation of dual velocity, commonly given by an expression of the form $\boldsymbol{\omega} = \omega + \epsilon v$, to perform kinematic control on a robotic arm, yielding a clever representation of the Jacobian matrix that uses dual quaternion screws. Their approach, however, has a fixed base and requires the use of base-frame coordinates—as opposed to body-frame coordinates, which are commonly used in the study of spacecraft motion—to describe the Plücker lines associated to the different joints of the system.

Given the significant interest that dual quaternions have garnered in the last decade in the realm of space applications, it is pertinent to contribute to the literature a straightforward treatment of kinematics with an emphasis on space-based robotic operations. In this paper, we aim to extend the study of robot kinematics using dual quaternions, mainly by lifting the condition that the robotic base must be fixed, allowing it instead to move freely in the three-dimensional space. Additionally, we consider the possibility of incorporating different types of joints, and provide the formulas for the dual velocity of each different type of joint. Along the way, we provide some important well-known results, such as the derivation of the famous quaternion kinematic law, and the aforementioned dual quaternion equivalent, as well as a collection of results that capture convex constraints using dual quaternions. While the latter expressions have been used in the field of Entry, Descent, and Landing (EDL), their incorporation in robotic manipulation for in-orbit servicing missions is also extremely beneficial in order to ensure safety and robustness.

This paper is structured as follows: Section 2 provides the mathematical tools necessary to use quaternion and dual quaternion algebras. Section 3 provides an overview of the most common kinematic tools in the robotics fields in dual quaternion form. Section 4 provides the development of the kinematic equations of motion using dual quaternions, and in Section 5 we provide a brief summary of some important constraint expressions for robotic manipulation cast using dual quaternions. Such constraints arise naturally in many in-orbit servicing missions. Addressing these constraints in a numerically efficient manner (e.g., casting them as convex constraints) leads to safe and elegant solutions of the in-orbit servicing problem.

2. Mathematical Preliminaries

In this section we give an introduction to quaternion and dual quaternion algebras, which provide convenient mathematical frameworks for attitude and pose representations respectively. Next, we provide the theoretical foundations required to study the kinematics of rigid bodies, and in particular how they pertain to serial manipulators.

2.1. Quaternions

The group of quaternions, as defined by Hamilton in 1843, extends the well-known imaginary unit j, which satisfies $j^2 = -1$. This non-abelian group is defined by $\mathbb{Q}_8 \triangleq \{-1, i, j, k : i^2 = j^2 = k^2 = ijk = -1\}$. The algebra constructed from $\mathbb{Q}_8$ over the field of real numbers is the quaternion algebra defined as $\mathbb{H} \triangleq \{q = q_0 + q_1 i + q_2 j + q_3 k : i^2 = j^2 = k^2 = ijk = -1, q_0, q_1, q_2, q_3 \in \mathbb{R}\}$. This defines an associative, non-commutative, division algebra.

In practice, quaternions are often referred to by their scalar and vectors parts as $q = (q_0, \bar{q})$, where $q_0 \in \mathbb{R}$ and $\bar{q} = [q_1, q_2, q_3]^\mathsf{T} \in \mathbb{R}^3$. The properties of the quaternion algebra are summarized in Table 1. Filipe and Tsiotras [7] also conveniently define a multiplication between real 4-by-4 matrices and quaternions, denoted by the $*$ operator, which resembles the well-known matrix-vector multiplication by simply representing the quaternion coefficients as a vector in $\mathbb{R}^4$. In other words, given $a = (a_0, \bar{a}) \in \mathbb{H}$ and a matrix $M \in \mathbb{R}^{4\times4}$ defined as

$$M = \begin{bmatrix} M_{11} & M_{12} \\ M_{21} & M_{22} \end{bmatrix}, \tag{1}$$

where $M_{11} \in \mathbb{R}, M_{12} \in \mathbb{R}^{1\times3}, M_{21} \in \mathbb{R}^{3\times1}$ and $M_{22} \in \mathbb{R}^{3\times3}$, then

$$M * a \triangleq (M_{11}a_0 + M_{12}\bar{a}, M_{21}a_0 + M_{22}\bar{a}) \in \mathbb{H}. \tag{2}$$

Table 1. Quaternion Operations.

Operation	Definition
Addition	$a + b = (a_0 + b_0, \bar{a} + \bar{b})$
Scalar multiplication	$\lambda a = (\lambda a_0, \lambda a)$
Multiplication	$ab = (a_0 b_0 - \bar{a} \cdot \bar{b}, a_0\bar{b} + b_0\bar{a} + \bar{a} \times \bar{b})$
Conjugate	$a^* = (a_0, -\bar{a})$
Dot product	$a \cdot b = (a_0 b_0 + \bar{a} \cdot \bar{b}, 0_{3\times1}) = \frac{1}{2}(a^*b + b^*a)$
Cross product	$a \times b = (0, a_0\bar{b} + b_0\bar{a} + \bar{a} \times \bar{b}) = \frac{1}{2}(ab - b^*a^*)$
Norm	$\|a\| = \sqrt{a \cdot a}$
Scalar part	$\mathrm{sc}(a) = (a_0, 0_{3\times1})$
Vector part	$\mathrm{vec}(a) = (0, \bar{a})$

Since any rotation can be described by three parameters, the unit norm constraint is imposed on quaternions for attitude representation. *Unit* quaternions are closed under multiplication, but not under addition. A quaternion describing the orientation of frame X with respect to frame Y, denoted by $q_{\mathsf{X/Y}}$, satisfies $q^*_{\mathsf{X/Y}}q_{\mathsf{X/Y}} = q_{\mathsf{X/Y}}q^*_{\mathsf{X/Y}} = 1$, where $1 \triangleq (1, \bar{0}_{3\times1})$. This quaternion can be constructed as $q_{\mathsf{X/Y}} = (\cos(\phi/2), \bar{n}\sin(\theta/2))$, where $\bar{n}$ and θ are the *unit* Euler axis, and Euler angle of the rotation respectively. It is worth emphasizing that $q^*_{\mathsf{Y/X}} = q_{\mathsf{X/Y}}$, and that $q_{\mathsf{X/Y}}$ and $-q_{\mathsf{X/Y}}$ represent the same rotation. Furthermore, given quaternions $q_{\mathsf{Y/X}}$ and $q_{\mathsf{Z/Y}}$, the quaternion describing the rotation from X to Z is given by $q_{\mathsf{Z/X}} = q_{\mathsf{Y/X}}q_{\mathsf{Z/Y}}$. For completeness purposes, we define $0 \triangleq (0, \bar{0}_{3\times1})$.

Three-dimensional vectors can also be interpreted as special cases of quaternions. Specifically, given $\bar{s}^\mathsf{X} \in \mathbb{R}^3$, the coordinates of a vector expressed in frame X, its quaternion representation is given by $s^{\times} = (0, \bar{s}^{\times}) \in \mathbb{H}^v$, where $\mathbb{H}^v$ is the set of *vector* quaternions defined as $\mathbb{H}^v \triangleq \{(q_0, \bar{q}) \in \mathbb{H} : q_0 = 0\}$ (see Reference [19] for further information). The change of the reference frame for a vector quaternion is achieved by the adjoint operation, and is given by $s^{\scriptscriptstyle Y} = q_{\scriptscriptstyle Y/X}^* s^{\times} q_{\scriptscriptstyle Y/X}$. Additionally, given $s \in \mathbb{H}^v$, we can define the operation $[\,\cdot\,]^{\times} : \mathbb{H}^v \to \mathbb{R}^{4 \times 4}$ as

$$[s]^{\times} = \begin{bmatrix} 0 & 0_{1 \times 3} \\ 0_{3 \times 1} & [\bar{s}]^{\times} \end{bmatrix}, \quad \text{where} \quad [\bar{s}]^{\times} = \begin{bmatrix} 0 & -s_3 & s_2 \\ s_3 & 0 & -s_1 \\ -s_2 & s_1 & 0 \end{bmatrix}. \tag{3}$$

For quaternions $a = (a_0, \bar{a})$ and $b = (b_0, \bar{b})$, the left and right quaternion multiplication operators $[\![\cdot]\!]_{\mathrm{L}}, [\![\cdot]\!]_{\mathrm{R}} : \mathbb{H} \to \mathbb{R}^{4 \times 4}$ will be defined as

$$[\![a]\!]_{\mathrm{L}} * b \triangleq [\![b]\!]_{\mathrm{R}} * a \triangleq ab, \tag{4}$$

where

$$[\![a]\!]_{\mathrm{L}} = \begin{bmatrix} a_0 & -a_1 & -a_2 & -a_3 \\ \hline a_1 & a_0 & -a_3 & a_2 \\ a_2 & a_3 & a_0 & -a_1 \\ a_3 & -a_2 & a_1 & a_0 \end{bmatrix} = \begin{bmatrix} a_0 & -\bar{a}^{\mathsf{T}} \\ \bar{a} & a_0 \mathbb{I}_3 + [\bar{a}]^{\times} \end{bmatrix}, \tag{5}$$

$$[\![b]\!]_{\mathrm{R}} = \begin{bmatrix} b_0 & -b_1 & -b_2 & -b_3 \\ \hline b_1 & b_0 & b_3 & -b_2 \\ b_2 & -b_3 & b_0 & b_1 \\ b_3 & b_2 & -b_1 & b_0 \end{bmatrix} = \begin{bmatrix} b_0 & -\bar{b}^{\mathsf{T}} \\ \bar{b} & b_0 \mathbb{I}_3 - [\bar{b}]^{\times} \end{bmatrix}. \tag{6}$$

2.2. Dual Quaternions

We define the dual quaternion group as

$$\mathbb{Q}_d := \{-1, i, j, k, \epsilon, \epsilon i, \epsilon j, \epsilon k : i^2 = j^2 = k^2 = ijk = -1, \tag{7}$$
$$\epsilon i = i\epsilon, \epsilon j = j\epsilon, \epsilon k = k\epsilon, \epsilon \neq 0, \epsilon^2 = 0\}.$$

The dual quaternion algebra arises as the algebra of the dual quaternion group $\mathbb{Q}_d$ over the field of real numbers, and is denoted as $\mathbb{H}_d$. When dealing with the modeling of mechanical systems, it is convenient to present this algebra as $\mathbb{H}_d = \{q = q_r + \epsilon q_d : q_r, q_d \in \mathbb{H}\}$, where ϵ is the dual unit. We call q_r the real part, and q_d the dual part of the dual quaternion q.

Filipe and Tsiotras [7,8,19,21] have laid out much of the groundwork in terms of the notation and basic properties of dual quaternions for spacecraft problems. The main properties of the dual quaternion algebra are listed in Table 2. Filipe and Tsiotras [7] also conveniently define a multiplication between matrices and dual quaternions, denoted by the $\star$ operator, that resembles the well-known real matrix-vector multiplication by simply representing the dual quaternion coefficients as a vector in $\mathbb{R}^8$. In other words, given $a = a_r + \epsilon a_d \in \mathbb{H}_d$ and a matrix $M \in \mathbb{R}^{8 \times 8}$ defined as

$$M = \begin{bmatrix} M_{11} & M_{12} \\ M_{21} & M_{22} \end{bmatrix}, \tag{8}$$

where $M_{11}, M_{12}, M_{21}, M_{22} \in \mathbb{R}^{4 \times 4}$, then

$$M \star a \triangleq (M_{11} * a_r + M_{12} * a_d) + \epsilon(M_{21} * a_r + M_{22} * a_d) \in \mathbb{H}_d. \tag{9}$$

Table 2. Dual Quaternion Operations.

Operation	Definition
Addition	$a + b = (a_r + b_r) + \epsilon(a_d + b_d)$
Scalar multiplication	$\lambda a = (\lambda a_r) + \epsilon(\lambda a_d)$
Multiplication	$ab = (a_r b_r) + \epsilon(a_d b_r + a_r b_d)$
Conjugate	$a^* = (a_r^*) + \epsilon(a_d^*)$
Dot product	$a \cdot b = (a_r \cdot b_r) + \epsilon(a_d \cdot b_r + a_r \cdot b_d) = \frac{1}{2}(a^* b + b^* a)$
Cross product	$a \times b = (a_r \times b_r) + \epsilon(a_d \times b_r + a_r \times b_d) = \frac{1}{2}(ab - b^* a^*)$
Circle product	$a \circ b = (a_r \cdot b_r + a_d \cdot b_d) + \epsilon 0$
Swap	$a^s = a_d + \epsilon a_r$
Norm	$\|a\| = \sqrt{a \circ a}$
Scalar part	$\mathrm{sc}(a) = \mathrm{sc}(a_r) + \epsilon \mathrm{sc}(a_d)$
Vector part	$\mathrm{vec}(a) = \mathrm{vec}(a_r) + \epsilon \mathrm{vec}(a_d)$

Analogous to the set of vector quaternions $\mathbb{H}^v$, we can define the set of vector dual quaternions as $\mathbb{H}_d^v \triangleq \{q = q_r + \epsilon q_d : q_r, q_d \in \mathbb{H}^v\}$. For vector dual quaternions we will define the skew-symmetric operator $[\,\cdot\,]^\times : \mathbb{H}_d^v \to \mathbb{R}^{8\times 8}$,

$$[s]^\times = \begin{bmatrix} [s_r]^\times & 0_{4\times 4} \\ [s_d]^\times & [s_r]^\times \end{bmatrix}. \tag{10}$$

For dual quaternions $a = a_r + \epsilon a_d$ and $b = b_r + \epsilon b_d \in \mathbb{H}_d$, the left and right dual quaternion multiplication operators $[\![\,\cdot\,]\!]_{\mathrm{L}}, [\![\,\cdot\,]\!]_{\mathrm{R}} : \mathbb{H}_d \to \mathbb{R}^{8\times 8}$ are defined as

$$ab \triangleq [\![a]\!]_{\mathrm{L}} \star b \triangleq [\![b]\!]_{\mathrm{R}} \star a, \tag{11}$$

where

$$[\![a]\!]_{\mathrm{L}} = \begin{bmatrix} [\![a_r]\!]_{\mathrm{L}} & 0_{4\times 4} \\ [\![a_d]\!]_{\mathrm{L}} & [\![a_r]\!]_{\mathrm{L}} \end{bmatrix} \quad \text{and} \quad [\![b]\!]_{\mathrm{R}} = \begin{bmatrix} [\![b_r]\!]_{\mathrm{R}} & 0_{4\times 4} \\ [\![b_d]\!]_{\mathrm{R}} & [\![b_r]\!]_{\mathrm{R}} \end{bmatrix}. \tag{12}$$

Since rigid body motion has six degrees of freedom, a dual quaternion needs two constraints to parameterize it. The dual quaternion describing the relative pose of frame B relative to frame I is given by $q_{\mathrm{B/I}} = q_{\mathrm{B/I},r} + \epsilon q_{\mathrm{B/I},d} = q_{\mathrm{B/I}} + \epsilon \frac{1}{2} q_{\mathrm{B/I}} r_{\mathrm{B/I}}^{\mathrm{B}}$, where $r_{\mathrm{B/I}}^{\mathrm{B}}$ is the position quaternion describing the location of the origin of frame B relative to that of frame I, expressed in B-frame coordinates. It can be easily observed that $q_{\mathrm{B/I},r} \cdot q_{\mathrm{B/I},r} = 1$ and $q_{\mathrm{B/I},r} \cdot q_{\mathrm{B/I},d} = 0$, where $0 = (0, \bar{0})$, providing the two necessary constraints. Thus, a dual quaternion representing a pose transformation is a *unit* dual quaternion, since it satisfies $q \cdot q = q^* q = 1$, where $1 \triangleq 1 + \epsilon 0$. Additionally, we also define $0 \triangleq 0 + \epsilon 0$.

Similar to the standard quaternion relationships, the frame transformations laid out in Table 3 can be easily verified.

Table 3. Unit Dual Quaternion Operations.

Composition of transformations	$q_{\mathrm{Z/X}} = q_{\mathrm{Y/X}} q_{\mathrm{Z/Y}}$
Inverse, Conjugate	$q_{\mathrm{Y/X}}^* = q_{\mathrm{X/Y}}$

In Reference [19] it was proven that for a dual unit quaternion $q \in \mathbb{H}_d$, q and $-q$ represent the same frame transformation, property inherited from the space of quaternions. Therefore, as is done in practice for quaternions, dual quaternions can be subjected to properization, which is the action of redefining a dual quaternion so that the scalar part of the quaternion is always positive. Formally, we can define the properization of a dual quaternion $q = q_r + \epsilon q_d$ as

$$q := -q \quad \text{if} \quad (q_r)_0 < 0, \tag{13}$$

where $(q_r)_0$ is the scalar part of q_r. Just like in the case of quaternions, dual quaternions also inherit the so-called *unwinding phenomenon*, first described in [22], which is most important in control applications.

A useful equation is the generalization of the velocity of a rigid body in dual form, which contains both the linear and angular velocity components. The dual velocity of the Y-frame with respect to the Z-frame, expressed in X-frame coordinates, is defined as

$$\omega_{Y/Z}^X = q_{X/Y}^* \omega_{Y/Z}^Y q_{X/Y} = \omega_{Y/Z}^X + \epsilon(v_{Y/Z}^X + \omega_{Y/Z}^X \times r_{X/Y}^X), \tag{14}$$

where $\omega_{Y/Z}^X = (0, \bar{\omega}_{Y/Z}^X)$ and $v_{Y/Z}^X = (0, \bar{v}_{Y/Z}^X)$, $\bar{\omega}_{Y/Z}^X$ and $\bar{v}_{Y/Z}^X \in \mathbb{R}^3$ are respectively the angular and linear velocity of the Y-frame with respect to the Z-frame expressed in X-frame coordinates, and $r_{X/Y}^X = (0, \bar{r}_{X/Y}^X)$, where $\bar{r}_{X/Y}^X \in \mathbb{R}^3$ is the position vector from the origin of the Y-frame to the origin of the X-frame expressed in X-frame coordinates. In particular, from Equation (14) we observe that the dual velocity of a rigid body assigned to frame B with respect to an inertial frame I, expressed in B-frame coordinates is given as $\omega_{B/I}^B = \omega_{B/I}^B + \epsilon v_{B/I}^B$. However, if we wanted to express this same dual velocity in inertial frame coordinates, as per Equation (14) we would get $\omega_{B/I}^I = \omega_{B/I}^I + \epsilon(v_{B/I}^I + \omega_{B/I}^I \times r_{I/B}^I)$. We will formally introduce frame transformations next.

2.3. Frame Transformations Using Dual Quaternions

As is common in the study of kinematics, frame transformations are vital for the determination of velocities and accelerations with respect to different frames. A dual velocity, or dual acceleration, can be described by a dual vector quaternion $s^X \in \mathbb{H}_d^v$ expressed in X-frame coordinates as $s^X \triangleq s_r^X + \epsilon s_d^X$, where s_r^X, $s_d^X \in \mathbb{H}^v$. As noted for Equation (14), frame transformations are given by the adjoint operation as

$$\begin{aligned}
s^Y &= q_{Y/X}^* s^X q_{Y/X} \\
&= (q_{Y/X} + \epsilon\tfrac{1}{2}r_{Y/X}^X q_{Y/X})^*(s_r^X + \epsilon s_d^X)(q_{Y/X} + \epsilon\tfrac{1}{2}r_{Y/X}^X q_{Y/X}) \\
&= (q_{Y/X}^* + \epsilon\tfrac{1}{2}q_{Y/X}^* r_{Y/X}^{X*})(s_r^X + \epsilon s_d^X)(q_{Y/X} + \epsilon\tfrac{1}{2}r_{Y/X}^X q_{Y/X}) \\
&= (q_{Y/X}^* - \epsilon\tfrac{1}{2}q_{Y/X}^* r_{Y/X}^X)(s_r^X + \epsilon s_d^X)(q_{Y/X} + \epsilon\tfrac{1}{2}r_{Y/X}^X q_{Y/X}) \\
&= (q_{Y/X}^* - \epsilon\tfrac{1}{2}q_{Y/X}^* r_{Y/X}^X)(s_r^X q_{Y/X} + \epsilon(s_d^X q_{Y/X} + s_r^X \tfrac{1}{2}r_{Y/X}^X q_{Y/X})) \\
&= q_{Y/X}^* s_r^X q_{Y/X} - \epsilon(\tfrac{1}{2}q_{Y/X}^* r_{Y/X}^X s_r^X q_{Y/X}) + \epsilon(q_{Y/X}^* s_d^X q_{Y/X} + q_{Y/X}^* s_r^X \tfrac{1}{2}r_{Y/X}^X q_{Y/X}) \\
&= s_r^Y + \epsilon(s_d^Y + \tfrac{1}{2}q_{Y/X}^* s_r^X q_{Y/X} q_{Y/X}^* r_{Y/X}^X q_{Y/X} - \tfrac{1}{2}q_{Y/X}^* r_{Y/X}^X q_{Y/X} q_{Y/X}^* s_r^X q_{Y/X}) \\
&= s_r^Y + \epsilon(s_d^Y + \tfrac{1}{2}s_r^Y r_{Y/X}^Y - \tfrac{1}{2}r_{Y/X}^Y s_r^Y) \\
&= s_r^Y + \epsilon(s_d^Y + \tfrac{1}{2}s_r^Y r_{Y/X}^Y - \tfrac{1}{2}(r_{Y/X}^Y)^*(s_r^Y)^*).
\end{aligned}$$

By the definition of the cross product of two quaternion quantities given in Table 1, we get that

$$\begin{aligned}
s^Y &= q_{Y/X}^* s^X q_{Y/X} \\
&= s_r^Y + \epsilon(s_d^Y + s_r^Y \times r_{Y/X}^Y) \\
&= s_r^Y + \epsilon(s_d^Y + r_{X/Y}^Y \times s_r^Y).
\end{aligned} \tag{15}$$

Analogously, the transformation of a dual vector $s^Y \triangleq s_r^Y + \epsilon s_d^Y$ can be easily derived using the procedure described above to be:

$$s^x = q_{Y/X} s^Y q_{Y/X}^*$$
$$= s_r^x + \epsilon(s_d^x + s_r^x \times r_{X/Y}^x) \tag{16}$$
$$= s_r^x + \epsilon(s_d^x + r_{Y/X}^x \times s_r^x).$$

As is standard notation, we can define the group adjoint operation for *unit* dual quaternions as

$$\mathrm{Ad}_q s \triangleq qsq^{-1} = qsq^*. \tag{17}$$

Therefore, using this notation, the frame transformations derived above can be cast as

$$s^x = \mathrm{Ad}_{q_{Y/X}} s^Y \tag{18}$$
$$s^Y = \mathrm{Ad}_{q_{Y/X}^*} s^x = \mathrm{Ad}_{q_{X/Y}} s^x \tag{19}$$

The power of dual quaternions goes beyond the ability to represent pose and transform dual velocities and accelerations. In fact, dual quaternions can natively—without constructs that fall outside the algebra—encode the most typical geometric objects such as points, lines and planes. The reader is referred to the literature to find such parameterizations and the correct dual quaternion transformation [15,16].

2.4. Derivation of Fundamental Kinematic Laws

In this section we will derive both the quaternion and dual quaternion kinematic laws. We will make the time dependence explicit only when necessary for clarity.

The three-dimensional attitude kinematics evolve as

$$\dot{q}_{X/Y} = \tfrac{1}{2} q_{X/Y} \omega_{X/Y}^x = \tfrac{1}{2} \omega_{X/Y}^Y q_{X/Y}, \tag{20}$$

where $\omega_{X/Y}^z \triangleq (0, \overline{\omega}_{X/Y}^z) \in \mathbb{H}^v$ and $\overline{\omega}_{X/Y}^z \in \mathbb{R}^3$ is the angular velocity of frame X with respect to frame Y expressed in Z-frame coordinates. On the other hand, the dual quaternion kinematics can be expressed as [7]

$$\dot{\boldsymbol{q}}_{X/Y} = \tfrac{1}{2} \boldsymbol{q}_{X/Y} \boldsymbol{\omega}_{X/Y}^x = \tfrac{1}{2} \boldsymbol{\omega}_{X/Y}^Y \boldsymbol{q}_{X/Y}. \tag{21}$$

Lemma 1. *The attitude of a rigid body evolves as* $\dot{q}_{X/Y} = \tfrac{1}{2} q_{X/Y} \omega_{X/Y}^x$, *as stated in Equation (20).*

Proof. Denote the infinitesimal rotation about axis $\hat{u}$ by $\Delta\theta$. The quaternion that represents this rotation is constructed as $\delta q_{X/Y}(\Delta t) \triangleq (\cos(\Delta\theta/2), \hat{u}\sin(\Delta\theta/2))$. Therefore, $q_{X/Y}(t+\Delta t) = q_{X/Y}(t)\delta q_{X/Y}(\Delta t)$. Then, for a small rotation angle, $\delta q_{X/Y}(\Delta t) = (1, \hat{u}\Delta\theta/2)$. Substituting into the previous expression for $q_{X/Y}(t+\Delta t)$, we obtain

$$q_{X/Y}(t+\Delta t) = q_{X/Y}(t)(1, \hat{u}\Delta\theta/2)$$
$$= q_{X/Y}(t)(1 + (0, \hat{u}\Delta\theta/2)) \tag{22}$$
$$= q_{X/Y}(t) + \tfrac{1}{2}q_{X/Y}(t)\hat{u}\Delta\theta.$$

Manipulating the expression and dividing by Δt, we obtain

$$\frac{q_{X/Y}(t+\Delta t) - q_{X/Y}(t)}{\Delta t} = \tfrac{1}{2}q_{X/Y}(t)\hat{u}\frac{\Delta\theta}{\Delta t}, \tag{23}$$

and invoking the limit as $\Delta t \to 0$ yields

$$\dot{q}_{X/Y}(t) = \tfrac{1}{2}q_{X/Y}(t)\omega^X_{X/Y}(t), \tag{24}$$

where we have defined the angular velocity as $\omega^X_{X/Y} \triangleq \hat{u}\dot{\theta}$, i.e., the rate of rotation about the instantaneous Euler axis. $\square$

Lemma 2. *The pose of a rigid body evolves as $\dot{\boldsymbol{q}}_{X/Y} = \tfrac{1}{2}\boldsymbol{q}_{X/Y}\boldsymbol{\omega}^X_{X/Y}$, as stated in Equation (21).*

Proof. Taking the derivative of $\boldsymbol{q}_{X/Y} = q_{X/Y} + \epsilon\tfrac{1}{2}q_{X/Y}r^X_{X/Y}$ we get

$$
\begin{aligned}
\dot{\boldsymbol{q}}_{X/Y} &= \dot{q}_{X/Y} + \epsilon(\tfrac{1}{2}\dot{q}_{X/Y}r^X_{X/Y} + \tfrac{1}{2}q_{X/Y}\dot{r}^X_{X/Y}) \\
&= \tfrac{1}{2}q_{X/Y}\omega^X_{X/Y} + \epsilon(\tfrac{1}{4}q_{X/Y}\omega^X_{X/Y}r^X_{X/Y} + \tfrac{1}{2}q_{X/Y}\dot{r}^X_{X/Y}) \\
&= \tfrac{1}{2}q_{X/Y}\omega^X_{X/Y} + \epsilon(\tfrac{1}{4}q_{X/Y}\omega^X_{X/Y}r^X_{X/Y} + \tfrac{1}{2}q_{X/Y}(v^X_{X/Y} - \omega^X_{X/Y} \times r^X_{X/Y})) \\
&= \tfrac{1}{2}q_{X/Y}\omega^X_{X/Y} + \epsilon(\tfrac{1}{2}q_{X/Y}v^X_{X/Y} + \tfrac{1}{4}q_{X/Y}\omega^X_{X/Y}r^X_{X/Y} - \tfrac{1}{2}q_{X/Y}(\omega^X_{X/Y} \times r^X_{X/Y})),
\end{aligned}
\tag{25}
$$

where we used the fact that $v^X_{X/Y} \triangleq dr^X_{X/Y}/dt = \dot{r}^X_{X/Y} + \omega^X_{X/Y} \times r^X_{X/Y}$. Using the definition of the cross product, we know that $\omega^X_{X/Y} \times r^X_{X/Y} = \tfrac{1}{2}(\omega^X_{X/Y}r^X_{X/Y} - (r^X_{X/Y})^*(\omega^X_{X/Y})^*) = \tfrac{1}{2}(\omega^X_{X/Y}r^X_{X/Y} - r^X_{X/Y}\omega^X_{X/Y})$. Evaluating this cross product into the above expression yields

$$
\begin{aligned}
&= \tfrac{1}{2}q_{X/Y}\omega^X_{X/Y} + \epsilon(\tfrac{1}{2}q_{X/Y}v^X_{X/Y} + \tfrac{1}{4}q_{X/Y}\omega^X_{X/Y}r^X_{X/Y} - \tfrac{1}{4}q_{X/Y}(\omega^X_{X/Y}r^X_{X/Y} - r^X_{X/Y}\omega^X_{X/Y})) \\
&= \tfrac{1}{2}q_{X/Y}\omega^X_{X/Y} + \epsilon\tfrac{1}{2}q_{X/Y}v^X_{X/Y} + \epsilon\tfrac{1}{4}q_{X/Y}r^X_{X/Y}\omega^X_{X/Y} \\
&= \tfrac{1}{2}q_{X/Y}\omega^X_{X/Y} + \epsilon\tfrac{1}{2}q_{X/Y}v^X_{X/Y} + \epsilon\tfrac{1}{4}q_{X/Y}r^X_{X/Y}\omega^X_{X/Y} + \epsilon^2\tfrac{1}{4}q_{X/Y}r^X_{X/Y}v^X_{X/Y}, \text{ since } \epsilon^2 = 0 \\
&= \tfrac{1}{2}q_{X/Y}(\omega^X_{X/Y} + \epsilon v^X_{X/Y}) + \tfrac{1}{2}\epsilon\tfrac{1}{2}q_{X/Y}r^X_{X/Y}(\omega^X_{X/Y} + \epsilon v^X_{X/Y}) \\
&= \tfrac{1}{2}(q_{X/Y} + \epsilon\tfrac{1}{2}q_{X/Y}r^X_{X/Y})(\omega^X_{X/Y} + \epsilon v^X_{X/Y}) \\
&= \tfrac{1}{2}\boldsymbol{q}_{X/Y}\boldsymbol{\omega}^X_{X/Y},
\end{aligned}
\tag{26}
$$

proving the desired result. $\square$

Remark 1. *The spatial kinematic equation $\dot{q}_{X/Y} = \tfrac{1}{2}\omega^Y_{X/Y}q_{X/Y}$ can be immediately derived as a direct consequence of the adjoint transformation equation $\omega^X_{X/Y} = q^*_{X/Y}\omega^Y_{X/Y}q_{X/Y}$, which implies $q_{X/Y}\omega^X_{X/Y} = \omega^Y_{X/Y}q_{X/Y}$.*

Remark 2. *The spatial kinematic equation $\dot{\boldsymbol{q}}_{X/Y} = \tfrac{1}{2}\boldsymbol{\omega}^Y_{X/Y}\boldsymbol{q}_{X/Y}$ can be immediately derived as a direct consequence of the adjoint transformation equation $\boldsymbol{\omega}^X_{X/Y} = \boldsymbol{q}^*_{X/Y}\boldsymbol{\omega}^Y_{X/Y}\boldsymbol{q}_{X/Y}$, which implies $\boldsymbol{q}_{X/Y}\boldsymbol{\omega}^X_{X/Y} = \boldsymbol{\omega}^Y_{X/Y}\boldsymbol{q}_{X/Y}$.*

3. Robot Kinematics Using Dual Quaternions

3.1. Dual Quaternion Notation

The forward kinematics of a robot can be easily laid out in dual quaternion form. In general, a dual quaternion encoding the relationship between two frames A and B is given as

$$\boldsymbol{q}_{B/A} = q_{B/A} + \epsilon\tfrac{1}{2}q_{B/A}r^B_{B/A}, \tag{27}$$

$$\boldsymbol{q}_{B/A} = q_{B/A} + \epsilon\tfrac{1}{2}r^A_{B/A}q_{B/A}, \tag{28}$$

where $q_{B/A}$ is the quaternion that represents the attitude change in going from reference frame A, to reference frame B. The position vectors $r^B_{B/A}$ and $r^A_{B/A}$ represent the position vector from the origin of frame A to the origin of frame B expressed in frame B, and frame A coordinates, respectively. Notice that Equations (27) and (28) can be equivalently expressed as follows:

$$\text{Rotation First:} \quad \boldsymbol{q}_{\text{B/A}} = (q_{\text{B/A}} + \epsilon 0)(1 + \epsilon \tfrac{1}{2} r^{\text{B}}_{\text{B/A}}), \tag{29}$$

$$\text{Translation First:} \quad \boldsymbol{q}_{\text{B/A}} = (1 + \epsilon \tfrac{1}{2} r^{\text{A}}_{\text{B/A}})(q_{\text{B/A}} + \epsilon 0), \tag{30}$$

leading to an intuitive decomposition of the underlying operations. In the forward kinematics, Equation (29) implies that the frame rotation is carried out first, and then a translation is carried out relative to the new frame. Equation (30) denotes a translation in the base frame, followed by an attitude change of the resulting frame. Throughout this work we will use the *translation first* approach.

3.2. Product of Exponentials Formula in Dual-Quaternion Form

The product of exponentials formula has been long used to study the forward kinematics of robots. Reference [23] has a thorough introduction to the topic, with many examples. In this section we lay out the main results that cast the product of exponentials (POE) formula in dual quaternion form. In particular, [20] has made use of the dual quaternion formalism to perform geometric control on a fixed-base robotic arm, where the forward kinematics of the robot are expressed using the POE formula.

As commonly used in robotics, the exponential operation takes an element of the Lie algebra for a given Lie group, and renders a group element. For the dual quaternion case, let the set of parameters $(\boldsymbol{\theta}, \boldsymbol{s}) \in D \times \mathbb{H}^v_d$, where $D = \{a + \epsilon a_d : a, a_d \in \mathbb{R} \text{ and } \epsilon^2 = 0\}$ is the set of dual numbers, parametrize a screw motion as shown in Figure 1. In particular, $\boldsymbol{\theta}$ and $\boldsymbol{s}$ are given by

$$\boldsymbol{\theta} = \theta + \epsilon d, \qquad \boldsymbol{\theta} \in D, \qquad \theta, d \in \mathbb{R}, \tag{31}$$

$$\boldsymbol{s} = \ell + \epsilon m, \qquad \boldsymbol{s} \in \mathbb{H}^v_d, \qquad \ell, m \in \mathbb{H}^v, \tag{32}$$

where θ is the angle of the screw motion, d is the translation along the screw axis, ℓ is the unit screw axis of the joint, and m is the moment vector of the screw axis of direction ℓ with respect to the origin of the local inertial frame. This implies that

$$m = r_{\text{P/I}} \times \ell, \tag{33}$$

where the point P lies on the screw axis. In robotic systems, the exponential mapping is commonly used to evaluate the forward kinematics of fixed-base robotic systems. We summarize the dual quaternion exponential mapping in the following lemma [14,20].

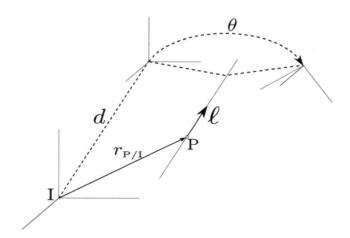

Figure 1. Screw motion parametrized by θ and $\boldsymbol{s}$.

Lemma 3. *The exponential operation,* $\exp : D \times \mathbb{H}_d^v \to \mathbb{H}_d$ *for a given pair* $(\boldsymbol{\theta}, s) \in D \times \mathbb{H}_d^v$ *defined as in Equations (31) and (32) is given as*

$$
\begin{aligned}
q &= \exp\left(\tfrac{1}{2}\boldsymbol{\theta}s\right), \quad q \in \mathbb{H}_d \\
&= \cos\left(\tfrac{1}{2}\boldsymbol{\theta}\right) + s\sin\left(\tfrac{1}{2}\boldsymbol{\theta}\right) \\
&= \left(\cos\left(\tfrac{1}{2}\theta\right), \ \ell\sin\left(\tfrac{1}{2}\theta\right)\right) + \epsilon\left(-\tfrac{1}{2}d\sin\left(\tfrac{1}{2}\theta\right), \ \tfrac{1}{2}d\ell\cos\left(\tfrac{1}{2}\theta\right) + m\sin\left(\tfrac{1}{2}\theta\right)\right).
\end{aligned}
\tag{34}
$$

Proof. Since $\boldsymbol{\theta} = \theta + \epsilon d \in D$, we have that

$$
\cos\left(\tfrac{1}{2}\boldsymbol{\theta}\right) = \cos\left(\tfrac{1}{2}\theta\right) + \epsilon\frac{d}{2}\left(-\sin(\tfrac{1}{2}\theta)\right)
\tag{35}
$$

$$
\sin\left(\tfrac{1}{2}\boldsymbol{\theta}\right) = \sin\left(\tfrac{1}{2}\theta\right) + \epsilon\frac{d}{2}\cos\left(\tfrac{1}{2}\theta\right).
\tag{36}
$$

It follows that

$$
q = \cos\left(\tfrac{1}{2}\boldsymbol{\theta}\right) + s\sin\left(\tfrac{1}{2}\boldsymbol{\theta}\right)
\tag{37}
$$

$$
= \cos\left(\tfrac{1}{2}\theta\right) - \epsilon\frac{d}{2}\sin\left(\tfrac{1}{2}\theta\right) + (\ell + \epsilon m)\left(\sin\left(\tfrac{1}{2}\theta\right) + \epsilon\frac{d}{2}\cos\left(\tfrac{1}{2}\theta\right)\right),
\tag{38}
$$

which yields the desired result upon expansion. $\square$

Remark 3. *By comparing Equations (28) and (34), it can be deduced that the effect of a joint motion can be characterized by an equivalent rotation and a translation. In particular, by equating the real parts of the dual quaternions, we have that*

$$
q_{\mathrm{B/A}} = \left(\cos\left(\tfrac{1}{2}\theta\right), \ \ell\sin\left(\tfrac{1}{2}\theta\right)\right),
\tag{39}
$$

and from the dual parts

$$
\tfrac{1}{2}r_{\mathrm{B/A}}^{\mathrm{A}} q_{\mathrm{B/A}} = \left(-\tfrac{1}{2}d\sin\left(\tfrac{1}{2}\theta\right), \ \tfrac{1}{2}d\ell\cos\left(\tfrac{1}{2}\theta\right) + m\sin\left(\tfrac{1}{2}\theta\right)\right).
\tag{40}
$$

Equivalently, $r_{\mathrm{B/A}}^{\mathrm{A}}$ *can be described as*

$$
r_{\mathrm{B/A}}^{\mathrm{A}} = (0, d\ell + m\sin(\theta) + (\cos(\theta) - 1)m \times \ell).
\tag{41}
$$

The inverse to the exponential mapping is the logarithmic mapping, $\ln : \mathbb{H}_d \to \mathbb{H}_d$, which is defined as

$$
\ln q = \tfrac{1}{2}\boldsymbol{\theta}s = \tfrac{1}{2}\theta\ell + \epsilon\tfrac{1}{2}(\theta m + d\ell).
\tag{42}
$$

Appendix A.6. of [20] explains how to retrieve $\{\theta, d, \ell, m\}$ given a dual quaternion, q.

Given the dual quaternion from the inertial (base) frame to the end effector, at the robots's home configuration, $q_{\mathrm{e,0/I}}$, and parameter s_i for each of the n joints of a robot at its home configuration, the product of exponentials formula yields

$$
q_{\mathrm{e/I}} = \exp\left(\tfrac{1}{2}\theta_1 s_1\right)\ldots\exp\left(\tfrac{1}{2}\theta_n s_n\right) q_{\mathrm{e,0/I}},
\tag{43}
$$

where joint 1 is closest to the base and joint n is closest to the end-effector. The exponential formula is effectively changing the spatial frame, as opposed to the body frame of the end-effector. Besides its simplicity to compute forward kinematics, the POE formula is straightforward to compute for a given configuration once the type of joint is known and the geometric properties of the robot are selected.

As this point, it is worth emphasizing that in regards to the moving frames used in space operations, the use of an inertial frame with respect to which one can perform spatial kinematics, as was done in [20], is impractical. Since the satellite base is constantly in motion, local-frame parameterizations of pose transformations across the links of the manipulator are preferred. Therefore, we favor the use of the forward-moving pose representations to express the location of the end-effector frame. We show next how to use the Denavit-Hartenberg parameterization in dual quaternions to capture such a transformation.

3.3. Denavit-Hartenberg Parameters in Dual Quaternion Form

The Denavit-Hartenberg parameters, commonly referred to as DH parameters, are four geometric quantities that allow identifying the relative pose of a joint with respect to another in a systematic manner. We will denote a set of DH parameters as $\{d_i, \theta_i, a_i, \alpha_i\}$ for joint i. The parameters d_i and θ_i are commonly referred to as *joint* parameters, while a_i and α_i are known as the *link* parameters. A complete description of the DH parameters for R and P joint types, and several examples of their use are provided in [24]. In [24] a thorough description of the orientation of the frames is also provided, to which the reader is referred. In [25], Gan et al. have used dual quaternions in combination with the DH parameter convention to capture the pose transformation between joints. For completeness, we provide these equations herein, making use of Figure 2.

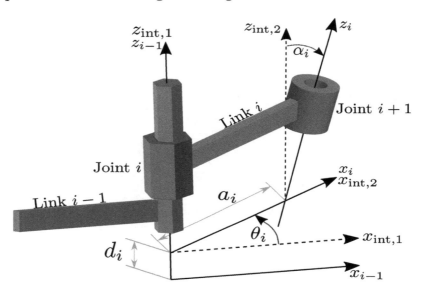

Figure 2. Denavit-Hartenberg parameters.

In words, the transformation from the reference frame assigned to the proximal joint (i.e., closer to the base of the robot) of a given link i, to the reference frame assigned to its distal joint (i.e., closer to the end effector), is described in terms of the DH parameters as:

1. From the origin O_{i-1}, displace along the Z_{i-1} (joint) axis by an amount d_i. Define this intermediate frame as $\{\text{int}, 1\}$.
2. Rotate about the Z_{i-1} axis by θ_i until axis X_{i-1} is superimposed to X_i
3. Translate along X_i by a distance of a_i. Define this intermediate frame as $\{\text{int}, 2\}$.
4. Rotate about the X_i axis by α_i

Mathematically, we can write this as the composition of four elementary dual quaternion operations, and summarize it further into two composite dual quaternions as

$$\boldsymbol{q}_{i/i-1} = \left(1 + \epsilon r^{\text{int},1}_{\text{int},1/i-1}\right)\left(q_{\text{int},2/\text{int},1} + \epsilon 0\right)\left(1 + \epsilon r^{\text{int},2}_{\text{int},2/\text{int},1}\right)\left(q_{i/\text{int},2} + \epsilon 0\right) \tag{44}$$

$$= \left(q_{\text{int},2/\text{int},1} + \epsilon r^{\text{int},1}_{\text{int},1/i-1} q_{\text{int},2/\text{int},1}\right)\left(q_{i/\text{int},2} + \epsilon r^{\text{int},2}_{\text{int},2/\text{int},1} q_{i/\text{int},2}\right) \tag{45}$$

where

$$r^{\text{int},1}_{\text{int},1/i\text{-}1} = (0, [0,0,d_i]^{\top}) \tag{46}$$

$$q_{\text{int},2/\text{int},1} = (\cos\theta_i/2, [0,0,\sin\theta_i/2]^{\top}) \tag{47}$$

$$r^{\text{int},2}_{\text{int},2/\text{int},1} = (0, [a_i,0,0]^{\top}) \tag{48}$$

$$q_{i/\text{int},2} = (\cos\alpha_i/2, [\sin\alpha_i/2,0,0]^{\top}) \tag{49}$$

Notice that while this is compact and readable up to multiplication of the dual quaternions, the same cannot be said about the end result compared to its homogeneous transformation matrix (HTM) counterpart. In fact, if we express $q_{i/i\text{-}1}$ component-wise, and cast it as a vector in $\mathbb{R}^8$ which is the typical representation of dual quaternions for numerical purposes, and compute the equivalent HTM, we get the following:

$$q_{i/i\text{-}1} = \begin{bmatrix} \cos(\alpha/2)\cos(\theta/2) \\ \sin(\alpha/2)\cos(\theta/2) \\ \sin(\alpha/2)\sin(\theta/2) \\ \cos(\alpha/2)\sin(\theta/2) \\ -\frac{1}{2}a_i\sin(\alpha_i/2)\cos(\theta_i/2) - \frac{1}{2}d_i\cos(\alpha_i/2)\sin(\theta_i/2) \\ \frac{1}{2}a_i\cos(\alpha_i/2)\cos(\theta_i/2) - \frac{1}{2}d_i\sin(\alpha_i/2)\sin(\theta_i/2) \\ \frac{1}{2}a_i\cos(\alpha_i/2)\sin(\theta_i/2) + \frac{1}{2}d_i\sin(\alpha_i/2)\cos(\theta_i/2) \\ \frac{1}{2}d_i\cos(\alpha_i/2)\cos(\theta_i/2) - \frac{1}{2}a_i\sin(\alpha_i/2)\sin(\theta_i/2) \end{bmatrix} \tag{50}$$

$$T_{i/i\text{-}1} = \left[\begin{array}{ccc:c} \cos\theta_i & \sin\theta_i & 0 & -a_i \\ -\cos\alpha_i\sin\theta_i & \cos\alpha_i\cos\theta_i & \sin\alpha_i & -d_i\sin\alpha_i \\ \sin\alpha_i\sin\theta_i & -\sin\alpha_i\cos\theta_i & \cos\alpha_i & -d_i\cos\alpha_i \\ \hdashline 0 & 0 & 0 & 1 \end{array} \right]. \tag{51}$$

While the HTM is more readable and faster to code, it uses 16 doubles and a multi-dimensional array to store the information and operate in the underlying algebra.

Remark 4. *Since the transformations associated to θ_i and d_i are about z_{i-1} and the operations associated to α_i and a_i happen about x_i, both stages of the DH transformation can be interpreted in the context of screw theory. Hence, the operation described by Equation (44) can be equivalently expressed as the composition of exponential operations given by*

$$q_{i/i\text{-}1} = \exp(\tfrac{1}{2}\boldsymbol{\theta}_1 s_1)\exp(\tfrac{1}{2}\boldsymbol{\theta}_2 s_2), \tag{52}$$

where $\boldsymbol{\theta}_1 = \theta_i + \epsilon d_i$ and $s_1 = (0, [0,0,1]^{\top}) + \epsilon 0$ and $\boldsymbol{\theta}_2 = \alpha_i + \epsilon a_i$ and $s_2 = (0, [1,0,0]^{\top}) + \epsilon 0$.

4. Manipulator Kinematics Using Dual Quaternions

In this section we provide examples to demonstrate how one can develop the kinematic equations for different types of serial manipulators using dual quaternions.

4.1. Example: Forward Kinematics with an Inertially Fixed Base

The serial RR configuration in Figure 3 will be used as an example of how to use dual quaternions for forward kinematics. Notice that the pose of the end effector with respect to the inertial frame is given by

$$q_{e/I} = q_{1/I}q_{2/1}q_{e/2}. \tag{53}$$

For the sake of exposition, these are given by

$$\boldsymbol{q}_{1/\text{I}} = \left(1 + \epsilon \tfrac{1}{2} r^{\text{I}}_{1/\text{I}}\right)\left(q_{1/\text{I}} + \epsilon 0\right), \tag{54}$$

$$\boldsymbol{q}_{2/1} = \left(1 + \epsilon \tfrac{1}{2} r^{1}_{2/1}\right)\left(q_{2/1} + \epsilon 0\right), \tag{55}$$

$$\boldsymbol{q}_{\text{e}/2} = \left(1 + \epsilon \tfrac{1}{2} r^{2}_{\text{e}/2}\right)\left(q_{\text{e}/2} + \epsilon 0\right), \tag{56}$$

where the *translation-first* approach has been used. Each of these quantities can be easily determined from the geometry of the problem. The position quaternions are given by $r^{\text{Y}}_{\text{X/Y}} = (0, \bar{r}^{\text{Y}}_{\text{X/Y}})$, and

$$\bar{r}^{\text{I}}_{1/\text{I}} = [0, 0, 0]^{\text{T}}, \tag{57}$$

$$\bar{r}^{1}_{2/1} = [l_1, 0, 0]^{\text{T}}, \tag{58}$$

$$\bar{r}^{2}_{\text{e}/2} = [l_2, 0, 0]^{\text{T}}, \tag{59}$$

while the quaternions are given by

$$q_{1/\text{I}} = \left(\cos \alpha_1 / 2, [0, 0, \sin \alpha_1 / 2]^{\text{T}}\right), \tag{60}$$

$$q_{1/\text{I}} = \left(\cos \alpha_2 / 2, [0, 0, \sin \alpha_2 / 2]^{\text{T}}\right), \tag{61}$$

$$q_{\text{e}/2} = 1. \tag{62}$$

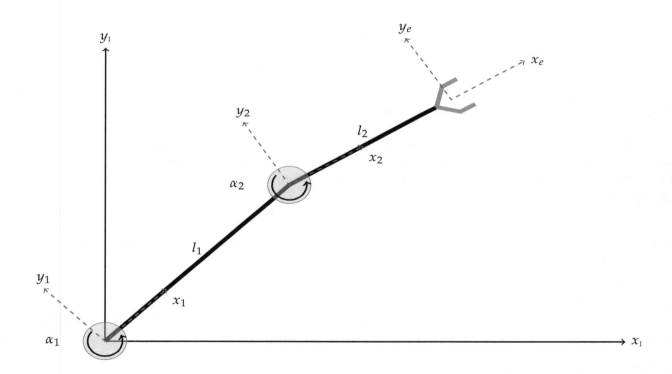

Figure 3. Robot arm configuration.

The time derivative of the dual quaternion yields information about the angular and linear velocity of the end-effector. In particular, we have that for a dual quaternion:

$$\dot{q}_{X/Y} = \tfrac{1}{2}q_{X/Y}\omega^X_{X/Y} = \tfrac{1}{2}\omega^Y_{X/Y}q_{X/Y},\tag{63}$$

where the first equality in Equation (63) is associated with a body-frame time derivative, and the second equality in Equation (63) is associated with a spatial-frame time derivative.

With these definitions in mind, we compute the time-rate of change of the pose of the end-effector as

$$\dot{q}_{e/I} = \dot{q}_{1/I}q_{2/1}q_{e/2} + q_{1/I}\dot{q}_{2/1}q_{e/2} + q_{1/I}q_{2/1}\dot{q}_{e/2}\tag{64}$$

$$= \tfrac{1}{2}q_{1/I}\omega^1_{1/I}q_{2/1}q_{e/2} + q_{1/I}\tfrac{1}{2}q_{2/1}\omega^2_{2/1}q_{e/2} + q_{1/I}q_{2/1}\tfrac{1}{2}q_{e/2}\omega^e_{e/2}.\tag{65}$$

Then, using Equation (63), we get that the dual velocity of the end effector with respect to the inertial frame is given by

$$\omega^e_{e/I} = 2q^*_{e/I}\dot{q}_{e/I}$$

$$= q^*_{e/I}q_{1/I}\omega^1_{1/I}q_{2/1}q_{e/2} + q^*_{e/I}q_{1/I}q_{2/1}\omega^2_{2/1}q_{e/2} + q^*_{e/I}q_{1/I}q_{2/1}q_{e/2}\overbrace{\omega^e_{e/2}}^{=0}$$

$$= q^*_{e/2}q^*_{2/1}q^*_{1/I}q_{1/I}\omega^1_{1/I}q_{2/1}q_{e/2} + q^*_{e/2}q^*_{2/1}q^*_{1/I}q_{1/I}q_{2/1}\omega^2_{2/1}q_{e/2}$$

$$= q^*_{e/2}q^*_{2/1}\omega^1_{1/I}q_{2/1}q_{e/2} + q^*_{e/2}\omega^2_{2/1}q_{e/2}$$

$$= \mathrm{Ad}_{q^*_{e/2}q^*_{2/1}}\omega^1_{1/I} + \mathrm{Ad}_{q^*_{e/2}}\omega^2_{2/1}$$

$$= \mathrm{Ad}_{(q_{2/1}q_{e/2})^*}\omega^1_{1/I} + \mathrm{Ad}_{(q_{e/2})^*}\omega^2_{2/1}$$

$$= \left[\mathrm{Ad}_{(q_{2/1}q_{e/2})^*}\xi^1_{1/I}, \quad \mathrm{Ad}_{(q_{e/2})^*}\xi^2_{2/1}\right]\dot{\bar{\alpha}}\tag{66}$$

$$= J^B(q,\xi)\dot{\bar{\alpha}},\tag{67}$$

where $J^B(q,\xi)$ is the Jacobian expressed in the body frame and

$$\bar{\alpha} = \begin{bmatrix}\alpha_1 \\ \alpha_2\end{bmatrix} \quad \text{and} \quad \dot{\bar{\alpha}} = \begin{bmatrix}\dot{\alpha}_1 \\ \dot{\alpha}_2\end{bmatrix}.\tag{68}$$

The elements ξ_i are the dual quaternion screws for each of the joints. In general, the screws for revolute and prismatic joints are listed in Table 4 for each of the three axes, and these are independent of the current robot configuration.

Table 4. Screw (ξ_i) for revolute and prismatic joints.

	Revolute Joint	Prismatic Joint
X-axis	$(0, [1,0,0]^T) + \epsilon 0$	$0 + \epsilon(0, [1,0,0]^T)$
Y-axis	$(0, [0,1,0]^T) + \epsilon 0$	$0 + \epsilon(0, [0,1,0]^T)$
Z-axis	$(0, [0,0,1]^T) + \epsilon 0$	$0 + \epsilon(0, [0,0,1]^T)$

4.2. Example: Forward Kinematics of a Floating Double Pendulum with End-Effector

Given the floating double pendulum shown in Figure 4, we want to model its kinematics. The difference with respect to the one shown in Figure 3 is that the first revolute joint is free to translate in 2D space.

The kinematic equations of motion can thus be derived as follows using a geometric description of the forward kinematics

$$q_{e/I} = q_{1/I}q_{2/1}q_{e/2},\tag{69}$$

where $q_{1/I}$, $q_{2/1}$, $q_{e/2}$ are given by Equations (54)–(56). However, $\bar{r}^I_{1/I} = [u, v, 0]^T$ determines the translation of the first revolute joint in 2D. It is clear that

$$\frac{\mathrm{d}}{\mathrm{d}t}\bar{r}_{1/I}^{I} = \dot{\bar{r}}_{1/I}^{I} = \bar{v}_{1/I}^{I} = [\dot{u}, \dot{v}, 0]^{\mathsf{T}} \tag{70}$$

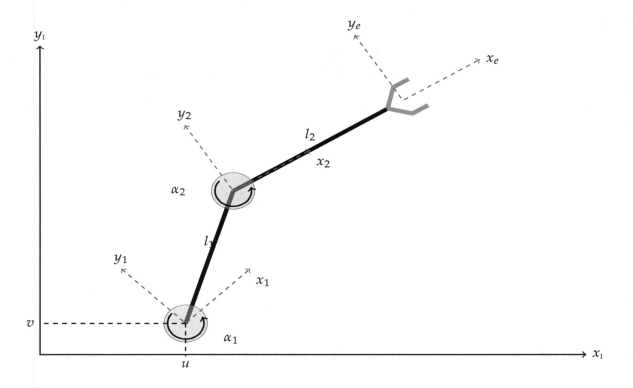

Figure 4. Robot arm configuration.

In this case, the time evolution of $\boldsymbol{q}_{1/I}$ is given by

$$\dot{\boldsymbol{q}}_{1/I} = \tfrac{1}{2}\boldsymbol{\omega}_{1/I}^{I}\boldsymbol{q}_{1/I} \tag{71}$$

as before, but we redefine the dual velocity as dictated by the definition in Equation (14) as

$$\boldsymbol{\omega}_{1/I}^{I} = \omega_{1/I}^{I} + \epsilon(v_{1/I}^{I} - \omega_{1/I}^{I} \times r_{1/I}^{I}). \tag{72}$$

The relationship derived earlier

$$\boldsymbol{\omega}_{e/I}^{e} = \mathrm{Ad}_{\boldsymbol{q}_{e/2}^{*}\boldsymbol{q}_{2/1}^{*}}\boldsymbol{\omega}_{1/I}^{1} + \mathrm{Ad}_{\boldsymbol{q}_{e/2}^{*}}\boldsymbol{\omega}_{2/1}^{2}$$

still holds. However, $\boldsymbol{\omega}_{1/I}^{1}$ must be computed from our knowledge of $\boldsymbol{\omega}_{1/I}^{I}$. While in quaternion and vector notation this might be troublesome, the expression using dual quaternions is simple and given by

$$\boldsymbol{\omega}_{1/I}^{1} = \boldsymbol{q}_{1/I}^{*}\boldsymbol{\omega}_{1/I}^{I}\boldsymbol{q}_{1/I} = \mathrm{Ad}_{\boldsymbol{q}_{1/I}^{*}}\boldsymbol{\omega}_{1/I}^{I}. \tag{73}$$

4.3. Manipulator on an Orbiting Spacecraft

For the general case, the robot base can move with six degrees of freedom, reinforcing the need for a convenient pose representation tool such as dual quaternions. Following an approach analogous to that proposed by Adorno in [26], in this section we will provide explicit expressions for the Jacobian matrix for different types of joints.

The kinematics of the robotic base, attached to frame B, would still be governed by the equation

$$\dot{q}_{B/I} = \tfrac{1}{2} q_{B/I} \omega^{B}_{B/I}, \tag{74}$$

while the kinematics of the joints depend on the type of joint. The dual velocities for the joints depend on frame positioning and on the selection of the generalized coordinates. In Table 5, we provide example generalized coordinates and their corresponding dual velocities. Simple numerical derivatives can yield the dual acceleration of the joint. It is worth emphasizing that for a $3 - 2 - 1/\psi - \theta - \phi$ rotation, the matrix $M(\phi_{i/i-1}, \theta_{i/i-1}, \psi_{i/i-1})$ will correspond to

$$M(\phi_{i/i-1}, \theta_{i/i-1}, \psi_{i/i-1}) = \begin{bmatrix} 1 & 0 & -\sin(\theta_{i/i-1}) \\ 0 & \cos(\phi_{i/i-1}) & \cos(\theta_{i/i-1})\sin(\phi_{i/i-1}) \\ 0 & -\sin(\phi_{i/i-1}) & \cos(\theta_{i/i-1})\cos(\phi_{i/i-1}) \end{bmatrix}. \tag{75}$$

In general, we can identify the pose of a satellite-mounted end-effector by

$$q_{e/I} = q_{B/I} q_{O/B} q_{0/O} \left(\prod_{i=1}^{n-1} q_{i/i-1} \right) q_{e/n-1}, \tag{76}$$

where $q_{O/B}$ represents the pose transformation from the body frame of the satellite, B, to the frame at the base of the satellite manipulator, denoted by O; frame i represents the joint frame attached to the i-th link, one of the n bodies composing the manipulator, at the location of the proximal joint; and e is the end-effector frame that is rigidly attached to the last link of the serial manipulator, $n - 1$. For clarity, an example of the product operator $\prod$ used on dual quaternions is given by

$$\prod_{i=1}^{2} q_{i/i-1} = q_{1/0} q_{2/1}. \tag{77}$$

Additionally, using the definition of frames described above, the first link of the manipulator is link 0 and its connecting joint to the satellite base is frame 0.

Then, since $q_{O/B}$ and $q_{e/n-1}$ are constant, the kinematics can be derived following the procedure of previous sections as

$$\begin{aligned} \dot{q}_{e/I} &= \dot{q}_{B/I} q_{O/B} q_{0/O} \left(\prod_{i=1}^{n-1} q_{i/i-1} \right) q_{e/n-1} + q_{B/I} q_{O/B} \dot{q}_{0/O} \left(\prod_{i=1}^{n-1} q_{i/i-1} \right) q_{e/n-1} \\ &\quad + q_{B/I} q_{O/B} q_{0/O} \sum_{k=1}^{n-1} \left(q_{k-1/0} \dot{q}_{k/k-1} q_{n-1/k} \right) q_{e/n-1} \\ &= \tfrac{1}{2} q_{B/I} \omega^{B}_{B/I} q_{O/B} q_{0/O} \left(\prod_{i=1}^{n-1} q_{i/i-1} \right) q_{e/n-1} + q_{B/I} q_{O/B} \tfrac{1}{2} q_{0/O} \omega^{0}_{0/O} \left(\prod_{i=1}^{n-1} q_{i/i-1} \right) q_{e/n-1} \\ &\quad + q_{B/I} q_{O/B} q_{0/O} \sum_{k=1}^{n-1} \left(q_{k-1/0} \tfrac{1}{2} q_{k/k-1} \omega^{k}_{k/k-1} q_{n-1/k} \right) q_{e/n-1}. \end{aligned} \tag{78}$$

Multiplying by $2q^{*}_{e/I}$ on the left, we get

$$\begin{aligned} \omega^{e}_{e/I} &= q^{*}_{e/I} q_{B/I} \omega^{B}_{B/I} q_{O/B} q_{0/O} \left(\prod_{i=1}^{n-1} q_{i/i-1} \right) q_{e/n-1} + q^{*}_{e/I} q_{B/I} q_{O/B} q_{0/O} \omega^{0}_{0/O} \left(\prod_{i=1}^{n-1} q_{i/i-1} \right) q_{e/n-1} \\ &\quad + q^{*}_{e/I} q_{B/I} q_{O/B} q_{0/O} \sum_{k=1}^{n-1} \left(q_{k-1/0} q_{k/k-1} \omega^{k}_{k/k-1} q_{n-1/k} \right) q_{e/n-1}, \end{aligned} \tag{79}$$

and carrying out the dual quaternion multiplications to simplify the expression yields

$$\omega_{e/I}^e = q_{e/B}^* \omega_{B/I}^B q_{e/B} + q_{e/0}^* \omega_{0/O}^0 q_{e/0} + \sum_{k=1}^{n-1} q_{e/k}^* \omega_{k/k\text{-}1}^k q_{e/k}. \tag{80}$$

Table 5. Generalized coordinates and dual velocities for different joint types.

Joint Type	Generalized Coordinate Parametrization, $\bar{\alpha}_i$	Dual Velocity, $\omega_{i/i\text{-}1}^i$
Revolute	$\theta_{i/i\text{-}1} \in \mathbb{R}^1$	$(0, [0,0,\dot\theta_{i/i\text{-}1}]^T) + \epsilon 0$
Prismatic	$z_{i/i\text{-}1} \in \mathbb{R}^1$	$0 + \epsilon(0, [0,0,\dot z_{i/i\text{-}1}]^T)$
Spherical	$[\phi_{i/i\text{-}1}, \theta_{i/i\text{-}1}, \psi_{i/i\text{-}1}]^T \in \mathbb{R}^3$	$(0, M(\phi_{i/i\text{-}1}, \theta_{i/i\text{-}1}, \psi_{i/i\text{-}1})[\dot\phi_{i/i\text{-}1}, \dot\theta_{i/i\text{-}1}, \dot\psi_{i/i\text{-}1}]^T) + \epsilon 0$
Cylindrical	$[\theta_{i/i\text{-}1}, z_{i/i\text{-}1}]^T \in \mathbb{R}^2$	$(0, [0,0,\dot\theta_{i/i\text{-}1}]^T) + \epsilon(0, [0,0,\dot z_{i/i\text{-}1}]^T)$
Cartesian	$[x_{i/i\text{-}1}, y_{i/i\text{-}1}, z_{i/i\text{-}1}]^T \in \mathbb{R}^3$	$0 + \epsilon(0, [\dot x_{i/i\text{-}1}, \dot y_{i/i\text{-}1}, \dot z_{i/i\text{-}1}]^T)$

In this form, it is straightforward to identify that in Equation (80), the first term yields the motion of the end-effector due to the motion of the base. The second and third terms provide the effect of the motion of the end-effector due to joint motion. We can now manipulate Equation (80) towards a more familiar structure

$$\omega_{e/I}^e = [\![q_{e/B}^*]\!]_L [\![q_{e/B}]\!]_R \star \omega_{B/I}^B + [\![q_{e/0}^*]\!]_L [\![q_{e/0}]\!]_R \star \omega_{0/O}^0 + \sum_{k=1}^{n-1} [\![q_{e/k}^*]\!]_L [\![q_{e/k}]\!]_R \star \omega_{k/k\text{-}1}^k. \tag{81}$$

Defining the vector of generalized coordinates as the vertical concatenation of the individual joint generalized coordinates, we can write

$$\omega_{e/I}^e = [\![q_{e/B}^*]\!]_L [\![q_{e/B}]\!]_R \star \omega_{B/I}^B + [\![q_{e/0}^*]\!]_L [\![q_{e/0}]\!]_R \zeta_0 \dot{\bar\alpha}_0 + \sum_{k=1}^{n-1} [\![q_{e/k}^*]\!]_L [\![q_{e/k}]\!]_R \zeta_k \dot{\bar\alpha}_k$$
$$= [\![q_{e/B}^*]\!]_L [\![q_{e/B}]\!]_R \star \omega_{B/I}^B + J(q,\zeta)\dot{\bar\alpha}. \tag{82}$$

Here, we have defined the body-frame Jacobian associated to joint motion as

$$J(q,\zeta) \triangleq \left[[\![q_{e/0}^*]\!]_L [\![q_{e/0}]\!]_R \zeta_0, \ \ldots, \ [\![q_{e/k}^*]\!]_L [\![q_{e/k}]\!]_R \zeta_k, \ \ldots, \ [\![q_{e/n\text{-}1}^*]\!]_L [\![q_{e/n\text{-}1}]\!]_R \zeta_{n\text{-}1} \right]. \tag{83}$$

The general term of the Jacobian mapping matrix, $[\![q_{e/k}^*]\!]_L [\![q_{e/k}]\!]_R \zeta_k$, where ζ_k is a screw matrix as defined in Table 6, is an improvement upon the more typical, adjoint-based methodology due to the ability of ζ_k to capture more than one degree of freedom in each of its different columns. For the case in which the adjoint formula $\mathrm{Ad}_{q_{e/k}^*}\xi_k$ is used, as in Equation (67), then ξ_k necessarily corresponds to one single generalized coordinate. In other words, the screws for the cylindrical, spherical and Cartesian joints would need to be separated into different columns, each of which has its adjoint operation applied independently.

For example, it would be easy to demonstrate that for a cylindrical ($d = 2$), spherical ($d = 3$) or Cartesian joints ($d = 3$),

$$\frac{\partial \omega_{i/i\text{-}1}^i}{\partial \dot{\bar\alpha}_k} \equiv [\![q_{e/k}^*]\!]_L [\![q_{e/k}]\!]_R \zeta_k \ \in \ \mathbb{R}^{8 \times d}, \tag{84}$$

but

$$\frac{\partial \omega_{i/i\text{-}1}^i}{\partial \dot{\bar\alpha}_k} \neq \mathrm{Ad}_{q_{e/k}^*}\xi_k \ \in \ \mathbb{H}_d, \tag{85}$$

for any physically intuitive $\xi_k \in \mathbb{H}_d^v$.

Table 6. Screw matrix for different joint types.

Joint Type	Screw Matrix, $\zeta_i = \partial \omega_{i/i\text{-}1}^i / \partial \dot{\alpha}_i$
Revolute	$[0,0,0,1,0,0,0,0]^\mathsf{T}$
Prismatic	$[0,0,0,0,0,0,0,1]^\mathsf{T}$
Spherical	$\begin{bmatrix} 0_{1\times3} \\ M(\phi_{i/i\text{-}1}, \theta_{i/i\text{-}1}, \psi_{i/i\text{-}1}) \\ 0_{4\times4} \end{bmatrix}$
Cylindrical	$\begin{bmatrix} 0,0,0,1,0,0,0,0 \\ 0,0,0,0,0,0,0,1 \end{bmatrix}^\mathsf{T}$
Cartesian	$\begin{bmatrix} 0_{5\times3} \\ \mathbb{I}_3 \end{bmatrix}$

5. Convex Constraints Using Dual Quaternions

When performing robotic operations, the incorporation of constraints is important where the safety of users, or a payload, is concerned. The dual quaternion framework is amenable to the incorporation of several convex constraints, which are of particular interest due to the availability of specialized codes to solve convex problems efficiently. The following presentation of convex constraints could be used in combination with a control approach such as the one proposed in [11], which is based on the differential dynamic programming algorithm.

In [27], the authors use dual quaternions as a pose parametrization representation to model convex state constraints for a powered landing scenario. In this section, we repurpose these same constraints for a space robotic servicing mission. The dual quaternion-based constraints will be provided without proof of convexity, since this is done in [27]. However, some properties of quaternions and some definitions are in order for a proper description of the results.

Lemma 4. *Given the quaternion $q \in \mathbb{H}$ and quaternions $r = (0, \bar{r}) \in \mathbb{H}^v$ and $y = (0, \bar{y}) \in \mathbb{H}^v$, the following equalities hold:*

$$(rq) \cdot (yq) = r \cdot y = (qr) \cdot (qy) \tag{86}$$

Proof. Using the definition of the quaternion dot product given in Table 1, the expression on the left becomes

$$
\begin{aligned}
(rq) \cdot (yq) &= \tfrac{1}{2}\left[(rq)^* yq + (yq)^* rq\right] \\
&= \tfrac{1}{2}\left[q^* r^* yq + q^* y^* rq\right] \\
&= \tfrac{1}{2} q^* \left[r^* y + y^* r\right] q \\
&= q^* (r \cdot y) q, \text{ and since } r \cdot y = (\bar{r} \cdot \bar{y}, 0_{3\times1}) = (\bar{r} \cdot \bar{y})\mathbf{1} \\
&= (\bar{r} \cdot \bar{y}) q^* q \\
&= (\bar{r} \cdot \bar{y})\mathbf{1} \\
&= r \cdot y.
\end{aligned} \tag{87}
$$

The second equality can be proven in the same manner. □

For the following facts, let us define

$$E_u \triangleq \begin{bmatrix} \mathbb{I}_4 & 0_{4\times4} \\ 0_{4\times4} & 0_{4\times4} \end{bmatrix} \tag{88}$$

and

$$E_d \triangleq \begin{bmatrix} 0_{4\times4} & 0_{4\times4} \\ 0_{4\times4} & \mathbb{I}_4 \end{bmatrix}. \tag{89}$$

Lemma 5. *Consider the dual quaternion $\boldsymbol{q}_{B/A} = q_{B/A} + \epsilon\frac{1}{2}q_{B/A}r^B_{B/A}$. Then, $\boldsymbol{q}_{B/A} \circ \boldsymbol{q}_{B/A} = (1 + \frac{1}{4}\|r^B_{B/A}\|^2, 0_{3\times1}) + \epsilon 0$*

Proof. By definition, $\boldsymbol{q}_{B/A} \circ \boldsymbol{q}_{B/A} = (q_{B/A} + \epsilon\frac{1}{2}q_{B/A}r^B_{B/A}) \circ (q_{B/A} + \epsilon\frac{1}{2}q_{B/A}r^B_{B/A}) = q_{B/A} \cdot q_{B/A} + (\frac{1}{2}q_{B/A}r^B_{B/A}) \cdot (\frac{1}{2}q_{B/A}r^B_{B/A}) + \epsilon 0$. By the unit norm constraint of the unit quaternions and applying Lemma 4 on the second summand, $\boldsymbol{q}_{B/A} \circ \boldsymbol{q}_{B/A} = (1 + \frac{1}{4}r^B_{B/A} \cdot r^B_{B/A}, 0_{3\times1}) + \epsilon 0$, from which the result follows. □

Lemma 6. *Consider the dual quaternion $\boldsymbol{q}_{B/A} = q_{B/A} + \epsilon\frac{1}{2}q_{B/A}r^B_{B/A}$. Then, $\boldsymbol{q}_{B/A} \circ (E_u \star \boldsymbol{q}_{B/A}) = 1$.*

Proof. Using the definition of E_u, we have $\boldsymbol{q}_{B/A} \circ (E_u \star \boldsymbol{q}_{B/A}) = \boldsymbol{q}_{B/A} \circ (q_{B/A} + \epsilon 0) = q_{B/A} \cdot q_{B/A} + \epsilon 0$. The result follows from the unit constraint of a unit quaternion. □

Lemma 7. *Consider the dual quaternion $\boldsymbol{q}_{B/A} = q_{B/A} + \epsilon\frac{1}{2}q_{B/A}r^B_{B/A}$. Then, $\boldsymbol{q}_{B/A} \circ (E_d \star \boldsymbol{q}_{B/A}) = \frac{1}{4}\|r^B_{B/A}\|^2 + \epsilon 0$.*

Proof. Using the definition of E_d, we have $\boldsymbol{q}_{B/A} \circ (E_d \star \boldsymbol{q}_{B/A}) = \boldsymbol{q}_{B/A} \circ (0 + \epsilon\frac{1}{2}q_{B/A}r^B_{B/A}) = (\frac{1}{2}q_{B/A}r^B_{B/A}) \cdot (\frac{1}{2}q_{B/A}r^B_{B/A}) + \epsilon 0$. The result follows from application of Lemma 4. □

Lemma 8. *Consider $\|r^B_{B/A}\| \leq \delta$. Then, $\boldsymbol{q}_{B/A} \circ \boldsymbol{q}_{B/A} \leq 1 + \frac{1}{4}\delta^2$.*

Proof. From Lemma 5, it follows that $\boldsymbol{q}_{B/A} \circ \boldsymbol{q}_{B/A} = 1 + \frac{1}{4}\|r^B_{B/A}\|^2 \leq 1 + \frac{1}{4}\delta^2$. □

Corollary 1. *Given the bound $\|r^B_{B/A}\| \leq \delta$, it follows that $\boldsymbol{q}_{B/A} \circ \boldsymbol{q}_{B/A} \in \left[1, 1 + \frac{1}{4}\delta^2\right]$, which is a closed and bounded set.*

It is worth emphasizing that in Lemmas 5 and 8 the bijective mapping between the circle product and the real-line is implied. In other words, since the circle product between two dual quaternions $\boldsymbol{a} \circ \boldsymbol{b} = s\mathbf{1}$ for some $s \in \mathbb{R}$, it will be commonly interpreted as $\boldsymbol{a} \circ \boldsymbol{b} = s$ for simplicity of exposition.

We are now ready to introduce three types of constraints in terms of dual quaternions:

1. Line-of-sight constraints.
2. Approach slope angle constraints, of which upper-and-lower bound constraints is a re-interpretation of the geometry.
3. Body attitude constraint with respect to an inertial direction.

For this, we will use notation consistent with [27]. Additionally, we require two auxiliary frames. We will define G as fixed on a gripper, and A as fixed on the target (say, an asteroid, or an object of interest) to be captured.

Proposition 1. *Consider the domain* $\mathcal{D} = \{\boldsymbol{q}_{\text{G/A}} \in \mathbb{H}_d : \boldsymbol{q}_{\text{G/A}} \circ \boldsymbol{q}_{\text{G/A}} \leq 1 + \frac{1}{4}\delta^2\}$. *The line of sight constraint depicted in Figure 5 can be encoded as*

$$r_{\text{A/G}}^{\text{G}} \cdot \hat{y}^{\text{G}} \geq \|r_{\text{A/G}}^{\text{G}}\| \cos\theta, \tag{90}$$

and it requires that the angle between $r_{\text{A/G}}^{\text{G}}$ *and* $\hat{y}^{\text{G}}$ *remains less than* θ. *Using dual quaternions, this constraint can be equivalently expressed as*

$$-\boldsymbol{q}_{\text{G/A}} \circ (M_H \star \boldsymbol{q}_{\text{G/A}}) + 2\|E_d \boldsymbol{q}_{\text{G/A}}\| \cos\theta \leq 0, \tag{91}$$

where

$$M_H = \begin{bmatrix} 0_{4\times4} & [\![\hat{y}^{\text{G}}]\!]_{\text{R}}^{\text{T}} \\ [\![\hat{y}^{\text{G}}]\!]_{\text{R}} & 0_{4\times4} \end{bmatrix}, \tag{92}$$

and it is convex over $\mathcal{D}$.

Proposition 2. *Consider the domain* $\mathcal{D} = \{\boldsymbol{q}_{\text{G/A}} \in \mathbb{H}_d : \boldsymbol{q}_{\text{G/A}} \circ \boldsymbol{q}_{\text{G/A}} \leq 1 + \frac{1}{4}\delta^2\}$. *The approach slope constraint depicted in Figure 6, and the upper-and-lower bounded approach constraint depicted in Figure 7, can be encoded as*

$$r_{\text{G/A}}^{\text{A}} \cdot \hat{z}^{\text{A}} \geq \|r_{\text{G/A}}^{\text{A}}\| \cos\phi, \tag{93}$$

and it requires that the angle between $r_{\text{G/A}}^{\text{A}}$ *and* $\hat{z}^{\text{A}}$ *remains less than* ϕ. *Using dual quaternions, this constraint can be equivalently expressed as*

$$-\boldsymbol{q}_{\text{G/A}} \circ (M_G \star \boldsymbol{q}_{\text{G/A}}) + 2\|E_d \boldsymbol{q}_{\text{G/A}}\| \cos\phi \leq 0, \tag{94}$$

where

$$M_G = \begin{bmatrix} 0_{4\times4} & [\![\hat{z}^{\text{A}}]\!]_{\text{L}}^{\text{T}} \\ [\![\hat{z}^{\text{A}}]\!]_{\text{L}} & 0_{4\times4} \end{bmatrix}, \tag{95}$$

and it is convex over $\mathcal{D}$.

Proposition 3. *Consider the domain* $\mathcal{D} = \{\boldsymbol{q}_{\text{B/I}} \in \mathbb{H}_d : \boldsymbol{q}_{\text{B/I}} \circ \boldsymbol{q}_{\text{B/I}} \leq 1 + \frac{1}{4}\delta^2\}$. *The attitude constraint depicted in Figure 8 can be encoded as*

$$\hat{n}^{\text{I}} \cdot (q_{\text{B/I}} \hat{n}^{\text{B}} q_{\text{B/I}}^*) \geq \cos\psi, \tag{96}$$

and it requires that the angle between the inertially fixed vector $\hat{n}^{\text{I}}$ *and the body fixed vector* $\hat{n}^{\text{B}}$ *remains less than* ψ. *Using dual quaternions, this constraint can be equivalently expressed as*

$$\boldsymbol{q}_{\text{B/I}} \circ (M_A \star \boldsymbol{q}_{\text{B/I}}) + \cos\psi \leq 0, \tag{97}$$

where

$$M_A = \begin{bmatrix} [\![\hat{z}^{\text{I}}]\!]_{\text{L}} [\![\hat{z}^{\text{B}}]\!]_{\text{R}} & 0_{4\times4} \\ 0_{4\times4} & 0_{4\times4} \end{bmatrix}, \tag{98}$$

and it is convex over $\mathcal{D}$.

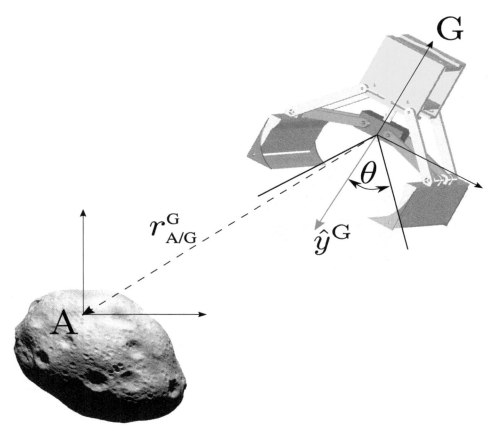

Figure 5. Line-of-sight constraint during grappling.

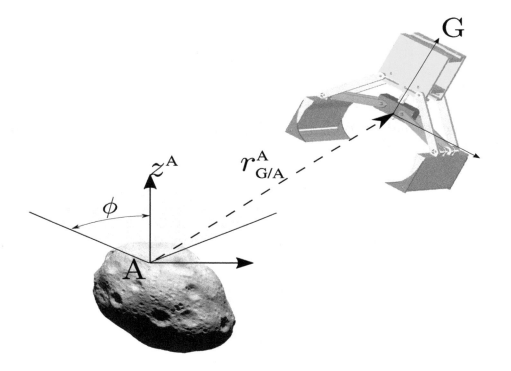

Figure 6. Approach slope constraint.

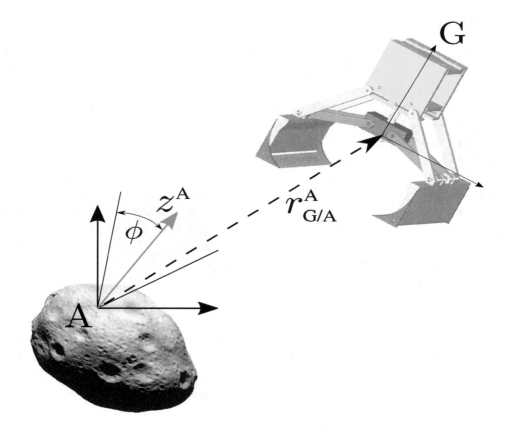

Figure 7. Upper-and-lower bounds constraint.

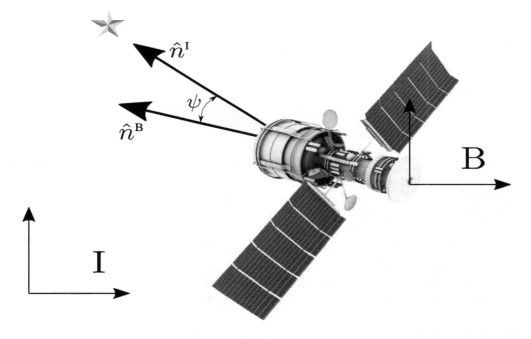

Figure 8. General attitude constraint with respect to inertial directions.

6. Conclusions

In this paper we have explored the use of dual quaternions for robot modeling. In particular, the main contribution of this paper is a generalizable framework to capture the kinematics of spacecraft-mounted robotic manipulators using a dual quaternion approach. We took a bare-bones approach that built up to a convenient description of the end-effector's dual velocity, making use of a more intuitive forward kinematics methodology than the existing methods in the literature. Previous works on robot kinematics using dual quaternions provided either strict geometry-dependent approaches or were only applicable to fixed-base robots. The work presented herein is highly relevant in combination with the latest literature in dynamic modeling of robot manipulators using dual quaternions. Additionally, in our study of kinematics, we developed a convenient and simple-to-implement representation of the body-frame Jacobian matrix. The proposed form of the Jacobian exploits a convenient matrix representation of the adjoint dual quaternion transformation so that, in combination with the newly proposed form of the screw matrix, it avoids the artificial separation of the contribution by the generalized speeds of a given joint. Finally, we have provided a summary, and re-interpretation, of several existing results on the topic of dual quaternions, emphasizing their applicability on spacecraft-mounted robots. These included results on the exponential and logarithmic maps, an exposition on the use of the DH parameters, and finally the casting of the dual quaternion representation of constraints (originally developed for EDL purposes) interpreted in the context of a gripper-target system on-board a spacecraft.

Future work in this area will aim at implementing kinematic control laws for end-effector pose control when the based is not fixed to an inertial reference frame. This should be possible by following the steps in Özgür and Mezouar [20], and through the use of the Generalized Jacobian Matrix [28].

Author Contributions: Both authors contributed equally in the preparation of the manuscript.

Acknowledgments: The authors would like to thank David S. Bayard, from the G&C section at the Jet Propulsion Laboratory for valuable conversations that improved the content of this article.

Abbreviations

The following abbreviations are used in this manuscript:

DQ   Dual quaternions
EDL   Entry, Descent and Landing

References

1.   Reed, B.B.; Smith, R.C.; Naasz, B.J.; Pellegrino, J.F.; Bacon, C.E. The Restore-L Servicing Mission. *AIAA Space Forum* **2016**. [CrossRef]
2.   NASA Goddard Space Flight Center. On-Orbit Satellite Servicing Study, Project Report. In *National Aeronautics and Space Administration, Goddard Space Flight Center*; Technical Report; NASA: Greenbelt, MD, USA, 2010.
3.   Saha, S.K.; Shah, S.V.; Nandihal, P.V. Evolution of the DeNOC-based dynamic modelling for multibody systems. *Mech. Sci.* **2013**, *4*, 1–20. [CrossRef]
4.   Todorov, E.; Erez, T.; Tassa, Y. MuJoCo: A physics engine for model-based control. In Proceedings of the 2012 IEEE/RSJ International Conference on Intelligent Robots and Systems, Vilamoura, Portugal, 7–12 October 2012.
5.   Featherstone, R. *Rigid Body Dynamics Algorithms*; Springer: Berlin/Heidelberg, Germany, 2008.
6.   Wang, J.; Liang, H.; Sun, Z.; Zhang, S.; Liu, M. Finite-Time Control for Spacecraft Formation with Dual-Number-Based Description. *J. Guid. Control Dyn.* **2012**, *35*, 950–962. [CrossRef]
7.   Filipe, N.; Tsiotras, P. Simultaneous Position and Attitude Control Without Linear and Angular Velocity Feedback Using Dual Quaternions. In Proceedings of the 2013 American Control Conference, Washington, DC, USA, 17–19 June 2013; pp. 4815–4820.

8. Filipe, N.; Tsiotras, P. Rigid Body Motion Tracking Without Linear and Angular Velocity Feedback Using Dual Quaternions. In Proceedings of the European Control Conference, Zurich, Switzerland, 17–19 July 2013; pp. 329–334.

9. Seo, D. Fast Adaptive Pose Tracking Control for Satellites via Dual Quaternion Upon Non-Certainty Equivalence Principle. *Acta Astronaut.* **2015**, *115*, 32–39. [CrossRef]

10. Tsiotras, P.; Valverde, A. Dual Quaternions as a Tool for Modeling, Control, and Estimation for Spacecraft Robotic Servicing Missions. In Proceedings of the Texas A&M University/AAS John L. Junkins Astrodynamics Symposium, College Station, TX, USA, 20–21 May 2018.

11. Valverde, A.; Tsiotras, P. Modeling of Spacecraft-Mounted Robot Dynamics and Control Using Dual Quaternions. In Proceedings of the 2018 American Control Conference, Milwaukee, WI, USA, 27–29 June 2018.

12. Perez, A.; McCarthy, J. Dual Quaternion Synthesis of Constrained Robotic Systems. *J. Mech. Des.* **2004**, *126*, 425–435. [CrossRef]

13. Perez, A. Dual Quaternion Synthesis of Constrained Robotic Systems. Ph.D. Thesis, University of California, Irvine, CA, USA, 2003.

14. Pennestrì, E.; Stefanelli, R. Linear algebra and numerical algorithms using dual numbers. *Multibody Syst. Dyn.* **2007**, *18*, 323–344. [CrossRef]

15. Radavelli, L.A.; De Pieri, E.R.; Martins, D.; Simoni, R. Points, Lines, Screws and Planes in Dual Quaternions Kinematics. In *Advances in Robot Kinematics*; Lenarcic, J., Khatib, O., Eds.; Springer: Berlin, Germany, 2014; pp. 285–293.

16. Leclercq, G.; Lefèvre, P.; Blohm, G. 3-D Kinematics Using Dual Quaternions: Theory and Applications in Neuroscience. *Front. Behav. Neurosci.* **2013**, *7*, 1–7. [CrossRef] [PubMed]

17. Quiroz-Omaña, J.J.; Adorno, B.V. Whole-Body Kinematic Control of Nonholonomic Mobile Manipulators Using Linear Programming. *J. Intell. Robot. Syst.* **2017**, *91*, 263–278. [CrossRef]

18. Brodsky, V.; Shoham, M. Dual numbers representation of rigid body dynamics. *Mech. Mach. Theory* **1999**, *34*, 693–718. [CrossRef]

19. Filipe, N. Nonlinear Pose Control and Estimation for Space Proximity Operations: An Approach Based on Dual Quaternions. Ph.D. Thesis, Georgia Institute of Technology, Atlanta, GA, USA, 2014.

20. Özgür, E.; Mezouar, Y. Kinematic Modeling and Control of a Robot Arm Using Unit Dual Quaternions. *Robot. Autom. Syst.* **2016**, *77*, 66–73. [CrossRef]

21. Filipe, N.; Tsiotras, P. Adaptive Position and Attitude-Tracking Controller for Satellite Proximity Operations Using Dual Quaternions. *J. Guid. Control Dyn.* **2014**, *38*, 566–577. [CrossRef]

22. Bhat, S.; Bernstein, D. A topological obstruction to global asymptotic stabilization of rotational motion and the unwinding phenomenon. In Proceedings of the 1998 American Control Conference, Philadelphia, PA, USA, 26 June 1998; doi:10.1109/acc.1998.688361.

23. Murray, R.M.; Li, Z.; Sastry, S.S. *A Mathematical Introduction to Robotic Manipulation*; CRC Press: Boca Raton, FL, USA, 1994.

24. Jazar, R.N. *Theory of Applied Robotics: Kinematics, Dynamics, and Control*; Springer: Berlin, Germany, 2010.

25. Gan, D.; Liao, Q.; Wei, S.; Dai, J.; Qiao, S. Dual Quaternion-Based Inverse Kinematics of the General Spatial 7R mechanism. *Proc. Inst. Mech. Eng. C J. Mech. Eng.* **2008**, *222*, 1593–1598. [CrossRef]

26. Adorno, B.V. Two-Arm Manipulation: From Manipulators to Enhanced Human-Robot Collaboration. Ph.D. Thesis, Université Montpellier II Sciences et Techniques du Languedoc, Montpellier, France, 2011.

27. Lee, U.; Mesbahi, M. Constrained Autonomous Precision Landing via Dual Quaternions and Model Predictive Control. *J. Guid. Control Dyn.* **2017**, *40*, 292–308. [CrossRef]

28. Umetani, Y.; Yoshida, K. Resolved motion rate control of space manipulators with generalized Jacobian matrix. *IEEE Trans. Robot. Autom.* **1989**, *5*, 303–314. [CrossRef]

7

Optimal Kinematic Design of a 6-UCU Kind Gough-Stewart Platform with a Guaranteed Given Accuracy

Guojun Liu

School of Mechanical Engineering, Hunan Institute of Science and Technology, Yueyang 414006, China; liuguojun_iest@163.com.

Abstract: The 6-UCU (U-universal joint; C-cylinder joint) kind Gough-Stewart platform is extensively employed in motion simulators due to its high accuracy, large payload, and high-speed capability. However, because of the manufacturing and assembling errors, the real geometry may be different from the nominal one. In the design process of the high-accuracy Gough-Stewart platform, one needs to consider these errors. The purpose of this paper is to propose an optimal design method for the 6-UCU kind Gough-Stewart platform with a guaranteed given accuracy. Accuracy analysis of the 6-UCU kind Gough-Stewart platform is presented by considering the limb length errors and joint position errors. An optimal design method is proposed by using a multi-objective evolutionary algorithm, the non-dominated sorting genetic algorithm II (NSGA-II). A set of Pareto-optimal parameters was found by applying the proposed optimal design method. An engineering design case was studied to verify the effectiveness of the proposed method.

Keywords: accuracy analysis; optimal design; multi-objective evolutionary algorithms; NSGA-II; 6-UCU kind Gough-Stewart platform

---

1. Introduction

With the advantages of high rigidity, high precision, and large carrying capacity, Gough-Stewart platforms (GSPs) are extensively used in virtual-reality motion simulators [1]. Many scholars studied the optimal design of the 6-UPS kind GSP [1], where U stands for the universal joint, P for the prismatic joint, and S for the spherical joint. When compared with spherical joints, universal joints can bear more tension [2]. Universal joints are extensively used as the passive joints of GSPs to connect hydraulic cylinders or electric cylinders to the moving platform and fixed base. Universal joints are used as the passive joints of the universal tyre test machine designed by Gough et al. [3–5], the motion simulator patented by Cappel [6], commercial flight simulators [7], the Ampelmann system [8], the Moog FCS 5000E motion base [9], the VARIAX machine tool [10], and the AMiBA hexapod telescope mount [11]. Universal joints are also used as the passive joints of the docking test system, and about 60 motion simulators designed by the team of Professor Junwei Han at the Harbin Institute of Technology, China. The cylinders of the hydraulic and electric actuators that can not only translate along the axis, but also rotate along the axis are cylindrical joints instead of prismatic ones. These GSPs are 6-UCU parallel manipulators [12,13], where C stands for the cylinder joint.

In highly accurate positioning applications, such as the docking test systems, high accuracy is required to achieve good simulation results. in China, the translational motion errors of the moving platform must be less than 1 mm in the whole workspace of the docking test system [14]. Many researchers studied the accuracy analysis of GSPs. Wang and Masory studied the effects of manufacturing errors on the accuracy of a GSP by modeling the GSP as serial legs based on the Denavit–Hartenberg method [15]. Ropponen and Arai presented the error model of a modified GSP by using the differentiation method [16]. Patel and Ehmann established an error model of a GSP, addressing all possible sources of errors [17]. Wang and Ehmann developed first- and second-order error models of a GSP [18]. Masory et al. presented an effective method of identifying kinematic parameters of a GSP using pose measurements [19]. Cong et al. developed a kinematic calibration method of a GSP by using a three-dimensional coordinate measuring machine containing actuator errors and passive joint errors [20]. Dai et al. proposed an accuracy analysis method for a docking test system [14]. Merlet and Daney proposed an optimal method of parallel manipulator considering manufacturing errors by using interval analysis [21–23]. Because of the dependence of the variables, a high level of expertise in interval analysis is needed for the proper usage of parallel manipulators [22]. To the best of our knowledge, there are few papers considering the errors of a GSP in optimal design processes.

In the optimal design process of a high-precision GSP, one needs to find the optimal solutions which meet the given accuracy requirement. An optimal design method for a GSP is proposed in this paper, so as to meet these accuracy requirements.

In this paper, the error model of the 6-UCU kind GSP is derived in Section 2. The optimal design method is proposed in Section 3. Section 4 presents a case study to illustrate the effectiveness of the proposed optimal design. The conclusions are given in Section 5.

2. Differential Error Model

In this section, the relationship between the structural errors and end-point errors of a 6-UCU kind GSP is discussed. The 6-UCU kind GSP consisted of a moving platform, a fixed base, and six limbs, as shown in Figures 1 and 2. The ith limb was connected to the fixed base via a universal joint, whose joint center point is denoted as B_i, and connected to the mobile platform via a universal joint with center P_i, as shown in Figure 2. For convenience, a Cartesian coordinate frame, $O_1 - X_1Y_1Z_1$, was attached to the moving platform with origin O_1. A Cartesian coordinate frame, $O - XYZ$, was attached to the fixed base with origin O.

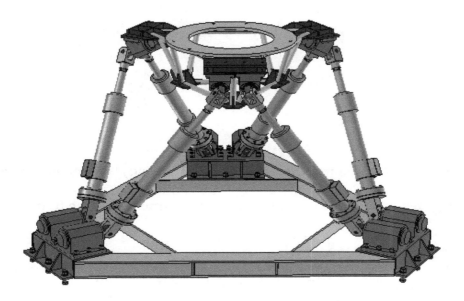

Figure 1. The 6-UCU (U—universal joint; C—cylinder joint) kind Gough-Stewart platform (GSP).

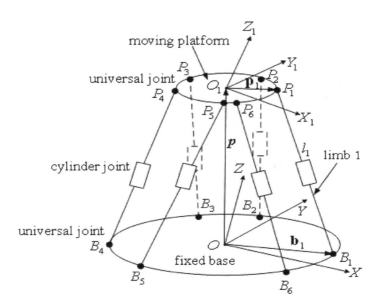

Figure 2. Schematic of the 6-UCU kind GSP.

The positions of the universal joints attached to the moving platform and the fixed base of the GSP formed semi-hexagons, as shown in Figure 3, where r_P is the radius of the platform-joint attachment circle, d_P is the distance between the shorter edges of the attachment point semi-hexagon on the moving platform, r_B is the radius of the base-joint attachment circle, and d_B is the distance between the shorter edges of the attachment point semi-hexagon on the fixed base.

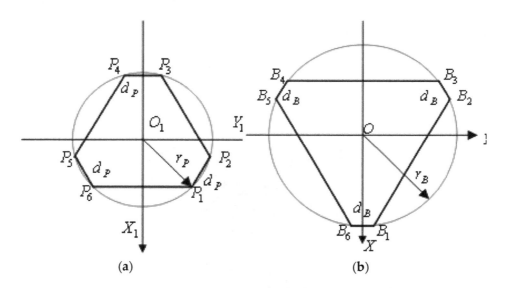

(a)   **(b)**

Figure 3. Schematic of the universal joint positions. **(a)** Moving platform; and **(b)** fixed base.

Referring to Figures 1 and 2, the position vector of P_i expressed in the frame $O - XYZ$ could be written as

$$\mathbf{p} + \mathbf{R}^{\mathbf{L}}\mathbf{p}_i = \mathbf{b}_i + l_{1i}\mathbf{n}_{1i} + l_{2i}\mathbf{n}_{2i}, \qquad (1)$$

where $\mathbf{R}$ is the rotation matrix from the frame $O_1 - X_1Y_1Z_1$ to $O - XYZ$, $\mathbf{p}$ is the position coordinate of the point O_1 measured in the frame $O - XYZ$, $\mathbf{b}_i$ is the position coordinate of the point B_i measured in the frame $O - XYZ$, $\mathbf{p}_i$ is expressed as $^{\mathbf{L}}\mathbf{p}_i$ in the frame $O - XYZ$, $^{\mathbf{L}}\mathbf{p}_i$ is the coordinate of the point P_i measured in the frame $O_1 - X_1Y_1Z_1$, l_{1i} is the distance from the point B_i to the lower plane of the

piston, l_{2i} is the distance from the point P_i to the lower plane of the piston, $\mathbf{n}_i$ is the unit vector of $\overrightarrow{B_i P_i}$, $\mathbf{n}_{1i}$ and $\mathbf{n}_{2i}$ are the unit vectors of $\mathbf{n}_i$ fixed to the cylinder and piston, respectively, and $\mathbf{n}_i = \mathbf{n}_{1i} = \mathbf{n}_{2i}$.
$\mathbf{^L p}_i$ and $\mathbf{b}_i$ are

$$\mathbf{^L p}_i = \begin{bmatrix} r_P \cos \theta_{Pi} & r_P \sin \theta_{Pi} & h_P \end{bmatrix}^T, \tag{2}$$

$$\mathbf{b}_i = \begin{bmatrix} r_B \cos \theta_{Bi} & r_B \sin \theta_{Bi} & h_B \end{bmatrix}^T, \tag{3}$$

where T is the matrix transpose, h_P is the z coordinate of the upper universal joint of the ith limb in the frame $O_1 - X_1 Y_1 Z_1$, h_B is the z coordinate of the lower universal joint of the ith limb in the frame $O - XYZ$, and θ_{Pi} and θ_{Bi} are as follows:

$$\begin{aligned} \mathbf{\theta}_P &= \begin{bmatrix} \theta_{P1} & \theta_{P2} & \theta_{P3} & \theta_{P4} & \theta_{P5} & \theta_{P6} \end{bmatrix}^T \\ &= \begin{bmatrix} \eta_A & \frac{2\pi}{3} - \eta_A & \frac{2\pi}{3} + \eta_A & \frac{4\pi}{3} - \eta_A & \frac{4\pi}{3} + \eta_A & -\eta_A \end{bmatrix}^T, \end{aligned} \tag{4}$$

$$\begin{aligned} \mathbf{\theta}_B &= \begin{bmatrix} \theta_{B1} & \theta_{B2} & \theta_{B3} & \theta_{B4} & \theta_{B5} & \theta_{B6} \end{bmatrix}^T \\ &= \begin{bmatrix} \eta_B & \frac{2\pi}{3} - \eta_B & \frac{2\pi}{3} + \eta_B & \frac{4\pi}{3} - \eta_B & \frac{4\pi}{3} + \eta_B & -\eta_B \end{bmatrix}^T, \end{aligned} \tag{5}$$

where

$$\eta_A = \frac{\pi}{3} - \arcsin\left(\frac{d_P}{2r_P}\right), \tag{6}$$

$$\eta_B = \arcsin\left(\frac{d_B}{2r_B}\right). \tag{7}$$

For the ith leg, the limb length l_i was derived as

$$l_i = l_{1i} + l_{2i}, \tag{8}$$

where l_i is the length of $\overrightarrow{B_i P_i}$.

Differentiating Equation (1), we obtained

$$\delta\mathbf{p} + \delta\mathbf{R}\mathbf{^L p}_i + \mathbf{R}\delta\mathbf{^L p}_i = \delta\mathbf{b}_i + \delta l_{1i}\mathbf{n}_{1i} + l_{1i}\delta\mathbf{n}_{1i} + \delta l_{2i}\mathbf{n}_{2i} + l_{2i}\delta\mathbf{n}_{2i}, \tag{9}$$

where $\delta\mathbf{R}$ could be rewritten as follows, according to Reference [16]:

$$\delta\mathbf{R} = \delta\mathbf{\theta} \times \mathbf{R} = \begin{bmatrix} 0 & -\delta\theta_Z & \delta\theta_Y \\ \delta\theta_Z & 0 & -\delta\theta_X \\ -\delta\theta_Y & \delta\theta_X & 0 \end{bmatrix} \mathbf{R} = \Omega\mathbf{R}, \tag{10}$$

where $\delta\mathbf{\theta} = \begin{bmatrix} \delta\theta_X & \delta\theta_Y & \delta\theta_Z \end{bmatrix}^T$ is the orientation error vector of the moving platform in the frame $O - XYZ$, and $\delta\mathbf{p} = \begin{bmatrix} \delta p_X & \delta p_Y & \delta p_Z \end{bmatrix}^T$ is the translational error vector of the moving platform in the frame $O - XYZ$.

Multiplication of $\mathbf{n}_i^T$ on both sides of Equation (10) yielded

$$\begin{aligned} &\mathbf{n}_i^T\delta\mathbf{p} + \mathbf{n}_i^T\delta\mathbf{\theta} \times \mathbf{R}\mathbf{^L p}_i + \mathbf{n}_i^T\mathbf{R}\delta\mathbf{^L p}_i \\ &= \mathbf{n}_i^T\delta\mathbf{b}_i + \mathbf{n}_i^T\delta l_{1i}\mathbf{n}_{1i} + \mathbf{n}_i^T l_{1i}\delta\mathbf{n}_{1i} + \mathbf{n}_i^T\delta l_{2i}\mathbf{n}_{2i} + \mathbf{n}_i^T l_{2i}\delta\mathbf{n}_{2i} \end{aligned}. \tag{11}$$

Since $\mathbf{n}_i^T \mathbf{n}_i = 1$, $\mathbf{n}_i = \mathbf{n}_{1i} = \mathbf{n}_{2i}$, $\delta l_{2i} = 0$, and $\delta l_i = \delta l_{1i}$, we had $\mathbf{n}_i^T \delta \mathbf{n}_i = 0$, $\mathbf{n}_i^T \delta \mathbf{n}_{1i} = 0$, and $\mathbf{n}_i^T \delta \mathbf{n}_{2i} = 0$. Therefore, Equation (11) could be rewritten as

$$\begin{aligned}
\mathbf{n}_i^T \delta \mathbf{p} + \mathbf{n}_i^T \delta \boldsymbol{\theta} \times \mathbf{R}^\mathbf{L} \mathbf{p}_i + \mathbf{n}_i^T \mathbf{R} \delta^\mathbf{L} \mathbf{p}_i & \\
= \mathbf{n}_i^T \delta \mathbf{b}_i + \mathbf{n}_i^T \delta l_{1i} \mathbf{n}_i & \\
= \mathbf{n}_i^T \delta \mathbf{b}_i + \delta l_{1i} & \\
= \mathbf{n}_i^T \delta \mathbf{b}_i + \delta l_i &
\end{aligned} \tag{12}$$

δl_i could be derived as

$$\begin{aligned}
\delta l_i = & \ \mathbf{n}_i^T \delta \mathbf{p} + \mathbf{n}_i^T \delta \boldsymbol{\theta} \times \mathbf{R}^\mathbf{L} \mathbf{p}_i + \mathbf{n}_i^T \mathbf{R} \delta^\mathbf{L} \mathbf{p}_i - \mathbf{n}_i^T \delta \mathbf{b}_i \\
= & \begin{bmatrix} \mathbf{n}_i^T & (\mathbf{R}^\mathbf{L} \mathbf{p}_i \times \mathbf{n}_i)^T \end{bmatrix} \begin{bmatrix} \delta \mathbf{p} \\ \delta \boldsymbol{\theta} \end{bmatrix} + \\
& \begin{bmatrix} \mathbf{n}_i^T \mathbf{R} & (-\mathbf{n}_i^T) \end{bmatrix} \begin{bmatrix} \delta^\mathbf{L} \mathbf{p}_i \\ \delta \mathbf{b}_i \end{bmatrix}
\end{aligned} \tag{13}$$

Once all six limbs were assembled, the errors could be expressed as

$$\delta \mathbf{l} = \mathbf{J}_P \delta \mathbf{x} + \mathbf{J}_s \delta \mathbf{s}, \tag{14}$$

where

$$\delta \mathbf{l} = \begin{bmatrix} \delta l_1 & \cdots & \delta l_6 \end{bmatrix}^T \in \mathfrak{R}^{6 \times 1}, \tag{15}$$

$$\delta \mathbf{x} = \begin{bmatrix} \delta \mathbf{p} \\ \delta \boldsymbol{\theta} \end{bmatrix} \in \mathfrak{R}^{6 \times 1}, \tag{16}$$

$$\delta \mathbf{s} = \begin{bmatrix} \delta^\mathbf{L} \mathbf{p}_i \\ \delta \mathbf{b}_1 \\ \vdots \\ \delta^\mathbf{L} \mathbf{p}_6 \\ \delta \mathbf{b}_6 \end{bmatrix} \in \mathfrak{R}^{36 \times 1}, \tag{17}$$

$$\mathbf{J}_P = \begin{bmatrix} \mathbf{n}_1^T & (\mathbf{R}^\mathbf{L} \mathbf{p}_i \times \mathbf{n}_1)^T \\ \vdots & \vdots \\ \mathbf{n}_6^T & (\mathbf{R}^\mathbf{L} \mathbf{p}_6 \times \mathbf{n}_6)^T \end{bmatrix} \in \mathfrak{R}^{6 \times 6}, \tag{18}$$

$$\mathbf{J}_s = \begin{bmatrix} \mathbf{n}_1^T \mathbf{R} & (-\mathbf{n}_1^T) & \cdots & \mathbf{0}_{1 \times 3} & \mathbf{0}_{1 \times 3} \\ \vdots & \vdots & \ddots & \vdots & \vdots \\ \mathbf{0}_{1 \times 3} & \mathbf{0}_{1 \times 3} & \cdots & \mathbf{n}_6^T \mathbf{R} & (-\mathbf{n}_6^T) \end{bmatrix} \in \mathfrak{R}^{6 \times 36}, \tag{19}$$

$$\mathbf{0}_{1 \times 3} = \begin{bmatrix} 0 & 0 & 0 \end{bmatrix}. \tag{20}$$

If the inverse of $\mathbf{J}_P$ existed, then $\delta \mathbf{x}$ could be derived as

$$\delta \mathbf{x} = \mathbf{J}_P^{-1} \delta \mathbf{l} - \mathbf{J}_P^{-1} \mathbf{J}_s \delta \mathbf{s}. \tag{21}$$

The first term on the right side represents the actuation-induced error, and the second term is the error caused by the position errors of the joints [16].

3. Optimal Design Method

In engineering, we may assume that the manufacturing tolerances on the geometrical parameters are bounded and the maximum values of $\delta \mathbf{l}$ and $\delta \mathbf{s}$ are known as a function of the manufacturing method. The maximum values of $\delta\mathbf{x}$ for the optimal design solutions had to be lower than the given accuracy by customers, before being dealt with as a constraint in the optimal design process in this paper.

Many scholars studied the optimal design of a GSP with only one optimal solution [24]. In practice, many different functional requirements of robots are intended to be satisfied [25]; thus, it is more appropriate to have multiple optimization solutions after the optimal kinematic design [24]. The condition number and determinant of the kinematic Jacobian matrix, $\mathbf{J}_P$, of a GSP were extensively used as optimal design objectives [24]; therefore, they were chosen as objective functions in the optimal design process in this paper. Accordingly, they were defined as

$$\text{cond}(\mathbf{J}_P) = \frac{\sigma_{\max}(\mathbf{J}_P)}{\sigma_{\min}(\mathbf{J}_P)}, \tag{22}$$

$$\omega = \sqrt{\det(\mathbf{J}_P\mathbf{J}_P^T)} = |\det(\mathbf{J}_P)|, \tag{23}$$

where $\sigma_{\max}(\mathbf{J}_P)$ and $\sigma_{\min}(\mathbf{J}_P)$ are the maximum and minimum singular values of $\mathbf{J}_P$ at one pose, respectively, and $\det(\mathbf{J}_P)$ is the determinant of $\mathbf{J}_P$.

In the optimal design process, evolutionary algorithms are extensively used to search the optimal solutions [26]. In this paper, the elitist non-dominated sorting genetic algorithm version II (NSGA-II), developed by Deb et al. [27], was employed to solve the multi-objective optimal design problems of the 6-UCU kind GSP, due to its good spread of solutions and convergence to obtain the Pareto front. Real-coded NSGA-II with a simulated binary crossover (SBX) operator [28] and a polynomial mutation operator [29] were adopted in this paper to search the minimum values of the objective functions. The parameters of the NSGA-II are shown in Table 1.

Table 1. Parameters of the non-dominated sorting genetic algorithm version II (NSGA-II).

Number of Iterations	1000
Population size	50
Crossover probability	0.9
Mutation probability	0.1
Distribution index for the simulated binary crossover (SBX)	20
Distribution index for the polynomial mutation	20

The optimal design procedures were proposed as described below.

Step 1. The customers usually provided the maximum motion requirements for each degree of freedom (DOF), as shown in Table 2.

Table 2. Typical requirements by customers.

	Payload		
Degree of Freedom (DOF)	Maximum Excursion	Speed	Acceleration
Roll	$\pm Ro_X(°)$	$\pm Rv_X(°/s)$	$\pm Ra_X(°/s^2)$
Pitch	$\pm Ro_Y(°)$	$\pm Rv_Y(°/s)$	$\pm Ra_Y(°/s^2)$
Yaw	$\pm Ro_Z(°)$	$\pm Rv_Z(°/s)$	$\pm Ra_Z(°/s^2)$
Surge	$\pm Tr_X(m)$	$\pm Tv_X(m/s)$	$\pm Ta_X(m/s^2)$
Sway	$\pm Tr_Y(m)$	$\pm Tv_Y(m/s)$	$\pm Ta_Y(m/s^2)$
Heave	$\pm Tr_Z(m)$	$\pm Tv_Z(m/s)$	$\pm Ta_Z(m/s^2)$

We transformed the requirements of the customer to 12 typical trajectories along six single axes [24]. Two typical travel trajectories of translation along the X axis were

$$S_X(t) = Tr_X \sin\left(\frac{Tv_X}{Tr_X}t\right), \tag{24}$$

$$S_X(t) = \frac{(Tv_X)^2}{Ta_X} \sin\left(\frac{Ta_X}{Tv_X}t\right), \tag{25}$$

where t is the run time.

Step 2. At each pose, $cond(\mathbf{J}_P)$ was calculated using Equation (22). When it was at or near the singularity, calculating $\delta\mathbf{x}$ using Equation (21) was potentially wrong. We used another method to solve this problem, as follows: if $cond(\mathbf{J}_P) > 10^6$, then $\delta\mathbf{x} = \begin{bmatrix} 10^7 & 10^7 & 10^7 & 10^7 & 10^7 & 10^7 \end{bmatrix}^T$; if $cond(\mathbf{J}_P) \leq 10^6$, $\delta\mathbf{x}$ was calculated using Equation (21). In the searching process, the guaranteed accuracy was given, and then it was dealt with as a constraint. The penalty function is the most often used technique in constrained optimization [26]; however, the penalty coefficients are very difficult to choose appropriately [30]. We handled the constraint as follows: if $\delta\mathbf{x}$ was in the required range, $cond(\mathbf{J}_P)$ was calculated using Equation (22), and ω was calculated using Equation (23). Otherwise, $cond(\mathbf{J}_P) = (10^7 + f_c)$ and $\omega = (-f_c)$. A subprogram was subsequently created to calculate the maximum value of $cond(\mathbf{J}_P)$ and the minimum value of ω for all 12 typical trajectories.

$$f_c = \|\frac{\delta p_X}{ac_{TX}}\| + \|\frac{\delta p_Y}{ac_{TY}}\| + \|\frac{\delta p_Z}{ac_{TZ}}\| + \|\frac{\delta\theta_X}{ac_{RX}}\| + \|\frac{\delta\theta_Y}{ac_{RY}}\| + \|\frac{\delta\theta_Z}{ac_{RZ}}\|, \tag{26}$$

where ac_{TX}, ac_{TY}, and ac_{TZ} are the required maximum linear displacement positioning errors along the X, Y, and Z axes, respectively; ac_{RX}, ac_{RY}, and ac_{RZ} are the required maximum angular displacement positioning errors along the X, Y, and Z axes, respectively.

Step 3. $cond(\mathbf{J}_P)_{max}$ was set as the maximum value of $cond(\mathbf{J}_P)$ calculated in step 2, and ω_{min} was set as the minimum value of ω calculated in step 2.

Step 4. The design variables were chosen as r_P, r_B, d_P, d_B, and H_0, where H_0 was the height of the upper universal joint plane to the lower universal joint plane at the home position. $f_1 = cond(\mathbf{J}_P)_{max}$ and $f_2 = -\omega_{min}$. NSGA-II was used to minimize the objective functions, f_1 and f_2, simultaneously.

For other requirements, the 12 typical trajectories were replaced by all the typical trajectories in the special applications. In the optimal kinematic design of a flight simulator, 31 typical trajectories of a Boeing 747 were considered in the design process by Advani [31]. If one desires a singularity-free workspace, as for motion simulators, the 12 typical trajectories in Step 2 would be replaced by all the typical trajectories and the 64 extreme positions [24,31].

4. Case Study

A case study is presented in this section to show the effectiveness of the proposed optimal design method. For this specific case, the requirements of the customer are shown in Table 3.

From a practical consideration, the ranges of the parameters were chosen as follows: $h_P = -0.45(m)$, $h_B = (-0.45 - H_0)(m)$, $0.75(m) \leq r_B \leq 3.0(m)$, $0.75(m) \leq r_P \leq 3.0(m)$, $r_P \leq r_B$, $1.0(m) \leq H_0 \leq 3.5(m)$, $0.25(m) \leq d_B \leq r_B$, and $0.22(m) \leq d_P \leq r_P$. The range of every element of $\delta\mathbf{l}$ and $\delta\mathbf{s}$ were in the range $[-0.3, 0.3](mm)$.

Table 3. Requirements of the customer.

Payload		10,000 kg	
Maximum Position Errors		Linear Travel	1 mm
		Angular Travel	0.1°
DOF	**Maximum Excursion**	**Speed**	**Acceleration**
Roll	$\pm25°$	$\pm20°/s$	$\pm210°/s^2$
Pitch	$\pm25°$	$\pm20°/s$	$\pm210°/s^2$
Yaw	$\pm30°$	$\pm20°/s$	$\pm210°/s^2$
Surge	±1 m	±0.7 m/s	±10 m/s^2
Sway	±1 m	±0.7 m/s	±10 m/s^2
Heave	±0.8 m	±0.6 m/s	±10 m/s^2

After applying the proposed optimization process, as outlined in Section 3, 50 optimal solutions and the Pareto-optimal front were found, as shown in Figure 4. Lastly, there were three optimum design points, labeled as a, b, and c, whose corresponding objective values and design parameters are shown in Table 4. As seen in Table 4, the maximum errors of all the linear displacements were lower than 1 mm, and the maximum errors of all the angular displacements were lower than 0.1°. Therefore, all the requirements by the customer, given in Table 3, were met. This showed the effectiveness of the proposed method for the optimal design purpose.

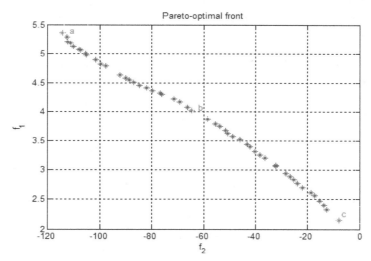

Figure 4. Pareto-optimal front of the solutions.

Table 4. Objective functions values from the Pareto sets and their corresponding design variables.

	a	b	c
d_B (m)	0.3409	0.3489	0.2500
d_P (m)	1.6915	0.5177	0.2200
r_B (m)	3.0000	2.9981	2.6552
r_P (m)	3.0000	2.4046	1.2128
H_0 (m)	2.4002	2.4044	2.4363
f_2	-114.2709	-64.5399	-8.0017
f_1	5.3615	4.0201	2.1464
δp_X (mm)	0.2	0.2	0.2
δp_Y (mm)	0.2	0.2	0.2
δp_Z (mm)	0.6	0.6	0.6
$\delta\theta_X$ (°)	0.0022	0.0025	0.0020
$\delta\theta_Y$ (°)	0.0018	0.0027	0.0026
$\delta\theta_Z$ (°)	0.0030	0.0029	0.0030

5. Conclusions

Motion simulators extensively use 6-UCU kind GSPs. A differential error model for a 6-UCU kind GSP was derived in this paper, and contained both the actuation-induced error and the error caused by the position errors of the joints. The guaranteed given accuracy was used as a constraint in the optimal design process of the high-accuracy motion simulators. An optimal kinematic design method was proposed by using the multi-objective evolutionary algorithm, NSGA-II. If $\mathbf{J}_P$ was near or at the singularity pose, the inverse of $\mathbf{J}_P$ potentially did not exist. $\delta\mathbf{x}$ could be wrongly calculated using Equation (21). In order to solve this problem, a method was presented where the condition number of $\mathbf{J}_P$ was compared with a large number. Another engineering method was proposed to handle the accuracy requirement constraints in the optimal design process. Multiple optimization solutions were found following implementation of the optimal kinematic design.

The effectiveness of the proposed method was verified through a practical optimal design case. The proposed optimal design method can be used as a guideline for the practical design of GSPs used in other applications with high-accuracy requirements.

References

1. Merlet, J.P. *Parallel Robots*, 2nd ed.; Springer: Dordrecht, The Netherlands, 2006; pp. 77–78. ISBN 978-1-4020-4132-7.
2. Huang, Z.; Kong, L.F.; Fang, Y.F. *Mechanism Theory of Parallel Robotic Manipulator and Control*; China Mechanical Press: Beijing, China, 1997; pp. 306–308. ISBN 7-111-05812-7. (In Chinese)
3. Fichter, E.; Kerr, D.; Rees-Jones, J. The Gough—Stewart platform parallel manipulator: A retrospective appreciation. *Proc. Inst. Mech. Eng. Part C J. Mech. Eng. Sci.* **2009**, *223*, 243–281. [CrossRef]
4. Gough, V.E.; Whitehall, S.G. Universal Tyre Testing Machine. In Proceedings of the 9th International Automobile Technical Congress, Bursa, Turkey, 7–8 May 2018.
5. Stewart, D. A Platform with six degrees of freedom. *Proc. Inst. Mech. Eng.* **1965**, *180*, 371–386. [CrossRef]
6. Cappel, K.L. Motion Simulator. U.S. Patent 3,295,224, 3 January 1967.
7. Ma, O. Mechanical Analysis of Parallel Manipulators with Simulation, Design and Control Applications. Ph.D. Thesis, McGill University, Montreal, QC, Canada, 1991.
8. Cerda Salzmann, D.J. Ampelmann: Development of the Access System for Offshore Wind Turbines. Ph.D. Thesis, Delft University of Technology, Delft, The Netherlands, 2010.
9. Blaise, J.; Bonev, I.; Monsarrat, B.; Briot, S.; Lambert, J.M.; Perron, C. Kinematic characterisation of hexapods for industry. *Ind. Robot Int. J.* **2010**, *37*, 79–88. [CrossRef]
10. *VARIAX: The Machine Tool of the Future-Today!* Giddings & Lewis, Inc.: Fond Du Lac, WI, USA, 1994.
11. Koch, P.M.; Kesteven, M.; Nishioka, H.; Jiang, H.; Lin, K.Y.; Umetsu, K.; Huang, Y.-D.; Raffin, P.; Chen, K.J.; Ibañez-Romano, F. The AMiBA hexapod telescope mount. *Astrophys. J.* **2009**, *694*, 1670–1684. [CrossRef]
12. Liu, G.; Zheng, S.; Ogbobe, P.; Han, J. Inverse kinematic and dynamic analyses of the 6-UCU parallel manipulator. *Appl. Mech. Mater.* **2012**, *127*, 172–180. [CrossRef]
13. Liu, G.; Qu, Z.; Liu, X.; Han, J. Singularity analysis and detection of 6-UCU parallel manipulator. *Robot Comput. Integr. Manuf.* **2014**, *30*, 172–179. [CrossRef]
14. Dai, X.; Huang, Q.; Han, J.; Li, H. Accuracy Synthesis of Stewart Platform used in Testing System for Spacecraft Docking Mechanism. In Proceedings of the International Conference on Measuring Technology and Mechatronics Automation, ICMTMA'09, Zhangjiajie, China, 11–12 April 2009; Volume 3, pp. 7–10. [CrossRef]
15. Wang, J.; Masory, O. On the Accuracy of a Stewart Platform-Part I: The Effect of Manufacturing Tolerances. In Proceedings of the IEEE International Conference on Robotics and Automation, Atlanta, GA, USA, 2–6 May 1993; pp. 114–120. [CrossRef]
16. Ropponen, T.; Arai, T. Accuracy Analysis of a Modified Stewart Platform Manipulator. In Proceedings of the IEEE International Conference on Robotics and Automation, Nagoya, Japan, 21–27 May 1995; pp. 521–525. [CrossRef]

17. Patel, A.J.; Ehmann, K.F. Volumetric error analysis of a Stewart platform-based machine tool. *CIRP Ann. Manuf. Technol.* **1997**, *46*, 287–290. [CrossRef]

18. Wang, S.M.; Ehmann, K.F. Error model and accuracy analysis of a six-DOF Stewart platform. *J. Manuf. Sci. Eng.* **2002**, *124*, 286–295. [CrossRef]

19. Masory, O.; Wang, J.; Zhuang, H. On the Accuracy of a Stewart Platform—Part II: Kinematic Calibration and Compensation. In Proceedings of the IEEE International Conference on Robotics and Automation, Atlanta, GA, USA, 2–6 May 1993; pp. 725–731. [CrossRef]

20. Cong, D.; Yu, D.; Han, J. Kinematics accuracy analysis and error compensation of Stewart platform. *J. Eng. Des.* **2006**, *13*, 162–165. (In Chinese)

21. Merlet, J.P.; Daney, D. Dimensional Synthesis of Parallel Robots with a Guaranteed Given Accuracy over a Specific Workspace. In Proceedings of the IEEE International Conference on Robotics and Automation, ICRA 2005, Barcelona, Spain, 18–22 April 2005; pp. 942–947. [CrossRef]

22. Merlet, J.-P.; Daney, D. Appropriate Design of Parallel Manipulators. In *Smart Devices and Machines for Advanced Manufacturing*; Wang, L., Xi, F., Eds.; Springer: London, UK, 2008; pp. 1–25. ISBN 978-1-84800-146-6.

23. Merlet, J.P. Interval analysis and reliability in robotics. *Int. J. Reliab. Saf.* **2009**, *3*, 104–130. [CrossRef]

24. Liu, G.; Qu, Z.; Han, J.; Liu, X. Systematic optimal design procedures for the Gough-Stewart platform used as motion simulators. *Ind. Robot* **2013**, *40*, 550–558. [CrossRef]

25. Angeles, J.; Park, F.C. Design and performance evaluation. In *Springer Handbook of Robotics*, 2nd ed.; Siciliano, B., Khatib, O., Eds.; Springer: Berlin, Germany, 2016; pp. 399–418. ISBN 978-3-319-32550-7.

26. Yu, X.; Gen, M. *Introduction to Evolutionary Algorithms*; Springer: London, UK, 2010; pp. 3–8. ISBN 978-1-84996-128-8.

27. Deb, K.; Pratap, A.; Agarwal, S.; Meyarivan, T.A.M.T. A fast and elitist multiobjective genetic algorithm: NSGA-II. *IEEE Trans. Evolut. Comput.* **2002**, *6*, 182–197. [CrossRef]

28. Deb, K.; Agrawal, R.B. Simulated binary crossover for continuous search space. *Complex Syst.* **1994**, *9*, 1–34.

29. Deb, K.; Goyal, M. A combined genetic adaptive search (GeneAS) for engineering design. *Comput. Sci. Inf. Syst.* **1996**, *26*, 30–45.

30. Deb, K. An efficient constraint handling method for genetic algorithms. *Comput. Method Appl. Mech.* **2000**, *186*, 311–338. [CrossRef]

31. Advani, S.K. The Kinematic Design of Flight Simulator Motion Bases. PhD. Dissertation, TU Delft, Delft, The Netherlands, 1998; pp. 103–191.

8

Nominal Stiffness of GT-2 Rubber-Fiberglass Timing Belts for Dynamic System Modeling and Design

Bozun Wang [1], Yefei Si [2], Charul Chadha [3], James T. Allison [1] and Albert E. Patterson [1,*]

[1]   Department of Industrial and Enterprise Systems Engineering, University of Illinois at Urbana-Champaign, 117 Transportation Building, 104 South Mathews Avenue, Urbana, IL 61801, USA; bozunw2@illinois.edu (B.W.); jtalliso@illinois.edu (J.T.A.)

[2]   Department of Mechanical Science and Engineering, University of Illinois at Urbana-Champaign, 144 Mechanical Engineering Building, 1206 West Green Street, Urbana, IL 61801, USA; yefeisi2@illinois.edu

[3]   Department of Aerospace Engineering, University of Illinois at Urbana-Champaign, 306 Talbot Laboratory, 104 South Wright Street, Urbana, IL 61801, USA; charulc2@illinois.edu

*   Correspondence: pttrsnv2@illinois.edu.

Abstract: GT-style rubber-fiberglass (RF) timing belts are designed to effectively transfer rotational motion from pulleys to linear motion in robots, small machines, and other important mechatronic systems. One of the characteristics of belts under this type of loading condition is that the length between load and pulleys changes during operation, thereby changing their effective stiffness. It has been shown that the effective stiffness of such a belt is a function of a "nominal stiffness" and the real-time belt section lengths. However, this nominal stiffness is not necessarily constant; it is common to assume linear proportional stiffness, but this often results in system modeling error. This technical note describes a brief study where the nominal stiffness of two lengths (400 mm and 760 mm) of GT-2 RF timing belt was tested up to breaking point; regression analysis was performed on the results to best model the observed stiffness. The experiments were performed three times, providing a total of six stiffness curves. It was found that cubic regression mod els ($R^2 > 0.999$) were the best fit, but that quadratic and linear models still provided acceptable representations of the whole dataset with R^2 values above 0.940.

Keywords: timing belt; belt stiffness; dynamic system modeling; mechatronic systems; 3D printers; robotics

1. Introduction

Timing belts are a common means of motion transfer between rotating motors/shafts in a machine or mechatronic system. Many small-to-medium sized mechatronic systems such as 3D printers [1], robots [2,3], desktop computer numerical control (CNC) machines [4], and positioners [5] use such belts, typically in the GT-style [6,7]. GT-style belts are specifically designed to effectively translate rotating motion from pulleys into linear motion with minimal deformation, slippage, and backlash. One of the fundamental characteristics of such a motion transfer system is that the length of the belts changes with time, causing time-variant stiffnesses in the belts which must be considered in dynamic system modeling and design. Note that the "stiffness" in the belt is considered only in the tension direction of the belt for this work, resulting in a stiffness that can be described as a single value or function instead of the full stiffness matrix [8,9].

When analyzing and designing any robotic and other mechatronic systems, it is vital that a good dynamic model of the system be developed and used. Since such systems often use some kind of flexible belts for motion transfer, the belt stiffness is a very important parameter in a system model. In cases where the length of the belt is constant (e.g., running between two fixed pulleys), the stiffness

k of the belt can be modeled as a spring where $f(x) = kx$; therefore, the stiffness of the belt is a function of its deflection x under load. In effect, this constant-length stiffness of the belt is its "nominal" design stiffness. However, in cases where the belt changes length during use, the effective length of these belt is a function of time and, therefore, its stiffness is also time-variant; this time-variant stiffness is the "effective" or apparent stiffness of the belt at some time t. It has been shown that the effective stiffness of the length-changing belts can be directly calculated as a function of the nominal stiffness value, the belt width, and the real-time length of the belt [10] such that:

$$k_i(t) = C_{sp}\frac{b}{L_i(t)} \tag{1}$$

where k_i is the effective stiffness as a function of time, C_{sp} is the nominal stiffness, b is the belt width, and $L_i(t)$ is the length of the belt section at time t. For any case where the length remains constant, the effective and nominal stiffnesses are equal since the value of $L_i(t)$ is a constant. Note that the value of C_{sp} may be a constant or function of material properties for different belt materials; it cannot be considered a function of time the way that the length of the belt is. The most commonly-used GT-style belt is the GT-2; Figure 1 shows the fundamental geometry and specifications for this type of belt.

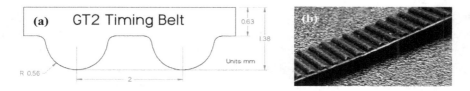

Figure 1. GT-2 belt (**a**) specifications and (**b**) basic geometry.

Figure 2 shows a common application, where a GT-style belt is used to transfer motion from a stepper motor to drive a linear positioning system. Also shown is a 2D dynamic model representation of such a system (Figure 2b), where the differences in effective stiffness, based on belt length, in the belt sections are clearly evident. The sections L_1 and L_2 change in effective stiffness as a function of time, while section L_3 stays constant during use [11,12] so the effective and nominal stiffnesses are equal.

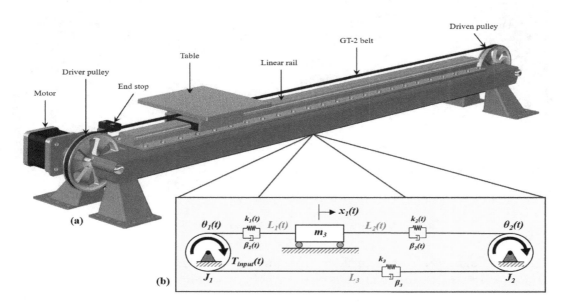

Figure 2. (**a**) simple positioning system that utilizes a GT-type belt to drive the table and (**b**) its representative dynamic model.

The work described in this note explored the nominal stiffness C_{sp} and the best way to model it in dynamic systems where belt length is not constant. Several previous studies have assumed that rubber-based timing belts have a linear nominal stiffness [5,11–18]. However, it is vital for designers and engineers working with dynamic systems which use belts for energy transfer to understand the true effects of the belt stiffness [19,20]. Therefore, experimental data was collected and used to derive conclusions on the true stiffness behavior of the GT-2 belts during use. The collected data was subjected to regression analysis to see which type of model best fit, allowing the comparison of models for the same dataset. The information in this study will prove useful, both in choosing k stiffness values for dynamic models and for judging expected model error if linear stiffness assumptions are used.

2. Procedure and Results

Two lengths of new GT-2 belts, 400 mm and 760 mm, were subjected to a simple tensile test until they ruptured. The test apparatus was a custom-built, screw-driven manual desktop test stand set up for tensile testing with 3000 N capability and a travel rate of $1/16$ in (1.6 mm) per screw revolution. The screw drive was rotated at a constant rate of 0.5 revolutions per second (0.8 mm/s), a reading being taken every revolution of the screw or every 1.6 mm. Since the length measurement was based on a count of the threads during travel, the uncertainty in length was too small to quantify; the digital readout for the unit used a load cell with a given uncertainty of 100 gram-force or 0.89 N. It was necessary to use this kind of manual tensile testing machine as none of the available standard machines were sensitive enough to measure the force-deflection behavior of these kinds of belts [21]. In addition, the discrete time measurement ensured a reasonably-sized dataset for curve-fitting. This was replicated twice to obtain a set of six different curves, three from each length. The ruptured belts were observed to fail suddenly and to show tearing of the glass fibers inside, as shown in Figure 3. The GT-2 belts used were a composite of neoprene (synthetic rubber) [22] and glass fibers, where the fibers appeared to drive the failure point of the belts.

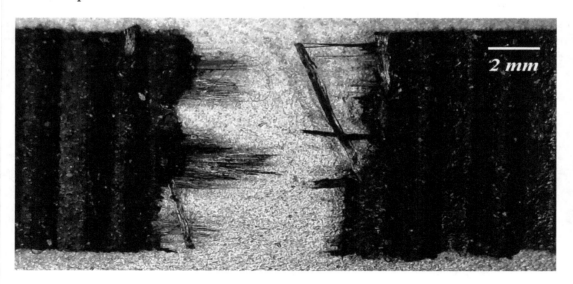

Figure 3. Belt break interface, showing broken fibers.

The collected data, in terms of force-deflection curves, are shown in Figure 4a, while the equivalent stress–strain curves for the tests are shown in Figure 4b. The length of the belts clearly had an effect on the force-deflection curves, but this largely disappeared when the length was accounted for in the stress–strain curves. Note that most of the curves show hyper-elastic behavior, i.e., there is no region in the curve where the stiffness is constant.

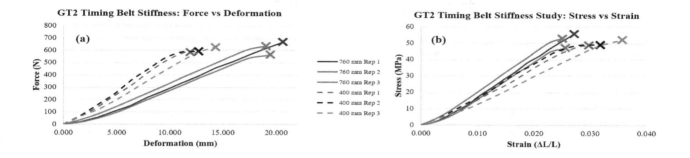

Figure 4. Belt stiffness curves: (**a**) force-deflection curve and (**b**) stress–strain curve.

As the nominal compliance of the belts was clearly found to be nonlinear, a regression analysis was performed to model the curves and find the level of unexplained variance in these curves. One of the most common polynomial regression models [23,24] used for hyper-elastic materials is the cubic polynomial. The basic model used for this study began with the following polynomial model:

$$\sigma_{belt} = A\varepsilon_{belt}^3 + B\varepsilon_{belt}^2 + C\varepsilon_{belt} + D \tag{2}$$

where a cubic model includes all of the variables, a quadratic model can be generated by setting $A = 0$, and a linear model can be used with $A = B = 0$. These curve fits, completed using Microsoft Excel™ (Microsoft Corp, Redmond, WA, USA) are shown in Figure 5a, and the fits for each of the variable and the resulting R^2 values are shown in the first six cases in Table 1.

After fitting the cubic models to each of the six sets of experimental data, the cubic model, a quadratic model, and a linear model were then fit to the entire set at once, as shown in Figure 5b. A significant drop in the R^2 value was noted for all of the models fit to the dataset, but differences between the cubic, quadratic, and linear models were observed to be small, as shown in Table 1.

It was observed that the low-strain region of the dataset (Figure 5b,c) conforms better to a linear model when the entire dataset is used. In actual use, it is most likely that the belts will not reach more than 20–30% of the belt breaking strength during normal use [14,25,26], so this is a valid assumption for many systems; this will, of course, need to be determined by the modeler or designer before using a linear belt model. If the low-strain assumption can be used, then the data fit a linear model with a slightly greater R^2 value than a quadratic model for the entire dataset and is certainly superior to a linear model for the entire dataset. The linear model for this case is shown in the last row of Table 1.

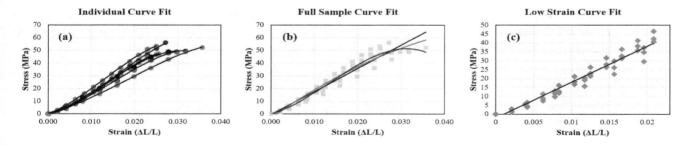

Figure 5. Curve fits for (**a**) individual belts (cubic model); (**b**) full sample curve fit (cubic, quadratic, and linear models); and (**c**) low-strain linear curve fit.

Table 1. Belt stiffness model curve fit data.

Case	Plot Reference	A	B	C	D	R^2
760 mm (cubic model) - $R1$	Figure 5a	-2.00×10^6	94,969	958.80	-0.5658	0.9996
760 mm (cubic model) - $R2$	Figure 5a	-3.00×10^6	144,204	445.86	-0.1163	0.9996
760 mm (cubic model) - $R3$	Figure 5a	-2.00×10^6	95,693	1281.60	-0.0723	0.9997
400 mm (cubic model) - $R1$	Figure 5a	-2.00×10^6	81,219	849.99	0.2296	0.9993
400 mm (cubic model) - $R2$	Figure 5a	-2.00×10^6	95,332	935.37	0.2840	0.9995
400 mm (cubic model) - $R3$	Figure 5a	$-922,283$	50,993	810.95	0.4965	0.9994
Full dataset (cubic model)	Figure 5b	-2.00×10^6	98,091	937.22	-0.1758	0.9672
Full dataset (quadratic model)	Figure 5b	-	$-19,340$	2408.9	-3.3656	0.9552
Full dataset (linear model)	Figure 5b	-	-	1821.1	-0.6140	0.9431
Low strain (linear model)	Figure 5c	-	-	2013.8	-2.3275	0.9573

3. Recommendations for Use and Applications

In cases where a time-variant belt length is used in a dynamic system model, the time-dependent stiffness of the belt must be considered, even when a mix of time-variant and time-invariant belt lengths are used. In practice, it is recommended that the modeler follow a three-step procedure:

1. Identify the nominal stiffness C_{sp} of each belt type used in the system (e.g., if two thicknesses of belts are used, two different nominal stiffnesses will be present). This information may be collected from manufacturer datasheets or from tests on each belt type, similar to the tests done in this technical report.

2. Decide if a linear or nonlinear nominal stiffness C_{sp} model will be used for each belt type. The primary driving force for this decision will be the computational cost for analyzing the system; for a simple system, it may be practical to use a nonlinear nominal stiffness model, but a linear model would be more feasible in a system with several elements. However, the importance of the model accuracy is a serious consideration and may justify a high computational cost if high accuracy is required.

3. Based on the configuration of the system and the decisions made in the first two steps, the effective stiffness k can take one of four forms:

    (a) If the belt length is constant and a linear model is used for C_{sp}, the effective stiffness in the equations of motion will be constant and described by

    $$k_i = C_{sp} \frac{b}{L} \tag{3}$$

    (b) If the belt length is constant and a nonlinear model is used to find C_{sp}, the nominal stiffness will be a function derived form a force-deflection curve. The effective stiffness in that belt section will be described by

    $$k_i = C_{sp}(x) \frac{b}{L} \tag{4}$$

    where $C_{sp}(x)$ is a continuous function of x.

    (c) If the belt length is time-variant and a linear model is used for C_{sp}, the effective stiffness in the equations of motion will be time-variant and described by

    $$k_i = C_{sp} \frac{b}{L(t)} \tag{5}$$

(d)   If the belt length is time-variant and a nonlinear model is used to find C_{sp}, the nominal stiffness will be a function derived form a force-deflection curve. In this case, the effective belt section stiffness will be described by

$$k_i = C_{sp}(x)\frac{b}{L(t)} \tag{6}$$

where $C_{sp}(x)$ is a continuous function of x and the belt length is a function of time. Therefore, the effective stiffness will be dependent on both the length of the belt and the amount of force placed on the belt.

When modeling these dynamic systems, it is recommended that the simplest model of the belt stiffness which gives acceptable accuracy be used in order to balance computational cost with extreme accuracy in the model. In most cases, the uncertainty in the material properties of the belt and the common use of linearization in dynamic models would erase any advantage to using an extremely high-fidelity belt model.

4. Conclusions

This short technical note presents the results of a brief exploratory study on modeling the nominal stiffness of GT-2 timing belts; this information can be used to more accurately model the true, time-variant, stiffness behavior of common GT-2 belts when the effective length of belt sections changes with time. It was observed that these belts do not behave in a linear way, as expected for belts with a hyper-elastic base material, but that a linear model can provide a reasonable approximation of the behavior under some conditions, particularly low-strain conditions. When possible, the cubic stiffness model should be used, but this would often be impractical for dynamic systems with many components, as it can cause a simple model to become nonlinear in more than one variable. When practical and necessary for problem tractability, a linear model may be used with a reasonable degree of accuracy. The modeler or designer should keep in mind that some uncertainty will exist with any belt model and should choose the model that best balances accuracy with computational cost.

Author Contributions: B.W. and A.E.P. conceived and designed the study. All authors helped to set up the experiments, collect data, perform the regression analyses, and write the report.

Nomenclature

b = Belt width (m)
β_i = Belt section i damping coefficient
C_{sp} = Nominal belt stiffness (N/m)
k_i = Effective (true) belt section i stiffness (N/m)
L_i = Belt section i length (m)
m_i = Mass of block i (kg)
θ_i = Pulley i angle (degrees)

References

1.   Laureto, J.; Pearce, J. Open Source Multi-Head 3D Printer for Polymer-Metal Composite Component Manufacturing. *Technologies* **2017**, *5*, 36. [CrossRef]

2.   Krahn, J.; Liu, Y.; Sadeghi, A.; Menon, C. A tailless timing belt climbing platform utilizing dry adhesives with mushroom caps. *Smart Mater. Struct.* **2011**, *20*, 115021. [CrossRef]

3.   Parietti, F.; Chan, K.; Asada, H.H. Bracing the human body with supernumerary Robotic Limbs for physical assistance and load reduction. In Proceedings of the IEEE International Conference on Robotics and Automation (ICRA), Hong Kong, China, 31 May–7 June 2014. [CrossRef]

4.   Choudhary, R.; Sambhav; Titus, S.D.; Akshaya, P.; Mathew, J.A.; Balaji, N. CNC PCB milling and wood engraving machine. In Proceedings of the International Conference On Smart Technologies For Smart Nation (SmartTechCon), Bangalore, India, 17–19 August 2017. [CrossRef]

5.   Sollmann, K.; Jouaneh, M.; Lavender, D. Dynamic Modeling of a Two-Axis, Parallel, H-Frame-Type XY Positioning System. *IEEE/ASME Trans. Mechatron.* **2010**, *15*, 280–290. [CrossRef]

6.   York Industries. York Timing Belt Catalog: 2mm GT2 Pitch (pp. 16). Available online: http://www.york-ind.com/print_cat/york_2mmGT2.pdf (accessed on 13 June 2018).

7.   SDP/SI. Handbook of Timing Belts, Pulleys, Chains, and Sprockets. Available online: www.sdp-si.com/PDFS/Technical-Section-Timing.pdf (accessed on 13 June 2018).

8.   Huang, J.L.; Clement, R.; Sun, Z.H.; Wang, J.Z.; Zhang, W.J. Global stiffness and natural frequency analysis of distributed compliant mechanisms with embedded actuators with a general-purpose finite element system. *Int. J. Adv. Manuf. Technol.* **2012**, *65*, 1111–1124. [CrossRef]

9.   Barker, C.R.; Oliver, L.R.; Breig, W.F. *Dynamic Analysis of Belt Drive Tension Forces During Rapid Engine Acceleration*; SAE Technical Paper Series; SAE International: Warrendale, PA, USA, 1991. [CrossRef]

10. Gates-Mectrol. Technical Manual: Timing Belt Theory. Available online: http://www.gatesmectrol.com/mectrol/downloads/download_common.cfm?file=Belt_Theory06sm.pdf&folder=brochure (accessed on 13 June 2018).

11. Hace, A.; Jezernik, K.; Sabanovic, A. SMC With Disturbance Observer for a Linear Belt Drive. *IEEE Trans. Ind. Electron.* **2007**, *54*, 3402–3412. [CrossRef]

12. Johannesson, T.; Distner, M. Dynamic Loading of Synchronous Belts. *J. Mech. Des.* **2002**, *124*, 79. [CrossRef]

13. Childs, T.H.C.; Dalgarno, K.W.; Hojjati, M.H.; Tutt, M.J.; Day, A.J. The meshing of timing belt teeth in pulley grooves. *Proc. Inst. Mech. Eng. D* **1997**, *211*, 205–218. [CrossRef]

14. Callegari, M.; Cannella, F.; Ferri, G. Multi-body modelling of timing belt dynamics. *Proc. Inst. Mech. Eng. K* **2003**, *217*, 63–75. [CrossRef]

15. Leamy, M.J.; Wasfy, T.M. Time-accurate finite element modelling of the transient, steady-state, and frequency responses of serpentine and timing belt-drives. *Int. J. Veh. Des.* **2005**, *39*, 272. [CrossRef]

16. Feng, X.; Shangguan, W.B.; Deng, J.; Jing, X.; Ahmed, W. Modelling of the rotational vibrations of the engine front-end accessory drive system: a generic method. *Proc. Inst. Mech. Eng. D* **2017**, *231*, 1780–1795. [CrossRef]

17. Rodriguez, J.; Keribar, R.; Wang, J. *A Comprehensive and Efficient Model of Belt-Drive Systems*; SAE Technical Paper Series; SAE International: Warrendale, PA, USA, 2010. [CrossRef]

18. Cepon, G.; Boltezar, M. *An Advanced Numerical Model for Dynamic Simulations of Automotive Belt-Drives*; SAE Technical Paper Series; SAE International: Warrendale, PA, USA, 2010. [CrossRef]

19. Tai, H.M.; Sung, C.K. Effects of Belt Flexural Rigidity on the Transmission Error of a Carriage-driving System. *J. Mech. Des.* **2000**, *122*, 213. [CrossRef]

20. Zhang, L.; Zu, J.W.; Hou, Z. Complex Modal Analysis of Non-Self-Adjoint Hybrid Serpentine Belt Drive Systems. *J. Vib. Acoust.* **2001**, *123*, 150. [CrossRef]

21. *Materials Testing Guide*; ADMET: Norwood, MA, USA, 2013.

22. Kumar, D.; Sarangi, S. Data on the viscoelastic behavior of neoprene rubber. *Data Brief.* **2018**, *21*, 943–947. [CrossRef] [PubMed]

23. Mansouri, M.; Darijani, H. Constitutive modeling of isotropic hyperelastic materials in an exponential framework using a self-contained approach. *Int. J. Solids Struct.* **2014**, *51*, 4316–4326. [CrossRef]

24. Shahzad, M.; Kamran, A.; Siddiqui, M.Z.; Farhan, M. Mechanical Characterization and FE Modelling of a Hyperelastic Material. *Mater. Res.* **2015**, *18*, 918–924. [CrossRef]

25. Tokoro, H. Analysis of transverse vibration in engine timing belt. *JSAE Rev.* **1997**, *18*, 33–38. [CrossRef]

26. Gerbert, G.; Jnsson, H.; Persson, U.; Stensson, G. Load Distribution in Timing Belts. *J. Mech. Des.* **1978**, *100*, 208. [CrossRef]

9

Kinematics Analysis of a Class of Spherical PKMs by Projective Angles

Giovanni Legnani [1,2,*] and Irene Fassi [2]

[1] Department of Mechanical and Industrial Engineering, University of Brescia, 25123 Brescia, Italy
[2] Institute of Intelligent Industrial Technologies and Systems for Advanced Manufacturing, National Research Council, 20133 Milan, Italy; irene.fassi@cnr.it
[*] Correspondence: giovanni.legnani@unibs.it.

Abstract: This paper presents the kinematics analysis of a class of spherical PKMs Parallel Kinematics Machines exploiting a novel approach. The analysis takes advantage of the properties of the projective angles, which are a set of angular conventions of which their properties have only recently been presented. Direct, inverse kinematics and singular configurations are discussed. The analysis, which results in the solution of easy equations, is developed at position, velocity and acceleration level.

Keywords: projective angles; spherical PKM; direct kinematics; inverse kinematics

---

1. Introduction

Spherical PKMs Parallel Kinematics Machines are a class of parallel manipulators whose mobile platform may rotate around one fixed point. They typically have 3 DOF Degrees of Freedom. They have been studied for a long time [1,2].

It is possible to identify two big classes of spherical PKMs. In the first one (class A), the spherical motion is guaranteed by the convergence in one point of the rotation axes of all of the joints. In the PKM of the second class (class B), the mobile platform is connected to the ground by a spherical or a universal joint that is placed in the center of rotation and the motion is transmitted to it by external legs that may have revolute joints of which the axes do not pass for the center of the rotations.

Among the spherical PKMs belonging to the first class, one of the most popular is the agile eye, which was firstly proposed by Gosselin and was then also adopted by others [2–7]. Its structure is presented in Figure 1. It has three identical legs, each with three revolute joints which have axes that all converge in the same point around which the end-effector rotates. This PKM exhibits an excellent rotational ability and it can assume a compact structure (Figure 2). However, as discussed in the following sections, this PKM is over constrained and, thus, requires high manufacturing and assembly precision to ensure a correct motion without overloading the structure with undesired internal stresses.

The properties of the projective angles can be used to describe the 3D rotation motion of any rigid body and, in this sense, are general. It is a particular angular convention with pros and cons, as with any other convention (Euler angles, quaternions . . .).

The application to PKM is particularly convenient to the class of spherical manipulators based on rotational joints forming cardan sequences, because in this case, the angles of the angular convention coincide with the joint angles of the PKM.

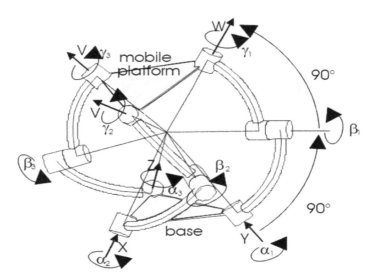

Figure 1. The kinematic structure of the agile eye.

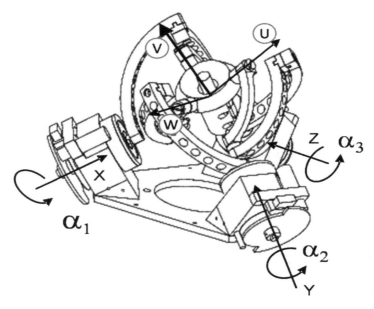

Figure 2. The agile eye (Adapted from [7]).

One example of the second class is the 3-DOF 3SPS/S that is presented in [8–11], which has three legs with spherical-prismatic-spherical-joints and one additional spherical joint connecting the mobile platform to the fixed base. Another example is described in [12,13], and it is shown in Figure 3. This PKM (better described in the following paragraphs of this paper) can be considered a non-overconstrained version of the agile-eye. The legs are serial kinematic chains for which standard methodologies for their analysis are available [14,15].

Several approaches have been developed for the synthesis of the different spherical PKMs. For example, Kong and Gosselin [16,17] proposed the adoption of the screw theory. Fang and Tsai [18] developed a family of spherical PKMs with legs of identical structure, while Karouia and Hervé [19] developed spherical PKMs with legs with different structures. A list with the classification of different spherical manipulators has been suggested by Hess-Coelho [20], which includes a methodology to evaluate their performances. Many other parallel orientation mechanisms have been described in numerous papers (e.g., [3,21–31]). Generally, the first joint of each leg is actuated, however some manipulators adopt transmissions based on parallelograms [31,32].

Analogous concepts were used to design non-redundant PKM with decoupled rotational and translational motion [33] and 2D orientating mechanisms [34,35].

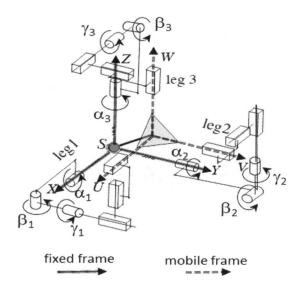

Figure 3. A non-overconstrained version of the agile-eye.

The concept of projective angles has been introduced by [36] and represents a useful approach for the kinematic analysis of PKMs. In [12], this concept has been applied to the angular position analysis of a non-overconstrained variation of the agile eye.

The present paper extends the abovementioned approach to the full kinematic analysis of some spherical PKMs, resulting in a set of easily solvable equations, at position, velocity and acceleration level. Specifically, the proposed approach will be used to solve the direct and inverse kinematics of spherical manipulators, belonging to the two mentioned classes (class A, B).

2. Materials and Methods

2.1. The Agile Eye

The 3RRR agile eye (Figures 1 and 2) is a particular version of a spherical PKM in which the axes of the rotational joints of each link form an angle of 90° and they all converge in one point, which is the centre of rotation.

The mobility of the agile-eye can be studied by the well know Grubler-Kutzbach formula. Having seven bodies and thus, a total of 42 DoF (degrees of freedom) and nine revolute joints for a total of 45 constrains, this means that there are six redundant constrains since the PKM clearly has 3 DoF.

The motion is possible because all of the joint axes converge in one point. The redundant constraints can be removed in several ways by changing the nature of the joints. Recently, a new non-overconstrained version of the agile-eye (Figure 3) was presented in [12].

The mobile platform is connected to the fixed base by three legs having a series of three revolute and three prismatic joints. The first revolute joint of each leg is actuated; in addition, a spherical joint is also present and connects the fixed base and the mobile platform. The PKM is of type 3RRRPPP/S. The joint actuator coordinates of this PKM are the rotations of the actuators that are connected to the first joint of each leg:

$$Q = \begin{bmatrix} \alpha_1 \\ \alpha_2 \\ \alpha_3 \end{bmatrix} \tag{1}$$

where the subscript i = 1, 2, 3 indicates the leg. The rotations of the non-actuated joints are represented by the angles β_i and γ_i. The choice of the name for the angles α_i and β_i, which is identical to those that we will use for the definitions of the projective angles (see the next section), is justified by the analogies that will be discussed further in this paper.

This PKM can be considered a variant of the agile-eye because the sequences of the revolute joints are the same. The prismatic joints remove the over constraints by leaving three additional translation degrees of freedom which are removed by the spherical joint.

The direct and the inverse kinematics studies the relation between the attitude of the mobile platform and the rotation of the joint actuators, and since the sequence of the joints is the same in the two platforms, their equation may be written in a unified way. Recently, a solution of the position analysis of the PKM of Figure 3 which was based on the concept of the projective angles was presented [12]. Moreover, the possibility to also use projective angles for velocity and acceleration analysis has been proposed in [13]. By joining these results, it is possible to propose a methodology for a full kinematic analysis of the presented class of spherical PKM.

2.2. The Projective Angles

The angular position (attitude) of one body may be represented by the rotation matrix expressing the angular position of a frame that is attached on the body with respect to a fixed frame.

We indicate by X, Y and Z the fixed axes, and by U, V and W the mobile axes (Figures 4–6). The rotation matrix is then:

$$\mathbf{R} = [\mathbf{u}, \mathbf{v}, \mathbf{w}] = \begin{bmatrix} u_x & v_x & w_x \\ u_y & v_y & w_y \\ u_z & v_z & w_z \end{bmatrix} \tag{2}$$

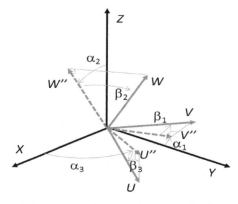

Figure 4. Definition of the projective angles α_i and of the auxiliary angles β_i.

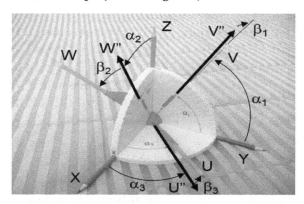

Figure 5. A 3D model to show the projective angles α_i and of the auxiliary angles β_i.

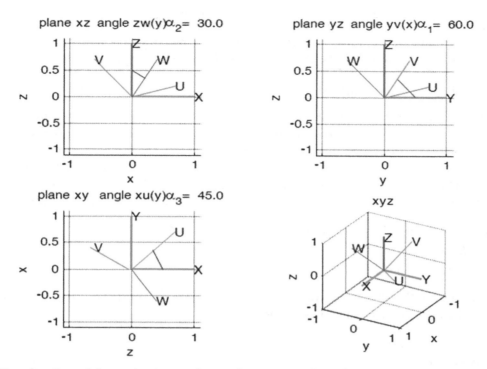

Figure 6. Visualization of the projective angles on the xy, yz and xz plane; in this example, the projective angles assume the value: 60°, 30°, 45°.

The projective angles α_1, α_2 and α_3 are defined as follows (Figures 5 and 6):

1. project the unit vector v of V on plane YZ obtaining vector v″, α_1 is the angle between v″ and Y;
2. project the unit vector w of W on plane XZ obtaining vector w″, α_2 is the angle between w″ and Z;
3. project the unit vector u of U on plane XY obtaining vector u″, α_3 is the angle between u″ and X.

The definition of the following auxiliary angles β_1, β_2 and β_3 is also necessary for the discussion (Figure 4).

1. β_1 is the angle between V and the plane YZ
2. β_2 is the angle between W and the plane XZ
3. β_3 is the angle between U and the plane XY

and so, the angular position of the mobile frame is:

$$\mathbf{R} = \begin{bmatrix} \mathbf{U} & \mathbf{V} & \mathbf{W} \end{bmatrix} = \begin{bmatrix} \cos\alpha_3\cos\beta_3 & -\sin\beta_1 & \sin\alpha_2\cos\beta_2 \\ \sin\alpha_3\cos\beta_3 & \cos\alpha_1\cos\beta_1 & -\sin\beta_2 \\ -\sin\beta_3 & \sin\alpha_1\cos\beta_1 & \cos\alpha_2\cos\beta_2 \end{bmatrix} \quad (3)$$

It is important to note that, by construction, $\cos(\beta_i) > 0$.
The projective angles of the mobile frame may by extracted from the rotation matrix $\mathbf{R}$ as:

$$\mathbf{A} = \begin{bmatrix} \alpha_1 \\ \alpha_2 \\ \alpha_3 \end{bmatrix} = \begin{bmatrix} \mathrm{atan2}(v_z, v_y) \\ \mathrm{atan2}(w_x, w_z) \\ \mathrm{atan2}(u_y, u_x) \end{bmatrix} \quad (4)$$

where atan2(a,b) is the four quadrant extension of the arctangent function atan(a/b). The possibility of using atan2 rather than other inverse trigonometric function (e.g., atan) eliminates the ambiguity of the identification of the correct quadrant and the risk of division by 0.

Therefore, knowing the angular position of the mobile frame (i.e., the rotation matrix R), the projective angles may be easily evaluated. The inverse operation (i.e., determining **R** when the projective angles are known) will be discussed further in this paper (Section 2.6).

2.3. Inverse Kinematics of the Spherical PKM

To study the mentioned spherical PKM, a fixed frame and a mobile frame were established, as shown in Figures 1–3. To simplify the analysis, the coordinate axes are chosen coincidently to the rotation axis of some revolute joints. The XYZ axis of the fixed base frame are parallel with the axes of the actuated joints, while the UVW axes of the mobile frame are parallel with the axis of the joints γ_i of the mobile frame. The inverse kinematics of the considered spherical PKM can be solved by realizing that the joints of each leg constitute a Cardanic sequence. With reference to Figures 3 and 4, we get:

leg 1: rotations $\alpha_1\beta_1\gamma_1$ around the following axis sequence XZY,
leg 2: rotations $\alpha_2\beta_2\gamma_2$ around YXZ,
leg 3: rotations $\alpha_3\beta_3\gamma_3$ around ZYX.

The rotation matrix **R** expressing the attitude of the mobile platform can easily be evaluated by considering each different leg. For leg 1, we got:

$$\mathbf{R} = \mathbf{R}_1 = \begin{bmatrix} \cos\beta_1\cos\gamma_1 & -\sin\beta_1 & \cos\beta_1\sin\gamma_1 \\ \cos\alpha_1\sin\beta_1\cos\gamma_1 + \sin\alpha_1\sin\gamma_1 & \cos\alpha_1\cos\beta_1 & \cos\alpha_1\sin\beta_1\sin\gamma_1 - \sin\alpha_1\cos\gamma_1 \\ \sin\alpha_1\sin\beta_1\cos\gamma_1 - \cos\alpha_1\sin\gamma_1 & \sin\alpha_1\cos\beta_1 & \sin\alpha_1\sin\beta_1\sin\gamma_1 + \cos\alpha_1\cos\gamma_1 \end{bmatrix} \tag{5}$$

and similar equations may be written for legs 2 and 3, which lead to the matrices $\mathbf{R}_2$ and $\mathbf{R}_3$. By indicating the rotation matrix as:

$$\mathbf{R} = \begin{bmatrix} u_x & v_x & w_x \\ u_y & v_y & w_y \\ u_z & v_z & w_z \end{bmatrix} = \begin{bmatrix} r_{11} & r_{12} & r_{13} \\ r_{21} & r_{22} & r_{23} \\ r_{31} & r_{32} & r_{33} \end{bmatrix} \tag{6}$$

the joint rotations can easily be determined, obtaining two solutions for each leg.
For leg 1 we got:

$$\begin{cases} \alpha_{11} = \text{atan2}(r_{32}, r_{22}) \\ \beta_{11} = \text{asin}(-r_{12}) \\ \gamma_{11} = \text{atan2}(r_{13}, r_{11}) \end{cases} , \quad \begin{cases} \alpha_{12} = \text{atan2}(-r_{32}, -r_{22}) \\ \beta_{12} = \pi - \text{asin}(-r_{12}) \\ \gamma_{12} = \text{atan2}(-r_{13}, -r_{11}) \end{cases} \tag{7}$$

and similar equations may be written for legs 2 and 3. The relations between the two solutions for the i-th leg is:

$$\begin{cases} \alpha_{i2} = \alpha_{i1} + \pi \\ \beta_{i2} = \pi - \beta_{i1} \\ \gamma_{i2} = \gamma_{i1} + \pi \end{cases} \tag{8}$$

and the inverse kinematics is therefore solved by selecting the equations that were associated with the first joint of each leg:

$$Q = \begin{bmatrix} \alpha_1 \\ \alpha_2 \\ \alpha_3 \end{bmatrix} = \begin{bmatrix} \text{atan2}(v_z, v_y) + k_1\pi \\ \text{atan2}(w_x, w_z) + k_2\pi \\ \text{atan2}(u_y, u_x) + k_3\pi \end{bmatrix} \quad k_i = 0, 1 \tag{9}$$

A total of 8 possible solutions were found. Kinematic singularities are present for the configurations for which the atan2 function is not defined, i.e., atan2(0,0) which happens when the second joint angle of one leg is ± 90 degrees and so $\cos(\beta_i) = 0$.

2.4. Direct Kinematics

To solve the direct kinematics (find the matrix $\mathbf{R}$ when the joint coordinates α_i are known), we can write the rotation matrix of the mobile platform by choosing one column for each of the matrix of each leg (Equation (5) for leg 1 and similar equations for the other legs). More precisely, we can bring the second column from R_1, the third column from R_2 and the first one from R_3.

$$\mathbf{R} = \begin{bmatrix} \mathbf{U} & \mathbf{V} & \mathbf{W} \end{bmatrix} = \begin{bmatrix} \cos\alpha_3\cos\beta_3 & -\sin\beta_1 & \sin\alpha_2\cos\beta_2 \\ \sin\alpha_3\cos\beta_3 & \cos\alpha_1\cos\beta_1 & -\sin\beta_2 \\ -\sin\beta_3 & \sin\alpha_1\cos\beta_1 & \cos\alpha_2\cos\beta_2 \end{bmatrix} \quad (10)$$

assuming $\cos(\beta_i) \neq 0$ and dividing each column by the corresponding $\cos(\beta_i)$, we got:

$$\mathbf{R}' = \begin{bmatrix} \mathbf{U}' & \mathbf{V}' & \mathbf{W}' \end{bmatrix} = \begin{bmatrix} \cos\alpha_3 & -a' & \sin\alpha_2 \\ \sin\alpha_3 & \cos\alpha_1 & -b' \\ -c' & \sin\alpha_1 & \cos\alpha_2 \end{bmatrix} \quad (11)$$

with $a' = \text{tg}\beta_1$, $b' = \text{tg}\beta_2$ and $c' = \text{tg}\beta_3$.

Since we are considering the direct kinematics, the angles α_i are given, while a', b' and c' can be computed by the conditions of orthogonality of the three columns:

$$\mathbf{V}' \cdot \mathbf{W}' = 0, \ \mathbf{W}' \cdot \mathbf{U}' = 0, \ \mathbf{U}' \cdot \mathbf{V}' = 0 \quad (12)$$

that can be expressed as:

$$\begin{cases} a'\sin\alpha_2 + b'\cos\alpha_1 - \cos\alpha_2\sin\alpha_1 = 0 \\ b'\sin\alpha_3 + c'\cos\alpha_2 - \cos\alpha_3\sin\alpha_2 = 0 \\ c'\sin\alpha_1 + a'\cos\alpha_3 - \cos\alpha_1\sin\alpha_3 = 0 \end{cases} \quad (13)$$

that in matrix form is:

$$\begin{bmatrix} \sin\alpha_2 & \cos\alpha_1 & 0 \\ 0 & \sin\alpha_3 & \cos\alpha_2 \\ \cos\alpha_3 & 0 & \sin\alpha_1 \end{bmatrix} \begin{bmatrix} a' \\ b' \\ c' \end{bmatrix} = \begin{bmatrix} \cos\alpha_2\sin\alpha_1 \\ \cos\alpha_3\sin\alpha_2 \\ \cos\alpha_1\sin\alpha_3 \end{bmatrix} \quad (14)$$

which is a linear system with respect to a', b' and c', of which the solution is:

$$\begin{cases} a' = \frac{\cos\alpha_2\sin\alpha_3 - \cos\alpha_1\cos\alpha_3\sin\alpha_2\sin\alpha_1}{\Delta} \\ b' = \frac{\cos\alpha_3\sin\alpha_1 - \sin\alpha_2\cos\alpha_1\sin\alpha_3\cos\alpha_2}{\Delta} \\ c' = \frac{\cos\alpha_1\sin\alpha_2 - \cos\alpha_3\cos\alpha_2\sin\alpha_1\sin\alpha_3}{\Delta} \end{cases} \quad (15)$$

where $\Delta = \cos\alpha_3\cos\alpha_2\cos\alpha_1 + \sin\alpha_3\sin\alpha_2\sin\alpha_1$, which is the determinant of the linear system that must not be zero for the invertibility of the matrix. Moreover, the determinant must be positive ($\Delta > 0$) because the mobile frame is right (for $\Delta < 0$ we get a left frame, while $\Delta = 0$ results in a singular configuration).

From Equation (15), the value of the angles β_i is immediately found:

$$\beta_1 = \text{atan}(a') + k_1\pi, \ \beta_2 = \text{atan}(b') + k_2\pi, \ \beta_3 = atan(c') + k_3\pi \quad (16)$$

with $k_i = 0, 1$ and so the direct kinematics has 8 different solutions and matrix $\mathbf{R}$ (Equation (10)) is determined.

By only considering the right frames, the singular configurations happened for the following angular position of the mobile platform:

$$\mathbf{R} = \mathbf{R_a} = \begin{bmatrix} 0 & 1 & 0 \\ 0 & 0 & 1 \\ 1 & 0 & 0 \end{bmatrix} \tag{17}$$

$$\mathbf{R} = \mathbf{R_b} = \begin{bmatrix} 0 & -1 & 0 \\ 0 & 0 & -1 \\ 1 & 0 & 0 \end{bmatrix} \tag{18}$$

$$\mathbf{R} = \mathbf{R_c} = \begin{bmatrix} 0 & 1 & 0 \\ 0 & 0 & -1 \\ -1 & 0 & 0 \end{bmatrix} \tag{19}$$

$$\mathbf{R} = \mathbf{R_d} = \begin{bmatrix} 0 & -1 & 0 \\ 0 & 0 & 1 \\ -1 & 0 & 0 \end{bmatrix} \tag{20}$$

which happens for $\cos(\beta_i) = 0$ and $\sin(\beta_i) = \pm 1$. For these configurations, the actuated joint angles α_i may assume any value.

2.5. Projective Angles and Spherical PKM

An analysis of the definition of the projective angles and of the kinematics of the considered spherical PKM highlights several analogies.

It is worth noting that Equations (4) and (9) are very similar and that they coincide for $k_i = 0$. It is therefore possible to conclude that if for the PKM we choose the solution with $\cos(\beta_i) > 0$, the joint coordinates of the PKM coincide with the projective angles of the mobile platform.

Similarly, the procedure that was adopted to solve the direct kinematics of the PKM can be adopted to compute the angular position of one frame which corresponds to an assigned value of the projective angles α_i. In this case, however, we must always assume $\cos(\beta_i) > 0$, and so the solution is unique.

Finally, if we describe the angular position of the mobile platform using the projective angles A, their relation to the joint coordinates Q is:

$$A = Q + \begin{bmatrix} k_1 \pi \\ k_2 \pi \\ k_3 \pi \end{bmatrix}, \; k_i = 0, \, 1 \tag{21}$$

and so

$$J = \frac{\partial A}{\partial Q} = \begin{bmatrix} 1 & 0 & 0 \\ 0 & 1 & 0 \\ 0 & 0 & 1 \end{bmatrix} \tag{22}$$

According to the above equations, and adopting the projective angle convention, the PKM can be considered 'decoupled', in the sense that each actuator influences just one of the end-effector coordinates [37]. Therefore, for $k_i = 0$, the joint coordinates of the PKM coincide with the projective angles of the mobile frame and also the angle β_i that was used in the description of the projective angles and in the spherical PKM (2nd joint of each leg) have the same values. If the 2nd solution is chosen for the PKM, its angles α_i and β_i differ from those of the projective angles by 180° (see Equation (21) for analogy). Of course, due to the non-integrability of the angular velocity vector Ω it is:

$$\dot{A} = \dot{Q} \text{ but } \Omega \neq \dot{Q} \tag{23}$$

because it is impossible to define a set of angular coordinates representing a 3D orientation of one body which has a time derivative that coincides with its angular velocity Ω [37–39].

A broader analysis of velocity and acceleration is found in the next sections. The analysis will be performed with reference to the abovementioned PKMs; by remembering Equation (21), the results are immediately extended to the projective angles.

We may conclude that the projective angles are a sort of "intrinsic" notation to study the considered spherical PKMs.

2.6. From Projective Angle to Rotation Matrix

Considering the analogies between the kinematic of the spherical PKM and the projective angles, the rotation matrix R that is associated with a set of projective angles A is found by Equations (10) and (16) assuming $k_1 = k_2 = k_3 = 0$ since, by definition of the projective angles, the cosine of the angles β_i are positive.

2.7. Velocity Analysis

Since the legs of the PKM form a Cardanic sequence, the velocity analysis is easily developed. The solution proposed in the following can be applied both to the spherical PKM under study and to the projective angles.

By considering the different legs, we have a first rotation α around one Cartesian axis, a second β rotation around an axis rotated by α and, finally, a third rotation γ around one axis which depends on α and β. For instance, for leg 1, the three rotation axes are defined by the following unit vectors $a_{11} = [1\ 0\ 0]^t$, $a_{12} = [0\ -\sin\alpha_1\ \cos\alpha_1]^t$ and $a_{13} = [-\sin\beta_i\ \cos\alpha_1\cos\beta_i\ \sin\alpha_1\cos\beta_i]^t$; therefore, for the three legs, and representing the angular velocity of the frame by $\Omega = \begin{bmatrix} \omega_x & \omega_y & \omega_z \end{bmatrix}^T$, we obtain the following results

$$\Omega = a_{i1}\dot\alpha_i + a_{i2}\dot\beta_i + a_{i3}\dot\gamma_{i3} \tag{24}$$

And so, for leg 1, it is:

$$\begin{bmatrix} \omega_x \\ \omega_y \\ \omega_z \end{bmatrix} = \begin{bmatrix} 1 & 0 & -\sin\beta_1 \\ 0 & -\sin\alpha_1 & \cos\alpha_1\cos\beta_1 \\ 0 & \cos\alpha_1 & \sin\alpha_1\cos\beta_1 \end{bmatrix} \begin{bmatrix} \dot\alpha_1 \\ \dot\beta_1 \\ \dot\gamma_1 \end{bmatrix} \tag{25}$$

and similar equations may be written for legs 2 and 3. The above presented equations may be inverted as:

$$\begin{bmatrix} \dot\alpha_1 \\ \dot\beta_1 \\ \dot\gamma_1 \end{bmatrix} = \begin{bmatrix} 1 & \cos\alpha_1\frac{\sin\beta_1}{\cos\beta_1} & \sin\alpha_1\frac{\sin\beta_1}{\cos\beta_1} \\ 0 & -\sin\alpha_1 & \cos\alpha_1 \\ 0 & \frac{\cos\alpha_1}{\cos\beta_1} & \frac{\sin\alpha_1}{\cos\beta_1} \end{bmatrix} \begin{bmatrix} \omega_x \\ \omega_y \\ \omega_z \end{bmatrix} \tag{26}$$

By considering the first row of the matrix of Equation (26) and the analogous equations for leg 2 and 3, the velocity of the actuated joints is then obtained by the angular velocity of the mobile platform:

$$Q = \begin{bmatrix} \dot\alpha_1 \\ \dot\alpha_2 \\ \dot\alpha_3 \end{bmatrix} = \begin{bmatrix} 1 & \cos\alpha_1\frac{\sin\beta_1}{\cos\beta_1} & \sin\alpha_1\frac{\sin\beta_1}{\cos\beta_1} \\ \sin\alpha_2\frac{\sin\beta_2}{\cos\beta_2} & 1 & \cos\alpha_2\frac{\sin\beta_2}{\cos\beta_2} \\ \cos\alpha_3\frac{\sin\beta_3}{\cos\beta_3} & \sin\alpha_3\frac{\sin\beta_3}{\cos\beta_3} & 1 \end{bmatrix} \begin{bmatrix} \omega_x \\ \omega_y \\ \omega_z \end{bmatrix} = C\Omega \tag{27}$$

while the angular velocity of the end-effector is obtained by the angular velocity of the actuators by inverting the equation:

$$
\Omega = \begin{bmatrix} \omega_x \\ \omega_y \\ \omega_z \end{bmatrix} = \begin{bmatrix} -\dfrac{\cos\alpha_2 \sin\alpha_3}{\Delta \tan\beta_1} & \dfrac{\sin\alpha_1 \sin\alpha_3}{\Delta \tan\beta_2} & \dfrac{\cos\alpha_1 \cos\alpha_2}{\Delta \tan\beta_3} \\[2ex] \dfrac{\cos\alpha_2 \cos\alpha_3}{\Delta \tan\beta_1} & -\dfrac{\cos\alpha_3 \sin\alpha_1}{\Delta \tan\beta_2} & \dfrac{\sin\alpha_1 \sin\alpha_2}{\Delta \tan\beta_3} \\[2ex] \dfrac{\sin\alpha_2 \sin\alpha_3}{\Delta \tan\beta_1} & \dfrac{\cos\alpha_1 \cos\alpha_3}{\Delta \tan\beta_2} & -\dfrac{\cos\alpha_1 \sin\alpha_2}{\Delta \tan\beta_3} \end{bmatrix} \begin{bmatrix} \dot{\alpha}_1 \\ \dot{\alpha}_2 \\ \dot{\alpha}_3 \end{bmatrix} = C^{-1}\dot{Q} \tag{28}
$$

where $\Delta = \cos\alpha_3 \cos\alpha_2 \cos\alpha_1 + \sin\alpha_3 \sin\alpha_2 \sin\alpha_1$.

Equations (27) and (28) are defined for all of the non-singular configurations (for direct or inverse kinematics), which implies:

$$
\Delta \neq 0, \ \cos\beta_i \neq 0 \text{ and } \tan\beta_i \neq 0 \tag{29}
$$

2.8. Acceleration Analysis

The acceleration analysis may be developed by starting from the velocity relation Equation (27) which synthetically reads:

$$
\dot{Q} = C\Omega \tag{30}
$$

by a derivation with respect to the time, we found that the joint actuators acceleration was necessary to give an angular acceleration $\dot{\Omega}$ to the end-effector:

$$
\ddot{Q} = \dot{C}\Omega + C\dot{\Omega} \tag{31}
$$

and inversely, the angular acceleration of the mobile platform is:

$$
\dot{\Omega} = C^{-1}\left(\ddot{Q} - \dot{C}\Omega\right) \tag{32}
$$

With

$$
\dot{C} = \sum_{i=1}^{3} \frac{\partial C}{\partial \alpha_i}\dot{\alpha}_i + \sum_{i=1}^{3} \frac{\partial C}{\partial \beta_i}\dot{\beta}_i = C_a + C_b \tag{33}
$$

where C_a and C_b are:

$$
\mathbf{C_a} = \begin{bmatrix} 0 & -\dot{\alpha}_1 \sin\alpha_1 \tan\beta_1 & \dot{\alpha}_1 \cos\alpha_1 \tan\beta_1 \\[1.5ex] \dot{\alpha}_2 \cos\alpha_2 \tan\beta_2 & 0 & -\dot{\alpha}_2 \sin\alpha_2 \tan\beta_2 \\[1.5ex] -\dot{\alpha}_3 \sin\alpha_3 \tan\beta_3 & \dot{\alpha}_3 \cos\alpha_3 \tan\beta_3 & 0 \end{bmatrix} \tag{34}
$$

The time derivatives of the angles β_i to be inserted in C_b are easily obtained from Equation (26) and those for legs 2 and 3 as:

$$
\begin{bmatrix} \dot{\beta}_1 \\ \dot{\beta}_2 \\ \dot{\beta}_3 \end{bmatrix} = \begin{bmatrix} 0 & -\sin\alpha_1 & \cos\alpha_1 \\ \cos\alpha_2 & 0 & -\sin\alpha_2 \\ -\sin\alpha_3 & \cos\alpha_3 & 0 \end{bmatrix} \begin{bmatrix} \omega_x \\ \omega_y \\ \omega_z \end{bmatrix} \tag{35}
$$

The proposed solution can be applied both to the spherical PKM under study and to the projective angles.

3. Results

This paper has presented a methodology for the full kinematics analysis of a class of spherical PKMs. This methodology takes advantage of the properties of the projective angles for which the analysis is extended to velocity and acceleration. All of the solutions are found and the singular cases are discussed.

4. Discussion

The properties of the projective angles can be used to describe the 3D rotation motion of any rigid body and, in this sense, this notation is general. It is a particular angular convention with pros and cons, as with any other convention (Euler angles, quaternions ...). The application to PKM is particularly convenient to the class of spherical manipulators based on rotational joints forming cardan sequences because, in this case, the angles of the angular convention coincide with the joint angles of the PKM. In this context, the paper highlights the numerous analogies between the direct and inverse kinematics of the PKM and the relations between the rotation matrix describing the orientation of the end-effector and the projective angles of the frame that were associated with it. We may conclude that the projective angles are a sort of "intrinsic" notation to study the considered spherical PKMs.

The paper solves this relation for angular position, velocity and acceleration.

The proposed methodology is an alternative solution to those that were proposed by classical papers.

The inverse kinematics analysis of the agile eye, as reported in [1], results, for any actuated joint, in a quadratic equation of the tangent of the angle of the corresponding joint rotation. With the proposed methodology, if we adopt the projection angles as angular convention, we do not need to perform calculations because the actuated joint rotation coincides with the projective angles of the mobile frame, therefore there is no need to perform calculations.

The direct analysis of the agile eye is reported in [2] and results in a polynomial of degree 8 which leads to 8 real solutions. When adopting the methodology that was proposed in this paper, it is necessary to solve a 3×3 linear system instead and optionally to add $\pm 180°$ to the angles to generate the different solutions.

The problem that was addressed in the paper has multiple solutions, both for direct and inverse kinematics. The chosen solution can be selected according to common robotic practice: each solution corresponds to a different assembly configuration. Different configurations can be reached by crossing a singular configuration. Since this operation may create control problems in the execution of trajectories generally, at each time, the "most close" configuration with respect to the current configuration is chosen. So, for direct kinematics, if $\mathbf{R}$ is the actual configuration at time T and $\mathbf{R}_1'$, $\mathbf{R}_2'$, ... are the different solutions for $t = T + dT$, the i-th solution chosen is the one that minimizes $\| \mathbf{R} - \mathbf{R}_i' \|$. A similar approach can be applied to the inverse kinematics minimizing the rotation of the motors to reach the next pose. This also ensures the absence of discontinuity in the joint motion.

If the angular position of the end-effector is represented by projective angles, the Jacobian matrix is the identity matrix and therefore, in the domain of the projective angles, the PKM can be defined as decoupled.

Author Contributions: Conceptualization, G.L. and I.F.; Writing—Original Draft Preparation, G.L. and I.F., Writing-Review & Editing G.L., Project Administration, I.F.

Acknowledgments: The 3D model of Figure 5 was designed and fabricated by Giulio Spagnuolo (ITIA-CNR). His cooperation is greatly appreciated.

References

1.   Gosselin, C.; Angeles, J. The Optimum Kinematic Design of a Spherical Three-Degrees-of-Freedom Parallel Manipulator. *ASME J. Mech. Trans. Autom. Des.* **1989**, *111*, 202–207. [CrossRef]
2.   Gosselin, C.M.; Sefrioui, J.; Richard, M.J. On the direct kinematics of spherical three-degree-of-freedom parallel manipulators of general architecture. *ASME J. Mech. Des.* **1994**, *116*, 594–598. [CrossRef]

3. Gosselin, C.M.; Lavoie, E. On the kinematic design of spherical three-degree-of-freedom parallel manipulators. *Int. J. Robot. Res.* **1993**, *12*, 394–402. [CrossRef]

4. Kong, X.; Gosselin, C.M. A formula that produces a unique solution to the forward displacement analysis of a quadratic spherical parallel manipulator: The Agile Eye. *ASME J. Mech. Robot.* **2010**, *2*, 044501. [CrossRef]

5. Malosio, M.; Pio Negri, S.; Pedrocchi, N.; Vicentini, F.; Caimmi, M.; Molinari Tosatti, L. A Spherical Parallel Three Degrees-of-Freedom Robot for Ankle-Foot Neuro-Rehabilitation. In Proceedings of the 34th Annual International Conference of the IEEE EMBS, San Diego, CA, USA, 28 August–1 September 2012.

6. Malosio, M.; Caimmi, M.; Ometto, M.; Molinari Tosatti, L. Ergonomics and kinematic compatibility of PKankle, a fully-parallel spherical robot for ankle-foot rehabilitation. In Proceedings of the 5th IEEE RAS & EMBS International Conference on Biomedical Robotics and Biomechatronics (BioRob), São Paulo, Brazil, 12–15 August 2014.

7. Gosselin, C.M.; St. Pierre, E.; Gagné, M. On the development of the Agile Eye. *IEEE Robot. Autom. Mag.* **1996**, *3*, 29–37. [CrossRef]

8. Innocenti, C.; Parenti-Castelli, V. Echelon Form Solution of Direct Kinematics for the General Fully-Parallel Spherical Wrist. *Mech. Mach. Theory* **1993**, *28*, 553–561. [CrossRef]

9. Wohlhart, K. Displacement Analysis of the General Spherical Stewart Platform. *Mech. Mach. Theory* **1994**, *29*, 581–589. [CrossRef]

10. Huang, Z.; Yao, Y.L. A New Closed-Form Kinematics of the Generalized 3-DOF Spherical Parallel Manipulator. *Robotica* **1999**, *17*, 475–485. [CrossRef]

11. Alici, G.; Shirinzadeh, B. Topology Optimisation and Singularity Analysis of a 3-SPS Parallel Manipulator with a Passive Constraining Spherical Joint. *Mech. Mach. Theory* **2004**, *39*, 215–235. [CrossRef]

12. Kuo, C.-H.; Dai, J.S.; Legnani, G. A non-overconstrained variant of the Agile Eye with a special decoupled kinematics. *Robotica* **2014**, *32*, 889–905. [CrossRef]

13. Legnani, G.; Fassi, I. Representation of 3D Motion by Projective Angles. In Proceedings of the ASME IDETC2015, Boston, MA, USA, 2–5 August 2015.

14. Legnani, G.; Casolo, F.; Righettini, P.; Zappa, B. A homogeneous matrix approach to 3D kinematics and dynamics—I. Theory. *Mech. Mach. Theory* **1996**, *31*, 573–587. [CrossRef]

15. Legnani, G.; Casolo, F.; Righettini, P.; Zappa, B. A homogeneous matrix approach to 3D kinematics and dynamics—II. Applications to chain of rigid bodies and serial manipulators. *Mech. Mach. Theory* **1996**, *31*, 589–605. [CrossRef]

16. Kong, X.; Gosselin, C.M. Type Synthesis of Three-Degree-of-Freedom Spherical Parallel Manipulators. *Int. J. Robot. Res.* **2004**, *23*, 237–245. [CrossRef]

17. Kong, X.; Gosselin, C.M. Type Synthesis of 3-DOF Spherical Parallel Manipulators Based on Screw Theory. *ASME J. Mech. Des.* **2004**, *126*, 101–126. [CrossRef]

18. Fang, Y.; Tsai, L.-W. Structure Synthesis of a Class of 3-DOF Rotational Parallel Manipulators. *IEEE Trans. Robot. Autom.* **2004**, *20*, 117–121. [CrossRef]

19. Karouia, M.; Hervé, J.M. Asymmetrical 3-Dof Spherical Parallel Mechanisms. *Eur. J. Mech. A Solids* **2005**, *24*, 47–57. [CrossRef]

20. Hess-Coelho, T.A. Topological Synthesis of a Parallel Wrist Mechanism. *ASME J. Mech. Des.* **2006**, *128*, 230–235. [CrossRef]

21. Karouia, M.; Hervé, J.M. A Three-DOF Tripod for Generating Spherical Rotation. In *Advances in Robot Kinematics*; Lenarčič, J., Stanišić, M.M., Eds.; Springer: Basel, Switzerland, 2000; pp. 395–402.

22. Di Gregorio, R. Kinematics of a New Spherical Parallel Manipulator with Three Equal Legs: The 3-URC Wrist. *J. Robot. Syst.* **2001**, *18*, 213–219. [CrossRef]

23. Di Gregorio, R. A New Parallel Wrist Using Only Revolute Pairs: The 3-RUU Wrist. *Robotica* **2001**, *19*, 305–309. [CrossRef]

24. Di Gregorio, R. A New Family of Spherical Parallel Manipulators. *Robotica* **2002**, *20*, 353–358. [CrossRef]

25. Di Gregorio, R. Kinematics of the 3-UPU Wrist. *Mech. Mach. Theory* **2003**, *38*, 253–263. [CrossRef]

26. Di Gregorio, R. The 3-RRS Wrist: A New, Simple and Non-Overconstrained Spherical Parallel Manipulator. *ASME J. Mech. Des.* **2004**, *126*, 850–855. [CrossRef]

27. Di Gregorio, R. Kinematics of the 3-RSR wrist. *IEEE Trans. Robot.* **2004**, *20*, 750–753. [CrossRef]

28. Karouia, M.; Hervé, J.M. Non-Overconstrained 3-Dof Spherical Parallel Manipulators of Type: 3-RCC, 3-CCR, 3-CRC. *Robotica* **2006**, *24*, 85–94. [CrossRef]

29. Gogu, G. Fully-Isotropic Three-Degree-of-Freedom Parallel Wrists. In Proceedings of the IEEE International Conference on Robotics and Automation, Rome, Italy, 10–14 April 2007; pp. 895–900.

30. Enferadi, J.; Tootoonchi, A.A. A Novel Spherical Parallel Manipulator: Forward Position Problem, Singularity Analysis, and Isotropy Design. *Robotica* **2009**, *27*, 663–676. [CrossRef]

31. Baumann, R.; Maeder, W.; Glauser, D.; Clavel, R. The PantoScope: A Spherical Remote-Center-of-Motion Parallel Manipulator for Force Reflection. In Proceedings of the IEEE International Conference on Robotics and Automation, Albuquerque, NM, USA, 20–25 April 1997; pp. 718–723.

32. Vischer, P.; Clavel, R. Argos: A Novel 3-DoF Parallel Wrist Mechanism. *Int. J. Robot. Res.* **2000**, *19*, 5–11. [CrossRef]

33. Kuo, C.-H.; Dai, J.S. Kinematics of a Fully-Decoupled Remote Center-of-Motion Parallel Manipulator for Minimally Invasive Surgery. *ASME J. Med. Devices* **2012**, *6*, 021008. [CrossRef]

34. Carricato, M.; Parenti-Castelli, V. A Novel Fully Decoupled Two-Degrees-of-Freedom Parallel Wrist. *Int. J. Robot. Res.* **2004**, *23*, 661–667. [CrossRef]

35. Gallardo, J.; Rodríguez, R.; Caudillo, M.; Rico, J.M. A Family of Spherical Parallel Manipulators with Two Legs. *Mech. Mach. Theory* **2008**, *43*, 201–216. [CrossRef]

36. Kuo, C.-H. Projective-Angle-Based Rotation Matrix and Its Applications. In Proceedings of the Second IFToMM Asian Conference on Mechanism and Machine Science, Tokyo, Japan, 7–10 November 2012.

37. Legnani, G.; Fassi, I.; Giberti, H.; Cinquemani, S.; Tosi, D. A new isotropic and decoupled 6-DoF parallel manipulator. *Mech. Mach. Theory* **2012**, *58*, 64–81. [CrossRef]

38. Angels, J. *Rational Kinematics*; Springer: London, UK, 1988.

39. Nappo, F. On the absence of a general link between the angular velocity and the characterization of a rigid displacement. *Rendiconti di Matematica* **1970**, *3*. Series VI.

10

Viability and Feasibility of Constrained Kinematic Control of Manipulators

Marco Faroni [1], Manuel Beschi [2], Nicola Pedrocchi [2] and Antonio Visioli [1],*

[1] Dipartimento di Ingegneria Meccanica e Industriale, University of Brescia, Via Branze 38, 25123 Brescia, Italy; m.faroni003@unibs.it

[2] Istituto di Sistemi e Tecnologie Industriali Intelligenti per il Manifatturiero Avanzato, National Research Council, Via Alfonso Corti 12, 20133 Milano, Italy; manuel.beschi@stiima.cnr.it (M.B.); nicola.pedrocchi@stiima.cnr.it (N.P.)

* Correspondence: antonio.visioli@unibs.it.

Abstract: Recent advances in planning and control of robot manipulators make an increasing use of optimization-based techniques, such as model predictive control. In this framework, ensuring the feasibility of the online optimal control problem is a key issue. In the case of manipulators with bounded joint positions, velocities, and accelerations, feasibility can be guaranteed by limiting the set of admissible velocities and positions to a viable set. However, this results in the imposition of nonlinear optimization constraints. In this paper, we analyze the feasibility of the optimal control problem and we propose a method to construct a viable convex polyhedral that ensures feasibility of the optimal control problem by means of a given number of linear constraints. Experimental and numerical results on an industrial manipulator show the validity of the proposed approach.

Keywords: manipulators; trajectory planning; kinematic constraints; optimization; viability; inverse kinematics

1. Introduction

Robotic systems typically present kinematic and/or dynamic limitations. For example, joint positions are generally bounded within an available range of motion, while actuators implicitly present velocity and acceleration/torque limits. Further constraints can be due to the environment where the robot has to operate (e.g., partial occupation of the robot workspace) or to safety reasons, which may determine velocity and acceleration limitations. Including such constraints in the development of planning and control methods is of utter importance, as their violation might lead to unrealizable motions (with consequent significant errors in the execution of the tasks) or to safety issues (for example, in the case workspace limits are not respected).

Constrained methods are widely diffused, for example, in the case of offline trajectory planning. The general approach consists in the formulation of a constrained optimization problem that should optimize a given objective, such as minimum execution time [1] or minimum energy consumption [2]. This is also the case of global redundancy resolution methods [3], where the robot has to perform an assigned task, and the redundancy can be exploited to optimize a desired objective, such as maximum manipulability [4], dexterity [5], or maximum joint range availability [6]. However, all these methods are typically performed offline as they need the prior knowledge of the task to be performed, and they lead to heavy computational burdens that do not permit their online execution. For this reason, they are not able to handle online trajectory generation and re-planning techniques and this is a relevant limitation, considering recent robotic applications such as collaborative robotics and robots in unstructured environments. To handle the robot constraints in online planning and control methods, optimization-based techniques are becoming more and more widespread as the increasing computing power of modern processors allows their real-time implementation with small sampling periods.

Quadratic programming, for example, is intensively exploited in the resolution of the Inverse Kinematic (IK) problem of redundant manipulators [7–11]. In brief, a manipulator is termed kinematically redundant with respect to a task when it has a number of degrees of freedom greater than the one of the desired task. The IK problem does not present a unique solution, as the transformation between the joint and the task space results in an undetermined system. Such problem is usually addressed at differential level in order to exploit the linear relation between joint and task velocities given by the Jacobian of the robot. This can be formulated as a chain of Quadratic Programs (QP) with decreasing priorities: the tasks are written as cost functions and the robot limits are represented as linear constraints. The possibility of including robot constraints in the QP represents a great advantage with respect to other typical methods (such as the unconstrained projection of secondary tasks in the null space of the Jacobian [12,13]). Further developments in this field explicitly face the problem of possible saturations and task deformations by introducing a scaling variable into the QP to slow down the execution of the tasks if needed [14].

A further improvement of these methods is represented by the use of Model Predictive Control (MPC) techniques, which aim at overcoming issues related to the fact that the previous methods only consider the current state of the robot. In fact, MPC-based methods are able to take into account the future evolution of the system, the tasks, and the constraints, improving the behavior of the robot in terms of task satisfaction and smoothness [15,16]. MPC is also applied to constrained motion planning [17,18] or low-level control applications [19].

Disregarding the particular field of application, a fundamental issue to address when using optimization-based control techniques is the feasibility of the online optimization problem. In fact, feasibility must be ensured in any possible state of the system. Otherwise, a solution to the problem might not exist, and this could cause the algorithm to stop (unless specific but sub-optimal strategies are activated when infeasibility occurs).

Feasibility of the online optimization problem is strictly related to the concept of set viability [20]. A set is said to be viable if, given an initial state within such set, the state trajectory can be kept within the set by means of a proper and realizable input function. In other words, keeping the state within a viable set ensures feasibility for all future time instants. On the contrary, if the state exits the viable set, infeasibility of the control problem will surely occur at a certain time in the future. This is particularly relevant in the case of state-constrained systems, as infeasibility might occur if no viability conditions are added to the control problem but only state and input limits are taken into account.

For robot manipulator control, this issue is common to any control strategy that takes into account both position and acceleration/torque constraints. In this case, the simplistic imposition of box constraints on the position and the acceleration/torque of the robot does not ensure the existence of a feasible solution to the optimization problem. For example, if a robot joint approaches its position limit with high velocity, the position bound will be exceeded due to the bounded admissible deceleration. Few works addressed this issue by means of manually-tuned heuristic strategies that aimed at reducing the velocity of the robot when it approached its position limits [21–23]. Many other control strategies did not take into account position bounds, assuming that the reference trajectories were implicitly feasible [24,25]. Notice that the viability guarantee does not ensure the feasibility of the reference trajectory, which could be only verified offline when the whole trajectory is known a priori. Such approach does not apply to online methods, which should be able to handle online trajectory generation and re-planning. In these cases, we can only ensure the feasibility of the optimal control control, which typically aims at minimizing the deviation of the performed motion with respect to the reference trajectory. This means that, although the desired motion might be infeasible, the optimal control problem remains feasible. In this case, a deformation with respect to the nominal trajectory could arise.

A formal viability guarantee was given in [26], which proposed an invariance control scheme for constrained robot manipulators. In particular, viability conditions were derived in order not to exceed workspace limitations with bounded joint accelerations. In [14], a local redundancy resolution technique was proposed: to respect a given deceleration limit in the breaking phase, the position-velocity state region was limited by imposing that the robot was able to stop within the given joint limit (in fact, limiting the state region to a viable set). This approach was similarly adopted in [27] in the context of robust constrained motion planning. Recently, viability for joint-constrained manipulators was explicitly addressed in [28], with a particular focus on the discrete-time implementation of the viability constraints.

All the above-mentioned approaches were based on the derivation of an analytical viability condition for a double integral system with bounded input and output. The resulting condition is quadratic in the velocity state. However, as all the above-mentioned works refer to local/feedback methods, such condition could be easily linearized around the current system state, resulting in variable box constraints on the control actions. However, such linearization is not possible for predictive strategies. In this case, a linear approximation of the viability set should be obtained. For example, Faroni et al.[29] gives a sufficient condition to approximate the quadratic viable set with a linear constraint and maintaining the viability property. Although the approximation technique is easy to implement and does not remarkably increase the computational complexity of the problem, the resulting allowed state region could result to be significantly smaller than the original one.

In this paper, we propose a method to derive a viable convex polyhedron for a robotic system with bounded joint positions, velocities, and accelerations. In particular, a simple optimization problem is set up to determine the maximum polyhedron that approximates the original viable set. The original quadratic condition is approximated by a polyhedron with a given number of sides, by maximizing the area of the allowed velocity-position state-region. Moreover, viability of the resulting set is ensured by imposing that, for all points on the polyhedron boundary, there exists a realizable input action that keeps the next state within the polyhedron itself, which is also demonstrated to be convex and can be therefore rewritten as a linear inequality constraint in optimization-based algorithms (such as linear MPC techniques). The paper shows that, by increasing the number of sides, the polyhedron gives a better approximation of the original viable set, as expected. This means that a larger admissible state region can be exploited by the controller when the resulting constraints are included in the optimization problem. Consequently, the controller can obtain smaller tracking errors when the desired task requires the robot joint to get close to the maximal viable set boundary. In particular, numerical results show the enlargement of the admissible state region as the number of sides increases. Such improvement is more and more evident as the maximum acceleration values gets smaller. Finally, experimental results on a Universal Robots UR10 manipulator demonstrates the validity of the proposed approach. In particular, an MPC algorithm is applied to the tracking problem of a given joint-space trajectory and the results with different viability conditions are compared. Firstly, an experimental example shows that the use of an invariance condition is of vital importance to ensure both the feasibility of the online optimization problem and the satisfaction of the manipulator limits. Secondly, the results show that the viable sets obtained by means of the proposed method permits to obtain smaller (or null) tracking errors in the case the required motion is close to the viability boundary, increasing the performance of the MPC controller.

The paper is organized as follows. Section 2 introduces the concept of set viability applied to constrained kinematic control of robot manipulators. Section 3 illustrates the proposed method for the computation of the optimal viable polyhedron and gives viability and convexity proves. Numerical results demonstrate the effectiveness of the proposed approach in Section 4, while experimental results on a Universal Robot UR10 manipulator are shown in Section 5. Section 6 concludes the paper.

2. Feasibility of the Constrained Kinematic Control Problem

Consider a generic robot joint and denote with q, $\dot{q}$, $\ddot{q}$ its position, velocity, and acceleration, respectively. The limits on q, $\dot{q}$, and $\ddot{q}$ can be therefore expressed as:

$$q_{min} \leq q \leq q_{max}, \tag{1}$$

$$\dot{q}_{min} \leq \dot{q} \leq \dot{q}_{max}, \tag{2}$$

$$\ddot{q}_{min} \leq \ddot{q} \leq \ddot{q}_{max}, \tag{3}$$

where q_{min}, $\dot{q}_{min}$, $\ddot{q}_{min}$, and q_{max}, $\dot{q}_{max}$, $\ddot{q}_{max}$ are the minimum and maximum joint position, velocity, and acceleration, respectively.

Assume a discrete-time implementation with sampling period T in which the acceleration is considered as constant along each sampling period. At time k, kinematic limits at the next sampling time $k + 1$ can be easily written as linear inequalities in the joint acceleration $\ddot{q}_k$ as:

$$q_{min} \leq q_k + T \dot{q}_k + \frac{1}{2} T^2 \ddot{q}_k \leq q_{max}, \tag{4}$$

$$\dot{q}_{min} \leq \dot{q}_k + T \ddot{q}_k \leq \dot{q}_{max}, \tag{5}$$

$$\ddot{q}_{min} \leq \ddot{q}_k \leq \ddot{q}_{max}. \tag{6}$$

However, the simplistic imposition of Equations (4)–(6) may result in an empty admissible set, as no feasible solution might exist that satisfies all constraints at the same time [28].

Indeed, the existence of a solution is guaranteed if Equations (4)–(6) are feasible for all the admissible states of the system. This is strictly correlated to the concept of set viability, as mentioned in Section 1. As each joint is modeled as a double integrator, the viability analysis traces back to the viability of a double integrator system with bounded input and output. By imposing the feasibility of Equations (4)–(6), the maximal viability set for the double integrator can be derived analytically [26]. Intuitively, it can be calculated by imposing that, applying the maximum deceleration $\ddot{q}_{min}$, the joint stops at $q = q_{max}$ with null velocity ($\dot{q} = 0$). The resulting condition (valid for the upper bound) is given by:

$$\begin{cases} q - \frac{\dot{q}^2}{2\ddot{q}_{min}} - q_{max} \leq 0 & \text{if } \dot{q} > 0 \\ q - q_{max} \leq 0 & \text{otherwise} \end{cases} \tag{7}$$

which expresses a quadratic condition in the system states. An analogous condition for the lower bound can be derived likewise.

Such condition can be easily linearized around the current velocity and position, in the case of local methods, as in [14,27]. The inclusion of Equation (7) in the QP ensures that the states of the system (i.e., the joint velocities and positions) remain within a viable set for which the problem is feasible. More details about the discrete implementation in robotic systems can be found in [28].

The necessity of deriving a linear approximation of the original quadratic equation (Equation (7)) comes from the advantages given by the linear formulation in optimization-based controllers (in particular, the significant decrement of computational time and the ease of implementation of linear MPC with respect to the nonlinear one). However, in the case of predictive strategies, the linearization adopted for local methods is not applicable.

To tackle this issue, Faroni et al. [29] proposed a linear viability condition based on a single constraint for each joint. Firstly, it showed that approximating the quadratic constraint in Equation (7) by means of a straight line passing through the extreme points of the maximal viable set (i.e., $(q_{max}, 0)$ and the intersection between Equation (7) and $\dot{q} = \dot{q}_{max}$ in Figure 1) leads to a non-viable set. Then, it derives a linear viability condition by imposing that the maximum deceleration along the linear constraint is exactly equal to the maximum admissible deceleration. In this way, the joint states can be always kept within the resulting set by means of a realizable acceleration. Such condition is given by:

$$-\frac{\ddot{q}_{\min}}{\dot{q}_{\max}}\, q + \dot{q} + \frac{\ddot{q}_{\min}}{\dot{q}_{\max}}\, q_{\max} \leq 0 \tag{8}$$

for the upper bound, and likewise for the lower one.

Inequalities (Equations (7) and (8)) are graphically represented in Figure 1. Linearity of Equation (8) is a great advantage, as it allows the straightforward inclusion in an optimization problem as a linear constraints. However, it is clear that Equation (8) implies a conservative reduction of the available state-space region compared to Equation (7), and such reduction worsens as the acceleration limits become smaller.

From a practical perspective, the reduction of the admissible state region shows its drawbacks when the robot is required to perform a motion that would violate Equation (8). In fact, the states laying between the linear and the quadratic constraints in Equations (7) and (8) are potentially realizable by the robot, but they are automatically excluded by the controller, with consequent deformation of the desired trajectory. Similarly, for redundant manipulators, shrinking the admissible state region could results in a worse satisfaction of the secondary objectives.

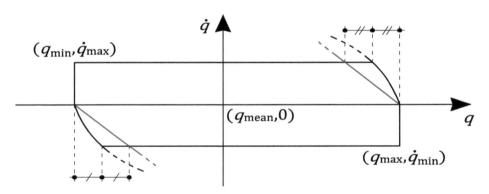

Figure 1. Viable admissible set for the double integrator system. Black: Maximal region, given by the quadratic inequality in Equation (7). Blue: Conservative linear inequality in Equation (8), as proposed in [29].

3. Proposed Method

As mentioned in the previous section, viability conditions are fundamental to ensure the feasibility of the control problem. The derivation of linear viability conditions (e.g., Equation (8)) allows their straightforward implementation in an optimization-based control framework, such as MPC algorithms. A strategy to enlarge the admissible region without waving its linearity is therefore proposed. It consists in increasing the number of linear constraints that compose the admissible set. We can therefore state the following problem.

Problem 1. *Given the acceleration, velocity, and configuration limits for each joint, find the maximal viable convex polyhedron determined by a given number of linear inequalities.*

We address this problem by converting it into an optimization problem. In particular, for each joint, two optimization problems are set up (one for the upper position bound and one for the lower position bound). Without loss of generality, consider the upper configuration bound for a generic joint (i.e., the first quarter in Figure 1). The optimization variables of the problem are given by the coordinates of the extremes of the segment composing the polyhedron shown in Figure 2. Given a number of segments h, such variables are denoted with the vectors:

$$p_x = (p_{x,0}, \ldots, p_{x,j}, \ldots, p_{x,h}), \tag{9}$$

$$p_y = (p_{y,0}, \ldots, p_{y,j}, \ldots, p_{y,h}), \tag{10}$$

whereas a generic point on the polyhedron is denoted with $P_j(p_{x,j}, p_{y,j})$.

Now, impose that $(p_{x,0}, p_{y,0}) = (q_{\text{mean}}, \dot{q}_{\text{max}})$, where $q_{\text{mean}} = \frac{q_{\text{max}} + q_{\text{min}}}{2}$ and $(p_{x,h}, p_{y,h}) = (q_{\text{max}}, 0)$. To maximize the area covered by the polyhedron, a cost function equivalent to the opposite of such area is defined as follows:

$$\psi = \frac{1}{2} \sum_{j=1}^{h} (p_{x,j-1} - p_{x,j})(p_{y,j-1} + p_{y,j}). \tag{11}$$

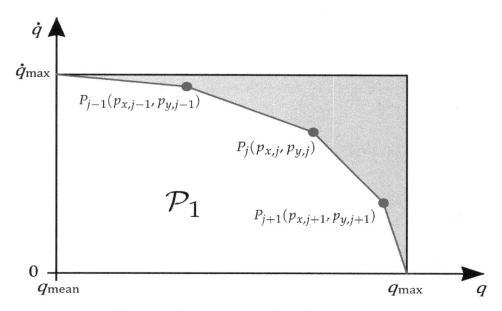

Figure 2. Construction of the polyhedron $\mathcal{P}_1$.

The extremes of the segments are required to lay in the first quarter and have to respect the maximum position and velocity bounds. This results in the following box constraints:

$$q_{\text{mean}} \leq p_x \leq q_{\text{max}}, \tag{12}$$

$$0 \leq p_y \leq \dot{q}_{\text{max}}. \tag{13}$$

Moreover, as a first necessary condition to the convexity of the polyhedron, we impose:

$$p_{x,j} - p_{x,j-1} \geq 0 \quad \forall j = 1, \ldots, h, \tag{14}$$

$$p_{y,j-1} - p_{y,j} \geq 0 \quad \forall j = 1, \ldots, h. \tag{15}$$

One last constraint needs to be imposed to ensure that the resulting polyhedron is viable and convex. To this purpose, we resort to a geometrical reasoning. Consider a generic segment $\overline{P_j P_{j+1}}$ whose extremes are the points $(p_{x,j}, p_{y,j})$ and $(p_{x,j+1}, p_{y,j+1})$ and denote with $\vec{n}_j$ the unitary vector normal to the segment and directed toward the inner region of the polyhedron. It results that:

$$\vec{n}_j = \left(\frac{p_{y,j+1} - p_{y,j}}{\|\overline{P_j P_{j+1}}\|}, \frac{p_{x,j} - p_{x,j+1}}{\|\overline{P_j P_{j+1}}\|} \right). \tag{16}$$

Consider then a generic point $P_\zeta(p_{x,\zeta}, p_{y,\zeta})$ on the segment. By applying the maximum deceleration $\ddot{q}_{\min}$ to the system, the states are forced to move on the curve given by:

$$\begin{cases} q = p_{x,\zeta} + p_{y,\zeta}t + \frac{1}{2}\ddot{q}_{\min}t^2 \\ \dot{q} = p_{y,\zeta} + \ddot{q}_{\min}t \end{cases} \tag{17}$$

The tangent vector to such curve in the $(q, \dot{q})$-plane is given by:

$$\vec{t} = \left(\frac{dq}{dt}, \frac{d\dot{q}}{dt} \right) = (p_{y,\zeta} + \ddot{q}_{\min}t, \ddot{q}_{\min}). \tag{18}$$

The vector $\vec{t}$ at $t = 0$ is therefore the tangent vector in the point P_ζ and results to be:

$$\vec{t}(p_{x,\zeta}, p_{y,\zeta}) = (p_{y,\zeta}, \ddot{q}_{\min}). \tag{19}$$

Note that $\vec{t}(p_{x,\zeta}, p_{y,\zeta})$ represents the direction of the state movement when the maximum realizable deceleration is applied. The state will therefore be able to remain below the considered segment if the following condition holds:

$$\vec{n}_j \cdot \vec{t}(p_{x,\zeta}, p_{y,\zeta}) \geq 0 \qquad \forall\, (p_{x,\zeta}, p_{y,\zeta}) \in \overline{P_j P_{j+1}} \tag{20}$$

where $(\cdot)$ denotes the scalar product between two vectors. As

$$\min_{p_{x,\zeta}, p_{y,\zeta}} \left(\vec{n}_j \cdot \vec{t}(p_{x,\zeta}, p_{y,\zeta}) \right) = \vec{n}_j \cdot \vec{t}(p_{x,j}, p_{y,j}), \tag{21}$$

condition in Equation (20) can be imposed on the sole extreme point P_j of each segment and becomes:

$$p_{y,j}(p_{y,j+1} - p_{y,j}) + \ddot{q}_{\min}(p_{x,j} - p_{x,j+1}) \geq 0. \tag{22}$$

Finally, the resulting optimization problem can be written as:

$$
\begin{aligned}
\underset{p_x, p_y}{\text{minimize}} \quad & \psi \\
\text{subject to} \quad & q_{\text{mean}} \leq p_x \leq q_{\max} \\
& 0 \leq p_y \leq \dot{q}_{\max} \\
& p_{x,j} - p_{x,j-1} \geq 0 \quad \forall j = 1, \ldots, h \\
& p_{y,j-1} - p_{y,j} \geq 0 \quad \forall j = 1, \ldots, h \\
& p_{y,j}(p_{y,j+1} - p_{y,j}) + \ddot{q}_{\min}(p_{x,j} - p_{x,j+1}) \geq 0 \quad \forall j = 1, \ldots, h-1 \\
& p_{x,0} = q_{\text{mean}} \\
& p_{y,0} = \dot{q}_{\max} \\
& p_{x,h} = q_{\max} \\
& p_{y,h} = 0
\end{aligned}
\tag{23}
$$

Following the same reasoning, an analogous optimization problem can be set up for the lower position bound. Note that, if the joint limits are symmetric, the solution for the lower bounds can be obtained by "mirroring" the solution of Equation (23) into the third quarter of the $(q, \dot{q})$-plane. Notice that, as Equation (23) is non-convex, global optimization such as genetic or multi-start algorithms should be adopted to solve the optimization problem (see Section 4).

Denoting with $\mathcal{P}_1$ the polyhedron determined by the segments obtained as solution of Equation (23), and with $\mathcal{P}_2$ the analogous polyhedron obtained for the lower bounds, the overall polyhedron for the single joint is given by:

$$\mathcal{P} = \mathcal{P}_1 \cup \mathcal{P}_2 \cup [q_{min}, q_{mean}] \times [0, \dot{q}_{max}] \cup [q_{mean}, q_{max}] \times [\dot{q}_{min}, 0]. \qquad (24)$$

Proposition 1. *Polyhedron $\mathcal{P}$ is a viable set with respect to the given joint position, velocity and acceleration limits. Moreover, such polyhedron is convex.*

Proof (Viability). Viability of the set is implicitly ensured by the imposition of Equation (22). $\qquad \square$

Proof (Convexity). To prove it by contradiction, denote the solution of Equation (23) with p_x^*, p_y^* and assume that the corresponding polyhedron $\mathcal{P}_1^*$ (i.e., the part of the polyhedron in the first quarter) is non-convex. Consider then a triple of consequent extremes $\{P_{j-1}, P_j, P_{j+1}\}$, which causes a non-convexity. Equations (20) and (21) imply:

$$\vec{n}_{j-1} \cdot \vec{t}(p_{x,j-1}^*, p_{y,j-1}^*) \geq 0, \qquad (25)$$
$$\vec{n}_j \cdot \vec{t}(p_{x,j}^*, p_{y,j}^*) \geq 0. \qquad (26)$$

Consider now the point P_c given by the projection of P_j onto $\overline{P_{j-1}P_{j+1}}$ and denote with $\vec{n}_c$ the vector normal to $\overline{P_{j-1}P_{j+1}}$. We want to prove that the polyhedron obtained by substituting P_c to P_j in $\mathcal{P}_1^*$ is a feasible and more efficient solution of Equation (23). First, constraints in Equations (12)–(15) are straightforwardly satisfied. Moreover,

$$\vec{n}_c \cdot \vec{t}(p_{x,j-1}^*, p_{y,j-1}^*) \geq \vec{n}_{j-1} \cdot \vec{t}(p_{x,j-1}^*, p_{y,j-1}^*) \geq 0 \qquad (27)$$

as $\vec{n}_{j-1} \cdot \vec{i} \leq \vec{n}_c \cdot \vec{i} \leq 0$ and $\vec{n}_c \cdot \vec{j} \leq \vec{n}_{j-1} \cdot \vec{j} \leq 0$ by construction (where $\vec{i}$ and $\vec{j}$ denote the unitary vectors directed as the horizontal and the vertical axis, respectively).

Moreover, as $\ddot{q}_{min}$ in Equation (19) is assumed to be negative, the following inequality holds:

$$\vec{n}_c \cdot \vec{t}(p_{x,c}, p_{y,c}) \geq \vec{n}_c \cdot \vec{t}(p_{x,j-1}^*, p_{y,j-1}^*) \geq 0. \qquad (28)$$

Note that Equations (27) and (28) implies that the new candidate solution satisfies Equation (22) and is therefore a feasible solution to Equation (23). Moreover, the candidate solution is more efficient than the assumed one, as

$$\psi(P_0, \dots, P_{j-1}, P_c, P_{j+1}, \dots, P_h) \leq \psi(P_0, \dots, P_{j-1}, P_j, P_{j+1}, \dots, P_h), \qquad (29)$$

which means that p_x^*, p_y^* are not the optimal solution of Equation (23), as a feasible and more efficient solution that eliminates such non-convexity exists. Applying this reasoning to any triples of extremes that cause a non-convexity implies that the optimal polyhedron $\mathcal{P}_1$ must be convex. The same demonstration can be applied to $\mathcal{P}_2$ leading to the same conclusion. $\qquad \square$

4. Numerical Results

Consider the first joint of the robot manipulator Universal Robot UR10. The joint configuration and velocity limits from the datasheet are given by:

$$q_{max} = -q_{min} = 3.14 \text{ rad}, \qquad \dot{q}_{max} = -\dot{q}_{min} = 2.16 \text{ rad/s}. \tag{30}$$

Typically, acceleration limits are not explicitly given in the datasheet. Indeed, the actual dynamic limit of the joint is usually determined by the available joint torque. However, handling viability directly in terms of torque limits is typically avoided due to the complexity of the problem. The common approach consists in estimating a (usually conservative) corresponding acceleration limit for each joint separately. A method for the estimation of such acceleration limit is given, for instance, in [28]. As a first example, assume that the acceleration limits are given by $\ddot{q}_{max} = -\ddot{q}_{min} = 6 \text{ rad/s}^2$.

We solve the optimization problem (Equation (23)) for different values of h to evaluate the resulting polyhedrons as the number of edges grows and to compare the results to the quadratic and the linear viability constraints (Equations (7) and (8), respectively). Table 1 shows the values of the inner area of polyhedron $\mathcal{P}_1$. The values are normalized with respect to the maximal viable set obtained by means of Equation (7), whose area is therefore equal to one (last column of Table 1). As expected, the smaller area is given by Equation (8) (first column of Table 1), while the extension of the polyhedron obtained by solving Equation (23) grows with the number of edges. In other words, the larger the number of edges, the closer the optimal polyhedron is to the maximal extension (given by Equation (7)). This is clearly shown in Figure 3, which depicts the different viable sets.

An analogous example is performed by using $\ddot{q}_{max} = -\ddot{q}_{min} = 3 \text{ rad/s}^2$. The values of the normalized area of $\mathcal{P}_1$ are shown in Table 2 and the resulting viable sets are depicted in Figure 4. The results lead to conclusions similar to the ones given by the previous example. However, it is clear that the magnitude of the phenomenon grows as the acceleration limits get smaller.

Of course the use of a larger value of h also gives some drawbacks. First, the computational complexity of Equation (23) rapidly grows with h as the number of variables and constraints is linear in h. As an example, the time needed to solve Equation (23) for $h = 1$ and $h = 11$ was in the order of 1 second and 15 s, respectively (the computation was performed in Matlab using a multi-start gradient-descent method on a standard laptop mounting a 2.5 GHz Intel Core i5-2520M processor).

Furthermore, the value of h determines the number of linear inequalities describing the polyhedron. In optimization-based control algorithms (such as linear MPC), this affects the computational complexity of the online optimal control problem, which is typically a critical issue in online methods, as mentioned in Section 1.

Table 1. Inner area of $\mathcal{P}_1$ for different values of h in case $\ddot{q}_{max} = -\ddot{q}_{min} = 6 \text{ rad/s}^2$.

Equation (8)	$h = 3$	$h = 4$	$h = 5$	$h = 6$	$h = 7$	$h = 8$	$h = 9$	$h = 10$	$h = 11$	Equation (7)
0.9139	0.9636	0.9773	0.9836	0.9871	0.9895	0.9911	0.9920	0.9923	0.9925	1

Table 2. Inner area of $\mathcal{P}_1$ for different values of h in case $\ddot{q}_{max} = -\ddot{q}_{min} = 3 \text{ rad/s}^2$.

Equation (8)	$h = 3$	$h = 4$	$h = 5$	$h = 6$	$h = 7$	$h = 8$	$h = 9$	$h = 10$	$h = 11$	Equation (7)
0.8200	0.9239	0.9525	0.9656	0.9731	0.9731	0.9813	0.9838	0.9857	0.9857	1

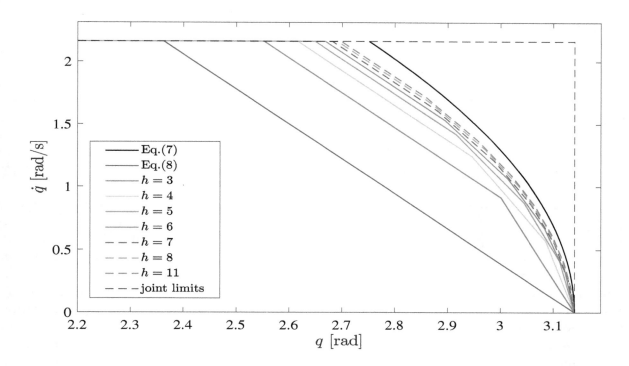

Figure 3. Comparison of the viable polyhedrons for different values of h (portion of interest of the first quarter) in the case $\ddot{q}_{max} = -\ddot{q}_{min} = 6 \, \text{rad/s}^2$.

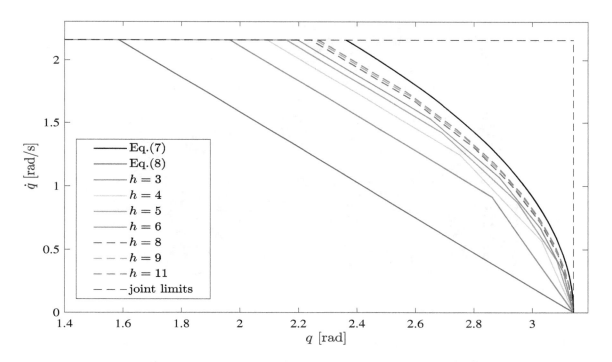

Figure 4. Comparison of the viable polyhedrons for different values of h (portion of interest of the first quarter) in the case $\ddot{q}_{max} = -\ddot{q}_{min} = 3 \, \text{rad/s}^2$.

5. Experimental Results

Experimental were also performed on a six-degrees-of-freedom Universal Robot UR10 manipulator to prove the validity of the proposed approach. For the purposes of the paper, the joint position and velocity limits have been set to slightly conservative values with respect to the ones given in the datasheet. In particular, they are set as:

$$Q_{max} = -Q_{min} = (\,3,\,3,\,3,\,3,\,3,\,3\,)\text{ rad,} \tag{31}$$

$$\dot{Q}_{max} = -\dot{Q}_{min} = (\,2,\,2,\,3,\,3,\,3,\,3\,)\text{ rad/s.} \tag{32}$$

The acceleration limits have been set as $\ddot{Q}_{max} = -\ddot{Q}_{min} = (\,3,\,3,\,3,\,3,\,3,\,3\,)\text{ rad/s}^2$. The experimental platform is shown in Figure 5. The objective of these experiments was the evaluation of the performance of an optimization-based control algorithm when different viability constraints are implemented. To this purpose, a simple MPC scheme was applied to a trajectory following problem. The MPC control scheme is in charge of following a position reference signal in the joint space. The model implemented in the MPC consists of a double integrator for each joint, as typical of robot kinematic control [15,30]. The input action is therefore represented by the joint accelerations, which then feed the low-level controller of the robot. Joint position, velocity, and acceleration limits can be implemented as linear constraints in the MPC (see [15] for details) and the online optimal control problem results to be a QP which minimizes the weighted sum of the tracking error and the control effort, as typical of linear MPC [31]. Notice that the choice of such a simple control scheme is due to the will of highlighting the behavior of a linear MPC controller in the case of different viability conditions. However, the proposed method could be straightforwardly applied to more complex MPC techniques such as [15,16,29].

Figure 5. Universal Robot UR10 used as experimental platform.

The MPC algorithm was implemented using a sampling period $T = 8$ ms and by setting a predictive and a control horizon $N = 20$ sampling periods. The robot trajectory is controlled by means of a ROS-based control architecture. Namely, a position controller runs in a ROS Kinetic Ubuntu 16.04. The controller communicates with the robot by means of a TCP connection. The controller takes the MPC position output as reference and receives the actual joint position. The controller output is the sum of a proportional action, with gain equal to seven, and a feedforward term equal to the MPC velocity output.

We considered a simple trajectory given by a straight line in the joint space, parameterized with respect to the normalized longitudinal length along the path, denoted with $r \in [0, 1]$. The trajectory $Q_{\text{des}}(t)$ is therefore defined as:

$$Q_{\text{des}}(t) = Q_{\text{start}} + (Q_{\text{end}} - Q_{\text{start}})\, r(t), \tag{33}$$

where $Q_{\text{start}} = (0, -1, 0.5, 0, 0, 0.5)$ rad and $Q_{\text{end}} = (2.9, -0.5, -1.5, 1, -0.5, 2.9)$ rad are the initial and the final points of the trajectory, respectively, and $r : [0, t_{\text{end}}] \to [0, 1]$, $t \mapsto r(t)$ is defined as a timing law with trapezoidal velocity profile, where t_{end} represents the total time of the trajectory, as depicted in Figure 6.

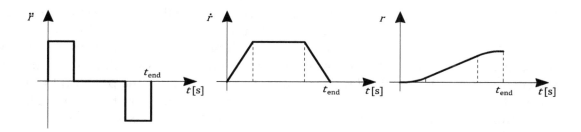

Figure 6. Timing law with trapezoidal velocity profile.

The control problem the MPC controller has to solve at each cycle k therefore results:

$$
\begin{aligned}
\underset{\ddot{Q}}{\text{minimize}} \quad & \sum_{i=1}^{N} \left\| Q_{\text{des}}(k+j) - Q(k+j) \right\|^2 + \lambda \sum_{i=0}^{N-1} \left\| \ddot{Q}(k+j) \right\|^2 \\
\text{subject to} \quad & \dot{Q}_{\min} \leq \dot{Q}(k+j) \leq \dot{Q}_{\max} && \forall j = 1, \ldots, N \\
& \ddot{Q}_{\min} \leq \ddot{Q}(k+j) \leq \ddot{Q}_{\max} && \forall j = 0, \ldots, N-1 \\
& Q(k+1) = Q(k) + T\dot{Q}(k) + 0.5T^2\ddot{Q}(k) \\
& \dot{Q}(k+1) = \dot{Q}(k) + T\ddot{Q}(k)
\end{aligned}
\tag{34}
$$

where Q, $\dot{Q}$, and $\ddot{Q} \in \mathbb{R}^6$ are the joint position, velocity, and acceleration vectors, respectively, and the controller effort weighting factor $\lambda = 1 \times 10^{-6}$ was tuned empirically. Different implementations of the position limits will then be added to the problem in order to evaluate the different behaviors of the system.

As a first example, the total time of the trajectory was chosen as $t_{\text{end}} = 2.07$ s and the behavior of the MPC controller was evaluated in three different scenarios:

- No position bounds are implemented.
- Box position constraints are implemented, which means that the following constraints are added to Equation (34):

$$Q_{\min} \leq Q(k+j) \leq Q_{\max} \qquad \forall j = 1, \ldots, N \tag{35}$$

- The linear viability condition in Equation (8) is applied, that is, Equation (35), and the following constraint are added to Equation (34):

$$-\frac{\ddot{Q}_{i,\min}}{\dot{Q}_{i,\max}} Q_i(k+j) + \dot{Q}_i(k+j) + \frac{\ddot{Q}_{i,\min}}{\dot{Q}_{i,\max}} Q_{i,\max} \leq 0 \qquad \forall j = 1, \ldots, N, \quad \forall i = 1, \ldots, 6; \tag{36}$$

where the subscript i denotes the ith element of the vector.

The viability constraints are imposed throughout the whole control horizon for the sake of clarity and because the small increment in the computational time (due to the larger number of linear constraints) does not represent a significant issue in the presented experimental tests. However, in the case of linear MPC, imposing such constraints only on the last time instant of the horizon would be enough to ensure feasibility of the control problem and would not give significant differences in terms of control performance. The imposition of the constraint throughout the whole horizon permits to obtain a more reliable prediction of the future trajectory. This can be helpful especially in the case the prediction is utilized in linearized methods such as [15], as a non-reliable prediction might worsen the performance of the method.

Notice that the trajectory is devised in such a way that the desired motions of the first and the sixth joints exceed the maximal viable set, as shown in Figure 7a,b. The figures also show the phase plot of the position/velocity variables computed by the MPC in the three above-mentioned cases. In the case no position bounds are implemented (dashed green line), the joint position obviously does not satisfy the position limit. However, when box position constraints are implemented in the MPC (dash-dotted purple line), infeasibility of the control problem occurs and, since that time, the joint is forced to decelerate with the maximum admissible deceleration, but the position limit cannot be satisfied anyway. This highlights the importance of the viability property. In fact, the implementation of the linear constraints in Equation (8) does ensure the feasibility of the control problem and permits to satisfy the position bound (with a deformation of the original trajectory) (dashed red line). These behaviors are clarified also in Figure 8a,b, where the ideal and the measured joint positions for Joint 1 and Joint 6 are shown.

A second experiment is performed by choosing a total trajectory time $t_{end} = 2.65$ s. In this case, the trajectory drives the first and the third joints in the region between Equations (7) and (8). The behavior of the MPC controller is evaluated in cases where three different viability constraints are implemented:

- The linear viability condition in Equation (8), that is, Equation (36) is added to Equation (34).
- The viable polyhedron obtained by solving Equation (23) with $h = 3$, that is, the following constraints are added to Equation (34):

$$(Q_i(k+j), \dot{Q}_i(k+j)) \in \mathcal{P}_{h=3}^i \qquad \forall j = 1, \ldots, N, \quad \forall i = 1, \ldots, 6; \qquad (37)$$

where the superscript i denotes the viable polyhedral of the ith joint.
- The viable polyhedron obtained by solving Equation (23) with $h = 4$, that is, the following constraints are added to Equation (34):

$$(Q_i(k+j), \dot{Q}_i(k+j)) \in \mathcal{P}_{h=4}^i \qquad \forall j = 1, \ldots, N, \quad \forall i = 1, \ldots, 6. \qquad (38)$$

Figure 9a,b shows the state trajectory for the first joint in the three different cases. Notice that the position limit is respected and the control problem remains feasible for all cases. However, the reduction of the admissible region determined by Equation (8) gives rise to a significant modification of the original trajectory, although the desired states are always potentially realizable by the robot (as the trajectory does not exceed Equation (7)). An improvement is obtained when the polyhedron with $h = 3$ is used.

Notice that the small velocity bumps visible in the figures are due to predictive nature of the controller. In fact, as the control scheme is based on the tracking of a position reference along the predictive horizon, the controller slightly increases the velocity when it realizes the viability constraint will be activated and a deformation of the task will arise (as the state will be forced to follow the constraint). By giving a small acceleration before the activation of the constraint, the controller minimizes the future deviation with respect to the given position reference.

Finally, the use of $h = 4$ permits to obtain a viable polyhedron that is large enough to enclose the whole desired trajectory. These behaviors are clarified also in Figure 10a,b, where the ideal and the measured joint positions for Joint 1 and Joint 6 are shown.

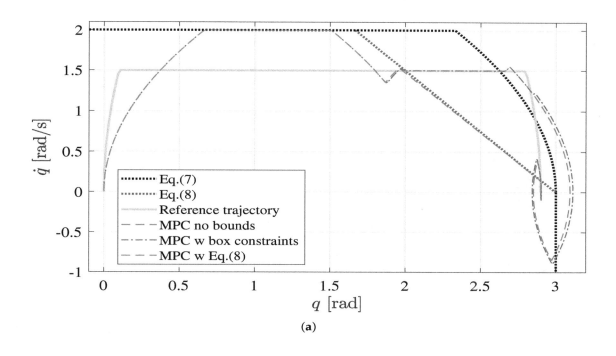

(a)

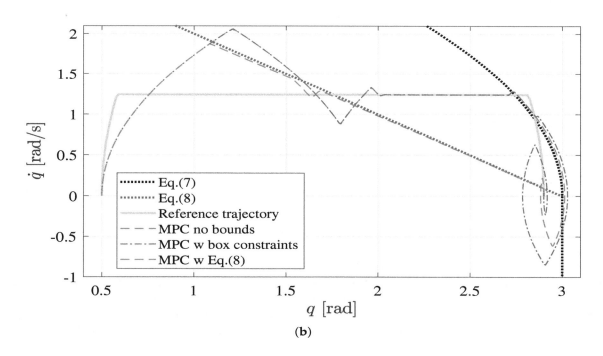

(b)

Figure 7. State trajectory in the $(q, \dot{q})$-plane for (**a**) Joint 1 and (**b**) Joint 3 for $T_{tot} = 2.07$ s: Reference trajectory (gray solid line); MPC without position bounds (green dashed line); MPC with box position bounds (purple dash-dot line); MPC with linear viability inequality (Equation (8)) (red dashed line); maximal viable set (black dotted line); and linear viability inequality ((Equation (8)) (blue dotted line).

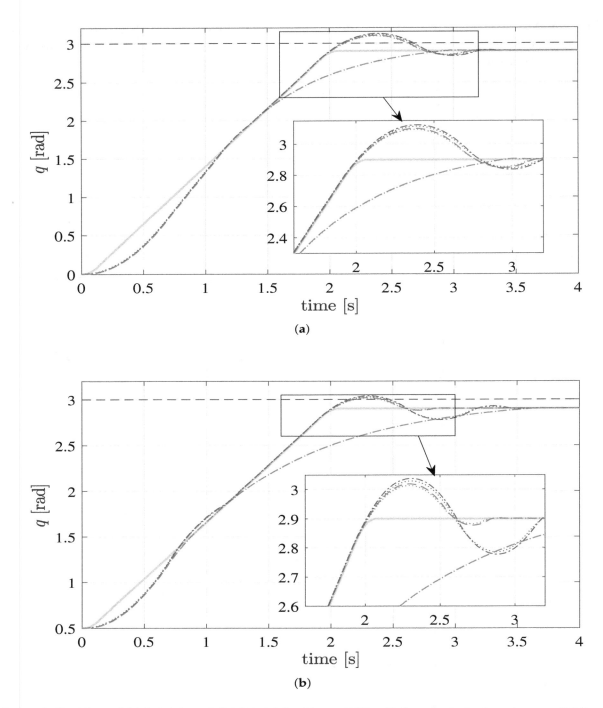

Figure 8. Position of (**a**) Joint 1; and (**b**) Joint 3 for $T_{\text{tot}} = 2.07$ s: Reference trajectory (gray solid line). Output of MPC given as reference signal to the low-level controller: MPC without position bounds (green dotted line); MPC with box position bounds (purple dotted line); and MPC with linear viability inequality (Equation (8)) (red dotted line). Measured position: MPC without position bounds (green dashed line); MPC with box position bounds (purple dash-dot line); and MPC with linear viability inequality (Equation (8)) (red dashed line).

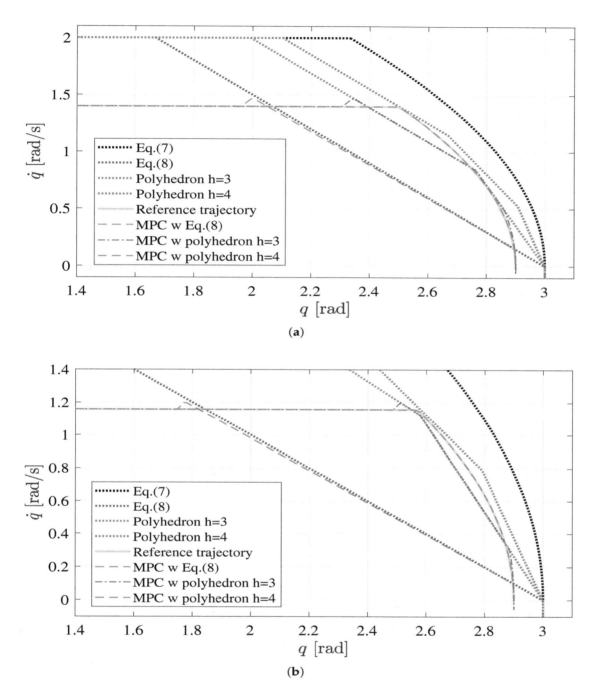

(a)

(b)

Figure 9. State trajectory in the $(q, \dot{q})$-plane for **(a)** Joint 1 and **(b)** Joint 3 for $T_{tot} = 2.65$ s: Reference trajectory (gray solid line); MPC with linear viability inequality (Equation (8)) (red dashed line); MPC implementing the viable polyhedron with $h = 3$ (purple dash-dot line); and MPC implementing the viable polyhedron with $h = 4$ (dashed green). Viable sets: Maximal (black dotted line); Linear viability inequality (Equation (8)) (blue dotted line); Polyhedron with $h = 3$ (light blue dotted line); and Polyhedron with $h = 4$ (gray dotted line).

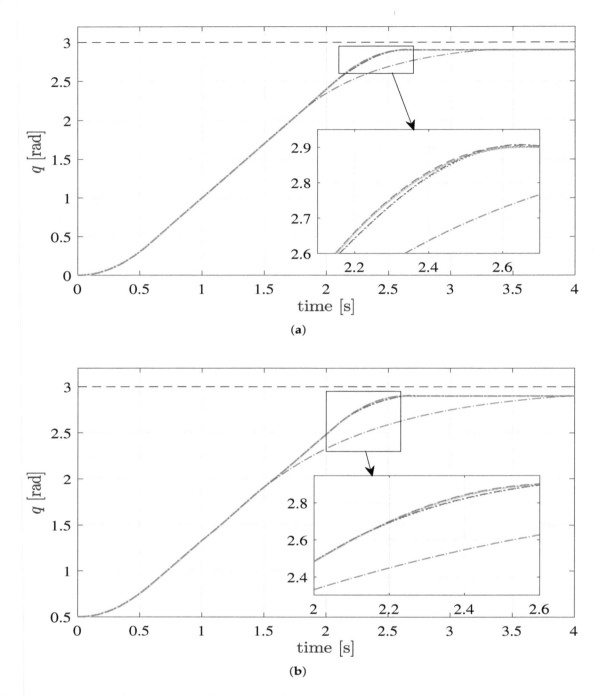

Figure 10. Position of (a) Joint 1 and (b) Joint 3 for $T_{\text{tot}} = 2.65$ s. Reference trajectory (gray solid line). Output of MPC given as reference signal to the low-level controller: MPC with linear viability inequality (Equation (8)) (red dotted line); MPC implementing the viable polyhedron with $h = 3$ (purple dotted line); and MPC implementing the viable polyhedron with $h = 4$ (green dotted line). Measured position: MPC with linear viability inequality (Equation (8)) (red dashed line); MPC implementing the viable polyhedron with $h = 3$ (purple dash-dot line); and MPC implementing the viable polyhedron with $h = 4$ (green dashed line).

6. Conclusions

In this paper, we have proposed a method to compute the maximal viable polyhedron for a robot manipulator with bounded joint positions, velocities, and accelerations. In the proposed approach, each joint is considered separately and two optimization problems are devised: one for the upper bounds and one for the lower ones. Given its number of sides, the resulting polyhedron maximizes the area of the admissible position/velocity region. Moreover, the set is proven to be viable and convex. In this way, it can be easily implemented as linear constraints in optimization-based control methods, such as linear MPC. Numerical results demonstrate that the percentage of area covered by the polyhedrons increases as the number of sides grows. This allows the controller to exploit a larger admissible state region and, thus, gives it a broader margin of maneuver in the case the desired motion drives the robot in proximity of the viability boundary. In this case, a better task following can be achieved, as demonstrated by means of experimental results on a six-degrees-of-freedom manipulator.

Author Contributions: M.F. devised the method and wrote the paper; M.B. and N.P. conceived and performed the experiments; and A.V. supervised the whole research work.

References

1. Verscheure, D.; Demeulenaere, B.; Swevers, J.; Schutter, J.D.; Diehl, M. Time-Optimal Path Tracking for Robots: A Convex Optimization Approach. *IEEE Trans. Autom. Control* **2009**, *54*, 2318–2327. [CrossRef]
2. Riazi, S.; Wigström, O.; Bengtsson, K.; Lennartson, B. Energy and Peak Power Optimization of Time-Bounded Robot Trajectories. *IEEE Trans. Autom. Sci. Eng.* **2017**, *14*, 646–657. [CrossRef]
3. Kazerounian, K.; Wang, Z. Global versus local optimization in redundancy resolution of robotic manipulators. *Int. J. Robot. Res.* **1988**, *7*, 3–12. [CrossRef]
4. Faroni, M.; Beschi, M.; Visioli, A.; Molinari Tosatti, L. A global approach to manipulability optimisation for a dual-arm manipulator. In Proceedings of the IEEE International Conference on Emerging Technologies and Factory Automation, Berlin, Germany, 6–9 September 2016; pp. 1–6.
5. Lee, J.; Chang, P.H. Redundancy resolution for dual-arm robots inspired by human asymmetric bimanual action: Formulation and experiments. In Proceedings of the IEEE International Conference on Robotics and Automation, Seattle, WA, USA, 26–30 May 2015; pp. 6058–6065.
6. Liegeois, A. Automatic Supervisory Control of the Configuration and Behavior of Multibody Mechanisms. *IEEE Trans. Syst. Man Cybern.* **1977**, *7*, 868–871.
7. Cheng, F.T.; Chen, T.H.; Sun, Y.Y. Resolving manipulator redundancy under inequality constraints. *IEEE Trans. Robot. Autom.* **1994**, *10*, 65–71. [CrossRef]
8. Kanoun, O.; Lamiraux, F.; Wieber, P.B. Kinematic control of redundant manipulators: Generalizing the task priority framework to inequality tasks. *IEEE Trans. Robot.* **2011**, *27*, 785–792. [CrossRef]
9. Escande, A.; Mansard, N.; Wieber, P.B. Hierarchical quadratic programming: Fast online humanoid-robot motion generation. *Int. J. Robot. Res.* **2014**, *33*, 1006–1028. [CrossRef]
10. Liu, M.; Tan, Y.; Padois, V. Generalized hierarchical control. *Auton. Robots* **2016**, *40*, 17–31. [CrossRef]
11. Rocchi, A.; Hoffman, E.M.; Caldwell, D.G.; Tsagarakis, N.G. Opensot: A whole-body control library for the compliant humanoid robot Coman. In Proceedings of the IEEE International Conference on Robotics and Automation, Seattle, WA, USA, 26–30 May 2015; pp. 6248–6253.
12. Siciliano, B.; Slotine, J.J.E. A general framework for managing multiple tasks in highly redundant robotic systems. In Proceedings of the International Conference on Advanced Robotics, Pisa, Italy, 19–22 June 1991; pp. 1211–1216.
13. Flacco, F.; De Luca, A. Discrete-time redundancy resolution at the velocity level with acceleration/torque optimization properties. *Robot. Auton. Syst.* **2015**, *70*, 191–201. [CrossRef]
14. Flacco, F.; De Luca, A.; Khatib, O. Control of Redundant Robots Under Hard Joint Constraints: Saturation in the Null Space. *IEEE Trans. Robot.* **2015**, *31*, 637–654. [CrossRef]
15. Faroni, M.; Beschi, M.; Visioli, A.; Molinari Tosatti, L. A Predictive Approach to Redundancy Resolution for Robot Manipulators. In Proceedings of the IFAC World Congress, Toulouse, France, 9–14 July 2017; pp. 8975–8980.

16. Faroni, M.; Beschi, M.; Visioli, A. Predictive Inverse Kinematics for Redundant Manipulators: Evaluation in Re-Planning Scenarios. In Proceedings of the IFAC Symposium on Robot Control, Budapest, Hungary, 27–30 August 2018.

17. Buizza Avanzini, G.; Zanchettin, A.M.; Rocco, P. Reactive Constrained Model Predictive Control for Redundant Mobile Manipulators. In *Intelligent Autonomous Systems 13, Advances in Intelligent Systems and Computing*; Springer: Berlin/Heidelberg, Germany, 2016; pp. 1301–1314.

18. Ghazaei Ardakani, M.M.; Olofsson, N.; Robertsson, A.; Johansson, R. Real-Time Trajectory Generation using Model Predictive Control. In Proceedings of the IEEE International Conference on Automation Science and Engineering, Gothenburg, Sweden, 24–28 August 2015; pp. 942–948.

19. Ferrara, A.; Incremona, G.P.; Magni, L. A robust MPC/ISM hierarchical multi-loop control scheme for robot manipulators. In Proceedings of the Annual Conference on Decision and Control, Florence, Italy, 10–13 December 2013; pp. 3560–3565.

20. Blanchini, F. Set invariance in control. *Automatica* **1999**, *35*, 1747–1767. [CrossRef]

21. Saab, L.; Ramos, O.E.; Keith, F.; Mansard, N.; Soueres, P.; Fourquet, J.Y. Dynamic whole-body motion generation under rigid contacts and other unilateral constraints. *IEEE Trans. Robot.* **2013**, *29*, 346–362. [CrossRef]

22. Park, K.C.; Chang, P.H.; Kim, S.H. The enhanced compact QP method for redundant manipulators using practical inequality constraints. In Proceedings of the IEEE International Conference on Robotics and Automation, Leuven, Belgium, 20 May 1998; pp. 107–114.

23. Kanehiro, F.; Morisawa, M.; Suleiman, W.; Kaneko, K.; Yoshida, E. Integrating geometric constraints into reactive leg motion generation. In Proceedings of the IEEE/RSJ International Conference on Intelligent Robots and Systems, Taipei, Taiwan, 18–22 October 2010; pp. 4069–4076.

24. Wensing, P.M.; Orin, D.E. Generation of dynamic humanoid behaviors through task-space control with conic optimization. In Proceedings of the IEEE International Conference on Robotics and Automation, Karlsruhe, Germany, 6–10 May 2013; pp. 3103–3109.

25. Guarino Lo Bianco, C.; Gerelli, O. Online trajectory scaling for manipulators subject to high-order kinematic and dynamic constraints. *IEEE Trans. Robot.* **2011**, *27*, 1144–1152. [CrossRef]

26. Scheint, M.; Wolff, J.; Buss, M. Invariance control in robotic applications: Trajectory supervision and haptic rendering. In Proceedings of the IEEE American Control Conference, Seattle, WA, USA, 11–13 June 2008; pp. 1436–1442.

27. Zanchettin, A.M.; Rocco, P. Motion planning for robotic manipulators using robust constrained control. *Control Eng. Pract.* **2017**, *59*, 127–136. [CrossRef]

28. Del Prete, A. Joint Position and Velocity Bounds in Discrete-Time Acceleration/Torque Control of Robot Manipulators. *IEEE Robot. Autom. Lett.* **2018**, *3*, 281–288. [CrossRef]

29. Faroni, M.; Beschi, M.; Pedrocchi, N.; Visioli, A. Predictive Inverse Kinematics for Redundant Manipulators with Task Scaling and Kinematic Constraints. *IEEE Trans. Robot*, **2018**, submitted.

30. Gerelli, O.; Guarino Lo Bianco, C. Nonlinear variable structure filter for the online trajectory scaling. *IEEE Trans. Ind. Electron.* **2009**, *56*, 3921–3930. [CrossRef]

31. Wang, L. *Model Predictive Control System Design and Implementation Using MATLAB®*; Springer Science & Business Media: Berlin/Heidelberg, Germany, 2009.

11

Design and Implementation of a Dual-Axis Tilting Quadcopter

Ali Bin Junaid [1], Alejandro Diaz De Cerio Sanchez [2], Javier Betancor Bosch [2], Nikolaos Vitzilaios [3] and Yahya Zweiri [2,4,*

[1] Department of Mechanical Engineering, KU Leuven, 3000 Leuven, Belgium; ali.binjunaid@kuleuven.be
[2] Faculty of Science, Engineering and Computing, Kingston University London, London SW15 3DW, UK; aldiazde@gmail.com (A.D.D.C.S.); J.Betancorbosch@kingston.ac.uk (J.B.B.)
[3] Department of Mechanical Engineering, University of South Carolina, Columbia, SC 29208, USA; VITZILAIOS@sc.edu
[4] Khalifa University Center for Autonomous Robotic Systems, Department of Aerospace Engineering, Khalifa University of Science and Technology, P.O. Box 127788 Abu Dhabi, UAE
* Correspondence: Y.Zweiri@Kingston.ac.uk.

Abstract: Standard quadcopters are popular largely because of their mechanical simplicity relative to other hovering aircraft, low cost and minimum operator involvement. However, this simplicity imposes fundamental limits on the types of maneuvers possible due to its under-actuation. The dexterity and fault tolerance required for flying in limited spaces like forests and industrial infrastructures dictate the use of a bespoke dual-tilting quadcopter that can launch vertically, performs autonomous flight between adjacent obstacles and is even capable of flying in the event of the failure of one or two motors. This paper proposes an actuation concept to enhance the performance characteristics of the conventional under-actuated quadcopter. The practical formation of this concept is followed by the design, modeling, simulation and prototyping of a dual-axis tilting quadcopter. Outdoor flight tests using tilting rotors, to follow a trajectory containing adjacent obstacles, were conducted in order to compare the flight of conventional quadcopter with the proposed over-actuated vehicle. The results show that the quadcopter with tilting rotors provides more agility and mobility to the vehicle especially in narrow indoor and outdoor infrastructures.

Keywords: aerial robotics; quadcopters; UAVs; dual-tilting; tilting rotors; over-actuation; flight control; rotorcraft

1. Introduction

Over the last few decades, the usage and deployment of UAVs (Unmanned Aerial Vehicles) have been growing, from hobby to military applications. Some of those applications include surveying, maintenance and surveillance tasks, transportation and manipulation, search and rescue [1,2]. Rotorcraft UAVs are of particular interest as they offer advanced capabilities such as Vertical Take Off and Landing (VTOL) and high agility. Quadcopters are the most researched and used platforms in this area.

A quadcopter's lift and thrust is generated by four propellers mounted on high-speed, high-power brushless DC motors. Quadcopters use an electronic control system and electronic sensors to stabilize themselves. With their small size and VTOL capability, quadcopters can fly indoors as well as outdoors. Similar to a conventional helicopters, quadcopters can hover but have significant other advantages such as ease of piloting and mechanical simplicity. Recently, there is a rapid growth in quadcopter development due to the high potential used in numerous commercial applications.

For real world applications, quadcopters require more payload capacity and should be more invulnerable and robust towards external disturbances. However, increased payload capacity demands up scaling the platforms which eventually results in decreased maneuverability and agility [3]. As stated in [3], the inertia of the platform is increased and requires larger control moments to achieve higher agility with the increase of the vehicle size. Secondly, the increased weight results in increased propeller size consequently increasing the inertia. The conventional quadcopter possesses such physical constraints with its control on larger scales.

Conventional quadcopters are under-actuated mechanical systems possessing less control inputs than Degrees Of Freedom (DOFs). Over the last decades, different control techniques have been proposed to deal with the quadcopter under-actuation for an effective and more controlled performance [3–5]. Still, the under-actuation of quadcopter has limitations on its flying ability in free or cluttered space.

In a conventional under-actuated quadcopter, actuators failure results in complete destabilization of the vehicle as its control is completely dependent on the symmetry of the lift. To alleviate this problem, several approaches have been proposed in previous studies. In [6], propellers with variable pitch and shifting of the Center of Gravity (COG) are suggested. These approaches enable to achieve the controllability of the vehicle in roll and pitch axes up to some extent. However, the yaw axis still remains uncontrollable. In [7], a bounded control law was proposed in order to have a safe landing for quadcopter in case of actuators failure.

One approach to overcome this issue is to increase the number of rotors. Typical example includes 4Y octocopters [8]. This approach has advantages such as the mechanical simplicity and reliability. However, this approach results in increased weight and increased inertia hence reducing the agility of the vehicle. Furthermore, increased number of rotors results in larger power consumption which immensely impacts the endurance of the aircraft. Nevertheless, standard hexarotors are not fail-safe multi-rotor platforms and cannot hover with five propellers [9].

Apart from the increasing number of rotors approach, other approaches include variation in the types of actuators keeping in consideration the key factors which are the vehicle's size and weight. Cutler et al. [10] showed an effective approach by using propellers with variable pitch. In this approach, while keeping the weight down, the bandwidth of its actuators was increased. However, actuator failure resulting in instability still remains an issue.

Other solutions to overcome the under-actuated problem include tilt-wing mechanisms [11], UAVs with non-parallel fixed thrust directions [12], or tilt-rotor actuations [13]. Similar approach with the dual-axis tilting of the rotors providing the broad range of control bandwidth for the same number of rotors is proposed in [14]. However, most of the platforms were developed and tested in indoor environments with low payload capacities. Furthermore, indoor navigation systems were used to develop the control of the tilting quadcopter. Similar concept of rotor tilting is used in [15,16], however, rotor tilting is limited to single axis only.

In [17], Control Moment Gyroscopes (CMG) were proposed to increase the control system bandwidth by merging a thrust vectoring approach with additional flywheels in order to be used as CMG, and a vane system for thrust vectoring. However, the extra weight of the flywheels and the thrust vectoring vane system results in increased weight of the aircraft and complicates the design of the vehicle. Gress [18] used Opposed Lateral Tilting (OLT) technique for using the gyroscopic effects for governing the pitch attitude of aircraft, using the propellers as gyroscopes. In [19], OLT proved to achieve higher controllability. The detailed model and control strategy for hovering, with experimental evidence is presented in [20]. To implement such control strategies and actuation mechanisms, it is important to keep in mind key factors for the development of such vehicles, which are weight, mechanical simplicity, cost-effectiveness and ability to manufacture the platform in less time.

Nowadays, advances in the fields of Computer Aided Design (CAD) and Rapid Prototyping (RP) have provided the tools to rapidly generate a prototype from a concept. RP technique allows to automatically construct physical models using additive manufacturing technology [21,22]. Mechanical parts or assembly can be quickly manufactured using 3D CAD design. This technique has emerged as an innovative tool to reduce the time and cost of manufacturing and fabrication by creating 3D product directly from computer aided design providing the ability to perform design validation and analysis [23,24].

This paper presents design and implementation of a novel actuation strategy which was proposed by co-author in [25] in order to increase the agility and control bandwidth of the conventional quadcopters in outdoor scenarios.

The introduction of dual-axis tilting to the propellers produces an over-actuated quadcopter with ability to have 12 control inputs with 6 control outputs. The tilting of propellers provides the necessary gyroscopic effects to provide fast control and action. The proposed tilting rotor solution for quadcopter uses additional 8 servomotors that allow the rotors to tilt in both axes, an over-actuated system can potentially track an arbitrary trajectory over time. As shown in [26,27], with single axis tilting, full controllability over the quad-rotor position and orientation provides possibility of hovering in a tilted configuration. The research here focuses on the design of such platform and conduct initial experiments to test the actuation modules in outdoor urban environment.

The development of the proposed concept is mainly based on arm design of the quadcopter in which each arm is able to generate three-actuation independently, including the rotors to achieve differential thrusting and dual tilting mechanism to provide broad range of control bandwidth. The computer-aided design (CAD) model was designed and analyzed using finite element analysis (FEA) for the structural rigidity and stability. The experiments were performed using the developed platform which achieved full controllability of the quadcopter thus transforming the system into an over-actuated machine. Flight tests were performed in a trajectory having sharp corners between adjacent obstacles in order to compare the behavior of conventional quadcopter configuration and the proposed actuation strategy. The results show that the quadcopter with tilting rotors provides more agility and mobility to the vehicle especially in narrow indoor and outdoor infrastructures.

The paper layout is as follows: first the design approach for the development of quadcopter with over-actuated mechanism along with its electronics to control the mechanisms is presented. Secondly, the modeling and simulation results are presented. Furthermore, the structural analysis of the rapid prototyped parts is presented ensuring the structural stability of the platform. The flight test results of the developed platform for the conventional and over-actuated configuration are presented. Finally, the results are discussed and some conclusions are drawn.

2. Design Approach

Two servomotors were used to achieve the dual tilting actuation with each rotor. The platform was designed in SolidWorks and manufactured using a ZORTRAX 3D printer [28]. The mechanical design of the quadcopter is shown in Figure 1.

Figure 1. CAD model of proposed quadcopter.

The proposed actuation concept of the quadcopter arms and motors is shown in Figures 2 and 3. The dual tilting mechanism of the arms is mainly based on the servomotor which is coupled with the arm through gears in order to compensate the high torque demand for rotation. This mechanism allows the movement of the whole arm around its axis as shown in Figure 2. Another servomotor is mounted in the arm which connected to the motor mount through push-pull mechanism which rotates parallel to the servo lever as shown in Figure 3. The two angles generated by the servomotors constitute towards the configuration of the rotation axes of the propellers.

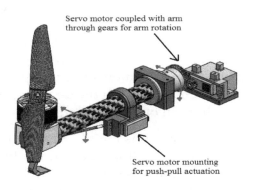

Figure 2. CAD model of proposed dual-tilting arm.

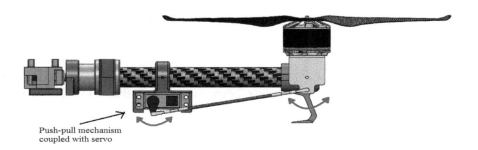

Figure 3. Push-pull mechanism for the tilting of rotor.

The gear mechanism for the coupling of servomotor with the arm of the quadcopter is installed with the gear ratio of 1:3 to fulfill the torque requirement for rotating the arm of the quadcopter. The coupling mechanism is shown in Figure 4. For the actuation of dual-axis tilting system, a total of eight servomotors are used for rotating the arms and motor mountings.

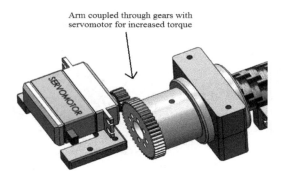

Figure 4. Gear coupling for the tilting of arm.

3. Modeling and Simulation of Over-Actuated Quadcopter

3.1. Modeling

For the development of the mathematical model which defines the dynamics of the dual-axis tilting quadcopter, Newton-Euler formulation has been used to derive the equations. The development process includes the definition of variables and axes reference, formulation using Newton-Euler equations, and including inertia variation [29]. The inertia of the quadcopter has been considered variable due to variation of the rotors position as a result of the tilting. These dynamic equations have been developed using Maple®.

To begin the analysis of the system, three frames have been selected in order to develop the system equations. The first frame is the world frame therefore we refer to it as the frame 1 (World Frame). The next frame is defined to be fixed in the center of gravity of the UAV and is referred as frame 2 (Body Frame). Finally, the third frame is defined to be fixed in each of the rotors and is referred as frame 3 (Rotor Frame), shown in Figure 5. The overall system model diagram is shown in Figure 6. In this paper, kinematics and dynamic modeling of the dual-axis titling system is developed (Propellers and Tilting block, shown in Figure 6), whereas the conventional quadcopter modeling details are standard text book material therefore omitted here.

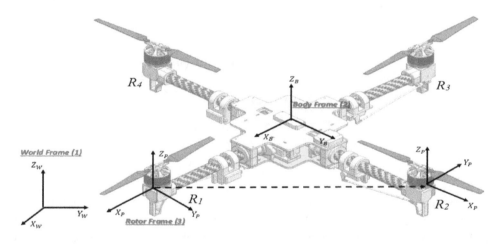

Figure 5. Rotor numbering and reference frames [29].

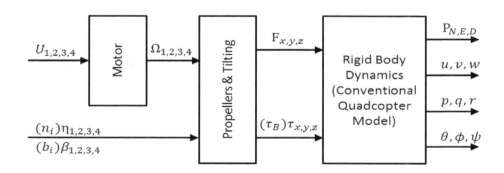

Figure 6. Dual-axis tilting quadcopter modeling diagram.

It is important to set reference frames in each rotor, to get the kinematic and dynamic equations for each rotor in the UAV. To simplify the equations, the four local frames of each rotor have been aligned with respect to the global axis following a rotational relationship with respect to the Z axis by

90° counter-clockwise. The expression which defines the rotation matrix is shown below as example of the rotation for the rotor 2.

$$
\begin{bmatrix} i_2 \\ j_2 \\ k_2 \end{bmatrix}_3 = \mathbf{R_Z}\left(\frac{\pi}{2}\right) \cdot \begin{bmatrix} i_1 \\ j_1 \\ k_1 \end{bmatrix}_3
$$

Similarly, the rotation matrix is applied for the rotors 3 and 4 with an angular difference of 90° counter-clockwise each, which represents 180° offset for the rotor 3 and 270° offset for the rotor 4. The general rotation matrix for each rotor therefore is defined as:

$$
\begin{bmatrix} i_i \\ j_i \\ k_i \end{bmatrix}_3 = \mathbf{R_Z}\left((i-1)\frac{\pi}{2}\right) \cdot \begin{bmatrix} i_1 \\ j_1 \\ k_1 \end{bmatrix}_3 \qquad for \quad i = 1 \cdots 4
$$

As frame 3 is fixed, every equation is translated from frame 3 to frame 2. As result of translating the local axis of each rotor with respect to frame 2, three matrices for the rotation of each axis are obtained as $\mathbf{R_X}$, $\mathbf{R_Y}$ and $\mathbf{R_Z}$.

$$
\mathbf{R_X} = \begin{bmatrix} 1 & 0 & 0 \\ 0 & \cos(\beta_i(t)) & -\sin(\beta_i(t)) \\ 0 & \sin(\beta_i(t)) & \cos(\beta_i(t)) \end{bmatrix} \tag{1}
$$

$$
\mathbf{R_Y} = \begin{bmatrix} \cos(\eta_i(t)) & 0 & \sin(\eta_i(t)) \\ 0 & 1 & 0 \\ -\sin(\eta_i(t)) & 0 & \cos(\eta_i(t)) \end{bmatrix} \tag{2}
$$

$$
\mathbf{R_Z} = \begin{bmatrix} \cos\left(\frac{\pi}{2}\cdot(i-1)\right) & -\sin\left(\frac{\pi}{2}\cdot(i-1)\right) & 0 \\ \sin\left(\frac{\pi}{2}\cdot(i-1)\right) & \cos\left(\frac{\pi}{2}\cdot(i-1)\right) & 0 \\ 0 & 0 & 1 \end{bmatrix}
$$

Finally, the rotation matrix from frame 2 to frame 1 is obtained by three rotation matrices, one in each axis of the frame 1. In this case, the Euler angles are used to identify the orientation of the aircraft. The angles are denoted by θ, φ, ψ corresponding to roll, pitch and yaw and representing the angular displacement of the UAV along the X, Y and Z axes respectively. The rotation matrix from the body to the fixed frame is shown in Equation (3).

$$
{}^2\mathbf{R_1} = \begin{bmatrix} c(\theta)c(\psi) & -c(\theta)s(\psi) & s(\theta) \\ s(\varphi)s(\theta)c(\psi)+c(\varphi)s(\psi) & -s(\varphi)s(\theta)s(\psi)+c(\varphi)c(\psi) & -s(\varphi)c(\theta) \\ -c(\varphi)s(\theta)c(\psi)+s(\varphi)s(\psi) & c(\varphi)s(\theta)s(\psi)+s(\varphi)c(\psi) & c(\varphi)c(\theta) \end{bmatrix} \tag{3}
$$

where c and s denote to the trigonometric functions *cosine* and *sine* respectively. With these equations the major problem for applying the Newton-Euler equations to our system is sorted. From this point, the calculation of the angular and linear accelerations for the UAV has been done by solving Equation (4). For the calculation of the angular accelerations, the general kinematic equation for the UAV has been used. The Equations (4) and (6) define the angular and linear accelerations of the UAV platform. The motors are rotating the propellers, this represents an angular velocity denoted by Ω.

$$
\vec{\alpha}_B = (I_B + 4I_P)^{-1} \cdot \left[\vec{M}_T - \sum_{i=1}^{4} ({}^2\mathbf{R_3})_i I_P \begin{pmatrix} \ddot{\beta}_i \\ \ddot{\eta}_i \\ \dot{\Omega}_i \end{pmatrix} + ({}^2\mathbf{R_3})_i \vec{\tau}_{ext} - \frac{d}{dt}(I_B)\vec{\omega}_B - \mathbf{\Theta} I_B \vec{\omega}_B \right] \tag{4}
$$

where Θ is the Skew matrix. This is formed by the angular velocity of the platform in three axes [29], it is given as:

$$\Theta = \begin{bmatrix} 0 & -R(t) & Q(t) \\ R(t) & 0 & -P(t) \\ -Q(t) & P(t) & 0 \end{bmatrix} \tag{5}$$

$$\begin{bmatrix} \dot{u} \\ \dot{v} \\ \dot{w} \end{bmatrix} = \begin{bmatrix} 0 \\ 0 \\ -g \end{bmatrix} + \frac{1}{m}\left[(^1\mathbf{R_2}) \cdot \sum_{i=1}^{4} (^2\mathbf{R_{3i}} \cdot \vec{T}_{P_i}) \right] \tag{6}$$

Expanding these equations and considering the rotation matrices, the dynamic equations for forces and torques can be obtained as follows:

$$\begin{bmatrix} F_x \\ F_y \\ F_z \end{bmatrix} = \begin{bmatrix} (k_{x1} + \lambda_{x1}) & (k_{x2}^* + \lambda_{x2}^*) & (-k_{x3} + \lambda_{x3}) & (-k_{x4}^* - \lambda_{x4}^*) \\ (k_{x1} + \lambda_{y1}) & (k_{x2}^* + \lambda_{y2}^*) & (-k_{x3} + \lambda_{y3}) & (-k_{y4}^* - \lambda_{y4}^*) \\ (k_{z1} + \lambda_{z1}) & (k_{z2}^* + \lambda_{z4}^*) & (-k_{z3} + \lambda_{z4}) & (-k_{z4}^* - \lambda_{z4}^*) \end{bmatrix} \begin{bmatrix} T_1 \\ T_2 \\ T_3 \\ T_4 \end{bmatrix}$$

$$\begin{bmatrix} \tau_x \\ \tau_y \\ \tau_z \end{bmatrix} = \begin{bmatrix} T_x \\ T_y \\ T_z \end{bmatrix} + \begin{bmatrix} l_{xx} & l_{xy} & l_{xz} \\ l_{yx} & l_{yy} & l_{yz} \\ l_{zx} & l_{zy} & l_{zz} \end{bmatrix} \begin{bmatrix} IP_x \\ IP_y \\ IP_z \end{bmatrix} + \begin{bmatrix} \rho_x \\ \rho_y \\ \rho_z \end{bmatrix}$$

The coefficients of the forces and the torques are given in Appendix A.

It is important to mention that the inertia of the system varies due to the tilting rotors and cannot be assumed as constant (time invariant). Therefore, the inertia matrix (Equation (7)) is modeled in this paper as:

$$I_B = \begin{bmatrix} I_{xx} & 0 & 0 \\ 0 & I_{yy} & 0 \\ 0 & 0 & I_{zz} \end{bmatrix} + \begin{bmatrix} I_{Px'x'} & 0 & 0 \\ 0 & I_{Py'y'} & 0 \\ 0 & 0 & I_{Pz''z''} \end{bmatrix} + m \cdot \begin{bmatrix} y_C^2 & 0 & 0 \\ 0 & x_C^2 & 0 \\ 0 & 0 & x_C^2 + y_C^2 \end{bmatrix} \tag{7}$$

Expanding Equation (7) yields the inertia values in the main axis which are given in Appendix B.

3.2. Simulation

The simulation of the dynamic equations is implemented using Matlab/Simulink© incorporating the motor dynamics, attitude controller for Roll, Pitch and Yaw, and the controller for tilting angles of the rotors (Figure 7). The quadcopter can be simulated to observe the flight behaviour under the influence of different control inputs i.e. attitude commands and tilting rotor angle inputs.

Rectangular path was simulated to observe the performance of UAV using tilting angle of rotors as inputs. In Figure 8, the simulation shows the movement of UAV in a rectangular path with only using the tilting capability of the rotors without changing its attitude. The platform was able to perform sharp cornering maneuvers.

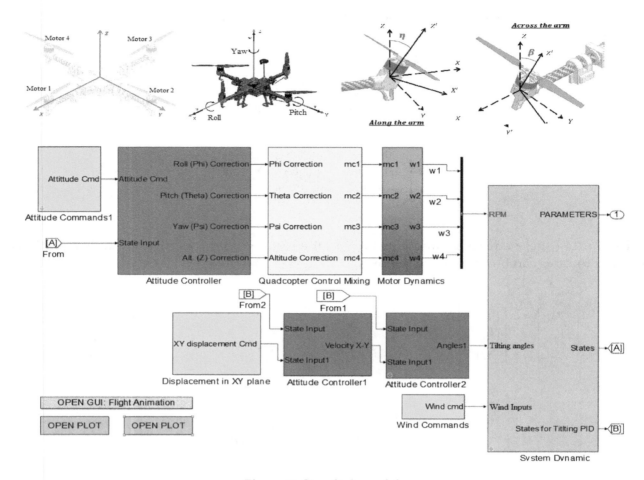

Figure 7. Simulink model.

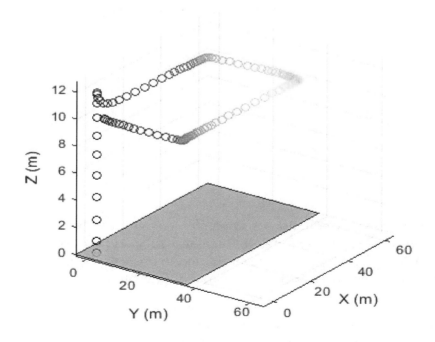

Figure 8. Rectangular path movement of UAV with tilting.

The trajectory was followed only using tilt rotor actuation. The rotor tilting angle along the arm (n) is used to follow a rectangular trajectory. Each rotor is coupled with the opposite rotor with tilting at angles shown in Figure 9. At each corner, the adjacent rotors are tilted to follow the path. Figure 10 shows the attitude of the UAV i.e., Roll and Pitch and it can be observed that the quadcopter performs the maneuver without changing its attitude.

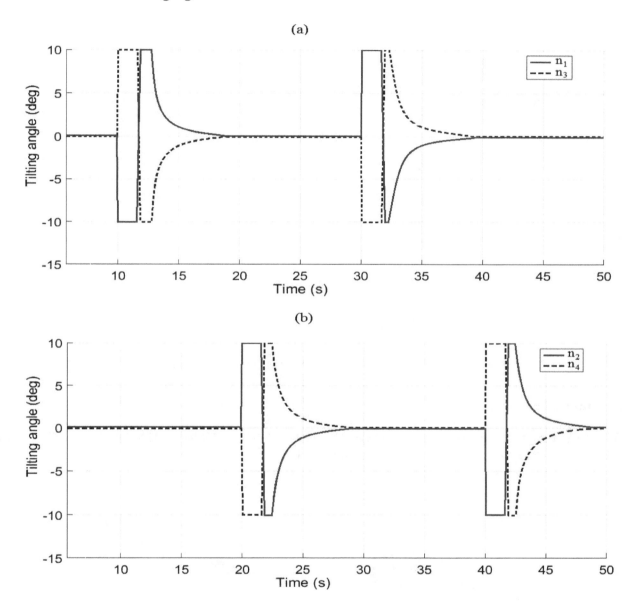

Figure 9. (a) Rotor tilting angles along the arm n_1 and n_3; (b) Rotor tilting angles along the arm n_2 and n_4.

From Figure 9, rotor angles along the arm n_1 and n_3 are tilted at 10 degrees without a change in attitude (as observed in Figure 10) and the quadcopter starts moving in that direction. On the corners, the adjacent rotors (n_2 and n_4) are tilted to follow the trajectory.

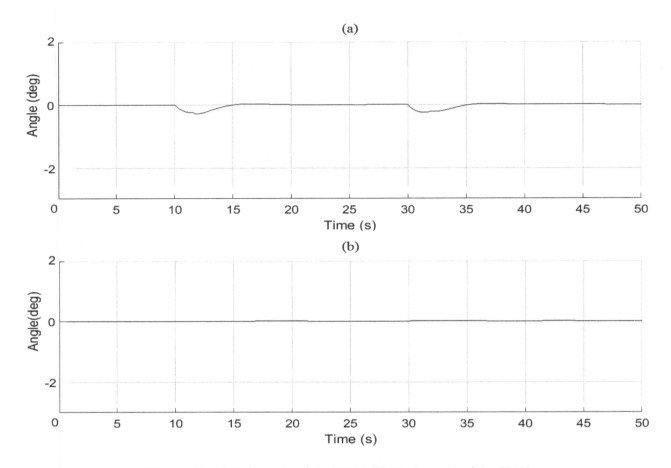

Figure 10. (a) Roll angle of the UAV; **(b)** Pitch angle of the UAV.

4. Experimental Setup

The four brushless motors selected were the Tiger T-MOTOR MN4014-9 400Kv with a 2-blade 15×5 Carbon Fiber propellers as shown in Figure 11.

Figure 11. (Left) Brushless motor; **(Right)** Propeller used for thrust generation.

For the Flight Control of the UAV, ArduPilotMega (APM) was used. APM is an open-source flight controller, able to control autonomous multicopters, fixed-wing aircraft, traditional helicopters and ground rovers. It is based on the Arduino electronics prototyping platform. Apart from the APM, which is mainly controlling the brushless motors and handling the inner loop control of the vehicle, the Arduino Leonardo microcontroller board is used to control the added servomotors. Figure 12 show the controllers used for the proposed vehicle.

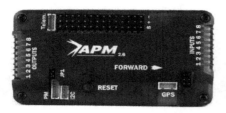 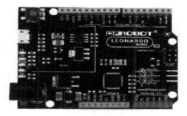

Figure 12. (Left) APM Flight controller; (Right) Arduino Leonardo.

The inner-loop control of the quadcopter is handled by the APM. This on-board inner-loop control device deals with control and stabilization required to hover and perform basic maneuvers. The APM board contains the inertial sensors required for the orientation and heading determination and its software includes the inner-loop control algorithm and the basic Graphical User Interface (GUI) for visualization. Generally, the APM input is provided by the RC remote control which allows the user to manually control and fly the multicopter. The TGY-i10 RC Controller was used as the input for the flight controller. GPS module was attached to the APM in order to navigate the aircraft and track its position for post-processing and analysis. The APM attitude controller was fused with the other controller which was aimed for controlling the inputs for the servo motors responsible for the tilting of the rotors. This controller mainly controls the tilting of the servomotors by combining the inputs for the attitude control of the UAV and the tilting commands for the rotors from the pilot. The commands generated by the human pilot using RC Controller are received by the Arduino Leonardo. The pilot has control of the conventional attitude of quadcopter and tilting servomotors. Based on the control inputs from the pilot, Arduino reads in Pulse Position Modulation (PPM) signals from the RC controller and generates commands (PPM signals) of roll, pitch, yaw and throttle to the APM and commands for servomotors simultaneously. This allows the pilot to have full control of all the actuators of the dual-tilting quadcopter. If the pilot wants to perform maneuvers using only tilting, the Arduino generates commands for the servo motors and proportionally generates attitude commands in order to compensate the tilting actuation.

Currently the tilting angle is limited to $10°$ for each servomotor. Unlike the conventional quadcopter, this gives the pilot additional control inputs i.e., tilting of the rotors along with conventional control for increased maneuverability. The altitude during all the flight tests was kept in altitude hold mode. The APM flight controller provides the feature of holding the altitude allowing to hover and maneuver at the desired altitude.

The control implementation here is rather basic since a more sophisticated development is out of the scope of this paper but remains in the future work agenda. The conventional inner loop controller of the APM [30] allows the quadcopter to respond to attitude commands of the pilot which are passed through Arduino Leonardo. Arduino Leonardo allows the control of the attitude and the tilting simultaneously. Figure 13 illustrates the experimental setup for control.

After assembling all the manufactured parts and integrating the related modules for the over-actuated mechanisms for the quadcopter, the final product is shown in Figure 14.

Table 1 presents some of the technical specifications of the developed quadcopter.

Table 1. UAV Technical Specifications.

Parameter	Specifications
UAV Dimensions	1048×1048 mm
Weight	4 kg
Endurance	20 min
Payload Capacity	2 kg

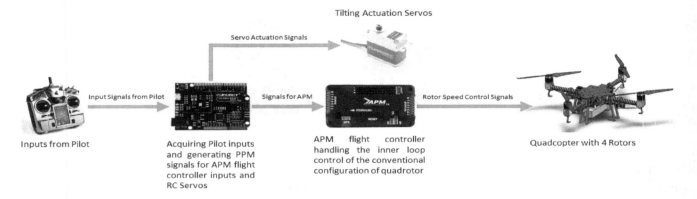

Figure 13. Experimental setup for control of the UAV.

Figure 14. Dual-tilting quadcopter prototype.

5. Results and Discussion

5.1. Flight Test with Conventional Actuation of Quadcopter

After the successful manufacturing and assembly of the vehicle, flight test was conducted for the conventional under-actuated configuration of the quadcopter in order to validate the design parameters and the dynamic model of the vehicle. Relevant flight variables of interest were analyzed in order to observe the behaviour and attitude for the designed vehicle. The results in Figures 15–17 show the control of each axis.

For each axis, the vehicle performs satisfactorily and flies according to the input angles given by the pilot through RC control. GPS is used for the position feedback as ground truth. The proposed system performs well for the conventional configuration and flight controls and validates the design of the platform and its flight stability for the conventional control. Therefore, it can be concluded that the vehicle is stable and able to maintain its attitude. Moreover, the platform can be used for developing and testing of the flight controls for the over-actuated configuration and performs well with proper control techniques for the over-actuated quadcopter design. With the development of over-actuated quadcopter controller, the manufactured vehicle gives plenty of control authority and high maneuverability due to its capability to incorporate the excess number of control inputs.

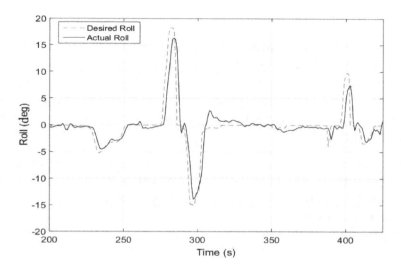

Figure 15. Roll command tracking of the vehicle.

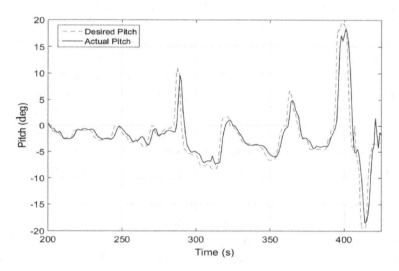

Figure 16. Pitch command tracking of the vehicle.

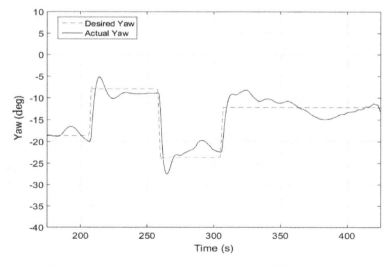

Figure 17. Yaw command tracking of the vehicle.

5.2. *Flight Test with Tilting Rotors*

The experimental scenario was created in order to validate the performance of the quadcopter using tilting rotors. As mentioned in Section 3, the quadcopter was simulated to imitate a rectangular trajectory which involves sharp cornering for the vehicle. Real corners (imitating trees in a dense forest) were created in an outdoor flying space, the quadcopter was flown to follow the rectangular trajectory along those corners with tilting and without tilting rotors. Figure 18 shows the track followed by the quadcopter using only tilting angles.

The attitude of the quadcopter during the flight can be observed in Figure 19 which shows that the quadcopter is stabilizing and maintaining its horizontal attitude without contributing in the movement in order to follow the trajectory. The rectangular trajectory including the cornering is achieved using only tilting of the rotors along the arms.

The same trajectory following was performed without using tilting of the rotors, with conventional configuration of the quadcopter. The results of trajectory followed by the quadcopter with conventional configuration is shown in Figure 20.

Comparing of the quadcopter attitude in both cases, i.e. with conventional actuation and with tilting rotor actuation for the same trajectory (Figures 19 and 21), it can be observed that the quadcopter with conventional actuation requires the whole frame to be tilted in order to maintain attitude. Whereas dual tilting actuation quadcopter provides the ability to maneuver in a way regardless of its attitude.

Furthermore, it is evident from the comparison of the trajectory followed by the quadcopter with two different actuation strategies that the quadcopter with tilting actuation of the rotors is able to maneuver efficiently through corners which minimizes the effort of the vehicle movement.

From Figures 18 and 20, it can be noticed that the conventional actuation of the quadcopter limits the motion of the vehicle around sharp corners up to certain extent, requiring a larger turning radius in order to do the cornering. The clearing distance through obstacles using tilting is d = 1.25 m, while with the conventional configuration, d = 2.65 m. It is clear that wider gap is required for the conventional quadcopter to fly through the obstacles whereas with tilting ability, the quadcopter is able to fly more precisely, reducing the clearing distance. The conventional quadcopter would not be able to execute such maneuvers as the under-actuation limits its ability. Tilting provides more controllability and ability to the vehicle as it can move without changing its attitude which helps the developed system to fly through narrow gaps and under trees canopy. The combination of tilting rotors with attitude control provide increased agility and control bandwidth in an urban outdoor scenario where the flying through narrow gaps and obstacles is challenging as compared to under-actuated quadcopter.

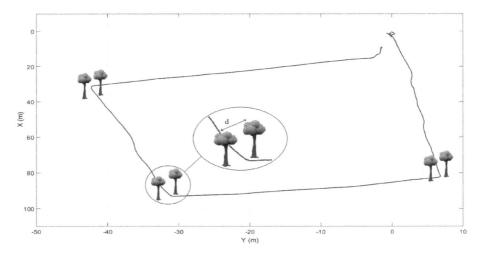

Figure 18. The real-time trajectory followed by UAV using tilting, d = 1.25 m.

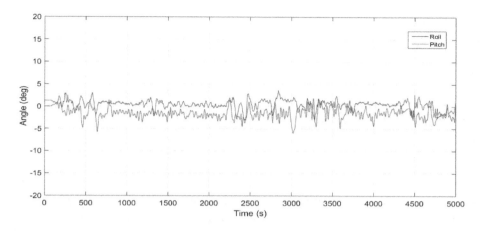

Figure 19. Roll and pitch angle of the UAV while maneuvering only with tilting.

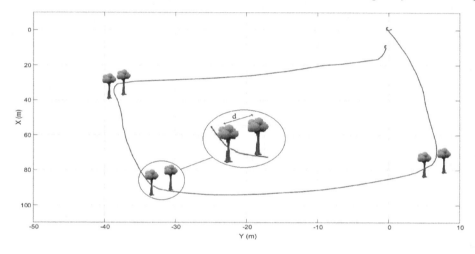

Figure 20. The real-time trajectory followed by UAV without using tilting, d = 2.65 m.

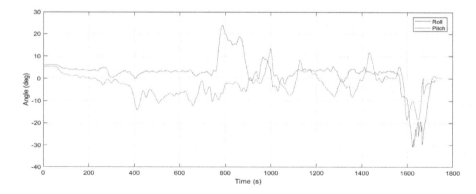

Figure 21. Roll and Pitch angle of the UAV while maneuvering without tilting.

6. Conclusions

This paper proposes a design and implementation of an over-actuated quadcopter with dual-axis tilting rotors. The CAD model was developed following the manufacturing of the system using rapid prototyping in order to minimize the manufacturing time and cost. The modeling and simulation of the over-actuated system allowed observing the behavior of the platform using different control inputs. The flight test results in outdoor conditions show satisfactory performance of the developed platform.

The experiments were performed to observe and compare the capability of over-actuated configuration against conventional configuration which showed that the proposed platform performs better when it comes to flying along corners and between adjacent obstacles, and gives better maneuverability compared to conventional quadcopter.

Integration of dual-axis tilting capability to quadcopters opens different research areas, including developing control and recovery strategies. Quadcopter with dual-tilting actuation will vastly expand its applications in search and rescue missions, detecting missing persons in a wide dense forest area with advantage of being able to fly under trees canopy.

Future work will focus on quantifying the energy consumption and the development of different fail-safe strategies in case of failure of one or two of the rotors. The development of failure strategies will use the full capability of over-actuation presented in the current research.

Author Contributions: All authors have made great contributions to the work. A.B.J., J.B.B. and Y.Z. designed and built the platform. A.D.D.C.S., J.B.B. and Y.Z. developed the model and the simulation. A.B.J., A.D.D.C.S. and N.V. conducted the experiments. A.B.J., N.V. and Y.Z. analyzed the data and revised the manuscript.

Acknowledgments: The authors would like to thank Miguel Guzman and Zineb Chelh for their help in the preliminary platform design and simulation.

Abbreviations

The following abbreviations are used in this manuscript:

θ	Pitch angle
φ	Roll angle
ψ	Yaw angle
ω_i	Angular velocity
L	Arm length
dIB_x	Time derivative of inertia along X axis
dIB_y	Time derivative of inertia along Y axis
dIB_z	Time derivative of inertia along Z axis
IB_{xx}	Body's inertia along X axis
IB_{yy}	Body's inertia along Y axis
IB_{zz}	Body's inertia along Z axis
IP_x	Propeller's inertia along X axis
IP_y	Propeller's inertia along Y axis
IP_z	Propeller's inertia along Z axis
i	number of rotor (1, 2, 3, 4)
$P_{N,E,D}$	Position in North, East and Down axis
P	Angular velocity in X axis
Q	Angular velocity in Y axis
R	Angular velocity in Z axis
n_i or η_i	Rotor tilting angle along the arm
b_i or β_i	Rotor tilting angle across the arm
x_W, y_W, z_W	Fixed frame (1)
x_B, y_B, z_B	Body frame (2)
x_P, y_P, z_P	Rotor frame (3)
R_x, R_y, R_z	Rotation matrices in x, y and z axis
α	Angular acceleration
u, v, w	Linear velocity in x, y and z axis
T_i	Thrust generated by the rotor i

Appendix A. Coefficients of the Forces and the Torques

The coefficients of the force dynamic equation in x-axis are as follows:

$$k_{xi} = \sin(\phi)\sin(\theta)\left[\cos(b_i)\left[\sin(n_i)\cos(\psi) - \cos(n_i)\right] - \sin(b_i)\sin(\psi)\right] \tag{A1}$$

$$k_{xi}^* = \sin(\phi)\sin(\theta)\left[\cos(b_i)\left[\sin(n_i)\cos(\psi) + \cos(n_i)\right] + \sin(b_i)\sin(\psi)\right] \tag{A2}$$

$$\lambda_{xi} = \cos(\theta)\left[\sin(b_i)\cos(\psi) + \sin(n_i)\cos(b_i) + \sin(\psi)\right] \tag{A3}$$

$$\lambda_{xi}^* = \cos(\theta)\left[\sin(b_i)\cos(\psi) - \sin(n_i)\cos(b_i) + \sin(\psi)\right] \tag{A4}$$

The coefficients of the force dynamic equation in y-axis are as follows:

$$k_{yi} = \sin(\phi)\sin(\theta)\left[\cos(b_i)\left[\sin(n_i)\cos(\psi) + \cos(n_i)\right] - \sin(b_i)\sin(\psi)\right] \tag{A5}$$

$$k_{yi}^* = \sin(\phi)\sin(\theta)\left[\cos(b_i)\left[\sin(n_i)\cos(\psi) - \cos(n_i)\right] + \sin(b_i)\sin(\psi)\right] \tag{A6}$$

$$\lambda_{yi} = \cos(\phi)\left[\sin(b_i)\cos(\psi) + \sin(n_i)\cos(b_i)\sin(\psi)\right] \tag{A7}$$

$$\lambda_{yi} = \cos(\phi)\left[\sin(B_i)\cos(\psi) + \sin(n_i)\cos(b_i)\sin(\psi)\right] \tag{A8}$$

The coefficients of the force dynamic equation in z-axis are as follows:

$$k_{z1} = \sin(\theta)\left[\sin(\psi)\left[\sin(n_1)\cos(b_1) - \sin(b_1)\cos(\phi)\right] - \sin(n_1)\cos(b_1)\cos(\phi)\cos(\psi)\right] \tag{A9}$$

$$k_{z2} = \sin(\theta)\left[\sin(\psi)\left[\sin(b_2) + \sin(n_2)\cos(b_2)\cos(\phi)\right] - \sin(b_2)\cos(\phi)\cos(\psi)\right] \tag{A10}$$

$$k_{z3} = -\sin(\theta)\left[\sin(\psi)\left[\sin(n_3)\cos(b_3) + \sin(b_3)\cos(\phi)\right] - \sin(n_3)\cos(b_3)\cos(\phi)\cos(\psi)\right] \tag{A11}$$

$$k_{z4} = \sin(\theta)\left[\sin(\psi)\left[\sin(b_4) + \sin(n_4)\cos(b_4)\cos(\phi)\right] - \sin(b_4)\cos(\phi)\cos(\psi)\right] \tag{A12}$$

$$\lambda_{z1} = \cos(n_1)\cos(b_1)\cos(\theta)\cos(\phi) - \sin(b_1)\sin(\phi)\cos(\psi) \tag{A13}$$

$$\lambda_{z2} = \cos(b_2)\left[\sin(n_2)\sin(\phi)\cos(\psi) - \cos(n_2)\cos(\phi)\cos(\theta)\right] \tag{A14}$$

$$\lambda_{z3} = -\cos(n_3)\cos(b_3)\cos(\theta)\cos(\phi) + \sin(b_1)\sin(\phi)\cos(\psi) \tag{A15}$$

$$\lambda_{z4} = -\cos(b_4)\left[\sin(n_4)\sin(\phi)\cos(\psi) - \cos(n_4)\cos(\phi)\cos(\theta)\right] \tag{A16}$$

The coefficients of the torque dynamic equation in x-axis are as follows:

$$l_{xx} = -\cos(n_1)IP_x\ddot{b}_1 - \sin(n_2)\sin(b_2)IP_x\ddot{b}_2 + \cos(n_3)IP_x\ddot{b}_3 + \sin(n_4) \\ \sin(b_4)IP_x\ddot{b}_4 \tag{A17}$$

$$l_{xy} = -\sin(n_1)\sin(b_1)IP_y\ddot{n}_1 + \cos(b_2)IP_y\ddot{n}_2 + \sin(n_3)\sin(b_3)IP_y\ddot{n}_3 - \\ \cos(b_4)IP_y \tag{A18}$$

$$l_{xz} = -\sin(n_1)\cos(b_1)IP_z\omega_1 - \sin(b_2)IP_z\omega_2 + \sin(n_3)\cos(b_3)IP_z\omega_3 + \\ \sin(b_4)IP_z\omega_4 \tag{A19}$$

$$\rho_x = QRIB_{yy} - QRIB_{zz} - PdIB_x \tag{A20}$$

The coefficients of the torque dynamic equation in y-axis are as follows:

$$l_{yx} = \sin(n_1)\sin(b_1)IP_x\ddot{b}_1 - \sin(n_2)IP_x\ddot{b}_2 - \sin(n_3)\sin(b_3)IP_x\ddot{b}_3 + \\ \sin(n_4)IP_x\ddot{b}_4 \tag{A21}$$

$$l_{yy} = -\cos(b_1)IP_y\ddot{n}_1 - \sin(n_2)\sin(b_2)IP_y\ddot{n}_2 + \cos(b_3)IP_y\ddot{n}_3 + \sin(n_4)$$
$$\sin(b_4)IP_y\ddot{n}_4 \tag{A22}$$

$$l_{yz} = \sin(b_1)IP_z\omega_1 - \sin(n_2)\cos(b_2)IP_z\omega_2 - \sin(b_3)IP_z\omega_3 + \sin(n_4)$$
$$\cos(b_4)IP_z\omega_4 \tag{A23}$$

$$\rho_y = PRIB_{zz} - PRIB_{xx} - QdIB_y \tag{A24}$$

The coefficients of the torque dynamic equation in z-axis are as follows:

$$l_{zx} = \sin(n_1)IP_x\ddot{b}_1 + \sin(n_2)IP_x\ddot{b}_2 + \sin(n_3)IP_x\ddot{b}_3 + \sin(n_4)IP_x\ddot{b}_4 \tag{A25}$$

$$l_{zy} = -\cos(n_1)\sin(b_1)IP_y\ddot{n}_1 - \cos(n_2)\sin(b_2)IP_y\ddot{n}_2 - \cos(n_3)\sin(b_3)IP_y\ddot{n}_3 -$$
$$\cos(n_4)\sin(b_4)IP_y\ddot{n}_4 \tag{A26}$$

$$l_{zz} = \cos(n_1)\cos(b_1)IP_z\omega_1 + \cos(n_2)\cos(b_2)IP_z\omega_2 + \cos(n_3)cos(b_3)IP_z\omega_3 +$$
$$\cos(n_4)\cos(b_4)IP_z\omega_4 \tag{A27}$$

$$\rho_z = QPIB_{xx} - PQIB_{yy} - RdIB_z \tag{A28}$$

Appendix B. Inertia Values

Expanding Equation (7) yields the inertia values in the main axis as follows:

$$I_{Bxx} = \quad I_{xx} + 2I_{Pxx} + 2I_{Pzz} + \lambda\left[\cos(2\eta_1) + \cos(2\eta_2) + \cos(2\eta_3) + \cos(2\eta_4)\right]$$
$$+2m_PL^2 \tag{A29}$$

$$I_{Byy} = \quad I_{yy} + 2I_{Pyy} + I_{Pxx} + I_{Pzz} - \tfrac{1}{2}\lambda\left[\cos(2\eta_1) + \cos(2\eta_2) + \cos(2\eta_3) + \cos(2\eta_4)\right]$$
$$+\zeta\left[\cos(2\beta_1) + \cos(2\beta_2) + \cos(2\beta_3) + \cos(2\beta_4)\right]$$
$$+\tfrac{1}{2}\lambda\left[\cos(2\eta_1)\cos(2\beta_1) + \cos(2\eta_2)\cos(2\beta_2) + \cos(2\eta_3)\cos(2\beta_3) + \cos(2\eta_4)\cos(2\beta_4)\right]$$
$$+2m_PL^2 \tag{A30}$$

$$I_{Bzz} = \quad I_{zz} + 2I_{Pyy} + I_{Pxx} + I_{Pzz} - \tfrac{1}{2}\lambda\left[\cos(2\eta_1) + \cos(2\eta_2) + \cos(2\eta_3) + \cos(2\eta_4)\right]$$
$$-\zeta\left[\cos(2\beta_1) + \cos(2\beta_2) + \cos(2\beta_3) + \cos(2\beta_4)\right]$$
$$-\tfrac{1}{2}\lambda\left[\cos(2\eta_1)\cos(2\beta_1) + \cos(2\eta_2)\cos(2\beta_2) + \cos(2\eta_3)\cos(2\beta_3) + \cos(2\eta_4)\cos(2\beta_4)\right]$$
$$+4m_PL^2 \tag{A31}$$

where the value of ζ and λ are defined as follows

$$\lambda = \tfrac{1}{2}(I_{Pxx} - I_{Pzz})$$
$$\zeta = \tfrac{1}{2}I_{Pyy} - \tfrac{1}{4}I_{Pxx} - \tfrac{1}{4}I_{Pzz} \tag{A32}$$

The last expression to be derived is the time derivative of the inertia matrix which is given by Equations (A29)–(A31). For each principal axis of the inertia, it yields:

$$\frac{d}{dt}(I_{Bxx}) = -2\lambda\left(\sin(2\eta_1)\left(\frac{d}{dt}(\eta_1)\right)\right) - 2\lambda\left(\sin(2\eta_2)\left(\frac{d}{dt}(\eta_2)\right)\right)$$
$$-2\lambda\left(\sin(2\eta_3)\left(\frac{d}{dt}(\eta_3)\right)\right) - 2\lambda\left(\sin(2\eta_4)\left(\frac{d}{dt}(\eta_4)\right)\right) \tag{A33}$$

$$\begin{aligned}
\frac{d}{dt}\left(I_B yy\right) =\ & \lambda \sin(2\eta_1)\tfrac{d}{dt}(\eta_1) + \lambda \sin(2\eta_2)\tfrac{d}{dt}(\eta_2) + \lambda \sin(2\eta_3)\tfrac{d}{dt}(\eta_3) \\
& + \lambda \sin(2\eta_4)\tfrac{d}{dt}(\eta_4) - \lambda \sin(2\eta_1)\cos(2\beta_1)\tfrac{d}{dt}(\eta_1) - \lambda \sin(2\eta_2)\cos(2\beta_2)\tfrac{d}{dt}(\eta_2) \\
& - \lambda \sin(2\eta_3)\cos(2\beta_3)\tfrac{d}{dt}(\eta_3) - \lambda \sin(2\eta_4)\cos(2\beta_4)\tfrac{d}{dt}(\eta_4) \\
& - 2\left(\zeta + \tfrac{1}{2}\lambda\cos(2\eta_1)\right)\sin(2\beta_1)\tfrac{d}{dt}(\beta_1) - 2\left(\zeta + \tfrac{1}{2}\lambda\cos(2\eta_2)\right)\sin(2\beta_2)\tfrac{d}{dt}(\beta_2) \\
& - 2\left(\zeta + \tfrac{1}{2}\lambda\cos(2\eta_3)\right)\sin(2\beta_3)\tfrac{d}{dt}(\beta_3) - 2\left(\zeta + \tfrac{1}{2}\lambda\cos(2\eta_4)\right)\sin(2\beta_4)\tfrac{d}{dt}(\beta_4)
\end{aligned}$$
(A34)

$$\begin{aligned}
\frac{d}{dt}\left(I_B zz\right) =\ & \lambda \sin(2\eta_1)\tfrac{d}{dt}(\eta_1) + \lambda \sin(2\eta_2)\tfrac{d}{dt}(\eta_2) + \lambda \sin(2\eta_3)\tfrac{d}{dt}(\eta_3) \\
& + \lambda \sin(2\eta_4)\tfrac{d}{dt}(\eta_4) + \lambda \sin(2\eta_1)\cos(2\beta_1)\tfrac{d}{dt}(\eta_1) + \lambda \sin(2\eta_2)\cos(2\beta_2)\tfrac{d}{dt}(\eta_2) \\
& + \lambda \sin(2\eta_3)\cos(2\beta_3)\tfrac{d}{dt}(\eta_3) + \lambda \sin(2\eta_4)\cos(2\beta_4)\tfrac{d}{dt}(\eta_4) \\
& + 2\left(\zeta + \tfrac{1}{2}\lambda\cos(2\eta_1)\right)\sin(2\beta_1)\tfrac{d}{dt}(\beta_1) + 2\left(\zeta + \tfrac{1}{2}\lambda\cos(2\eta_2)\right)\sin(2\beta_2)\tfrac{d}{dt}(\beta_2) \\
& + 2\left(\zeta + \tfrac{1}{2}\lambda\cos(2\eta_3)\right)\sin(2\beta_3)\tfrac{d}{dt}(\beta_3) + 2\left(\zeta + \tfrac{1}{2}\lambda\cos(2\eta_4)\right)\sin(2\beta_4)\tfrac{d}{dt}(\beta_4)
\end{aligned}$$
(A35)

References

1. Mahony, R.; Kumar, V. Aerial Robotics and the Quadrotor. *IEEE Robot. Autom. Mag.* **2012**, *19*, 19. [CrossRef]
2. Doherty, P.; Rudol, P. A UAV Search and Rescue Scenario with Human Body Detection and Geolocalization. In Proceedings of the Australian Conference on Artificial Intelligence, Gold Coast, Australia, 2–6 December 2007; pp. 1–13. [CrossRef]
3. Mahony, R.; Kumar, V.; Corke, P. Multirotor Aerial Vehicles: Modeling, Estimation, and Control of Quadrotor. *IEEE Robot. Autom. Mag.* **2012**, *19*, 20–32. [CrossRef]
4. Hua, M.D.; Hamel, T.; Morin, P.; Samson, C. A Control Approach for Thrust-Propelled Underactuated Vehicles and its Application to VTOL Drones. *IEEE Trans. Autom. Control* **2009**, *54*, 1837–1853.
5. Hua, M.D.; Hamel, T.; Morin, P.; Samson, C. Introduction to feedback control of underactuated VTOL vehicles: A review of basic control design ideas and principles. *IEEE Control Syst.* **2013**, *33*, 61–75. [CrossRef]
6. Lo, C.H.; Shin, H.S.; Tsourdos, A.; Kim, S. Modeling and Simulation of Fault Tolerant Strategies for A Quad Rotor UAV. In Proceedings of the AIAA Modeling and Simulation Technologies Conference, Minneapolis, MI, USA, 13–16 August 2012.
7. Morozov, Y.V. Emergency Control of a Quadrocopter in Case of Failure of Two Symmetric Propellers. *Autom. Remote Control* **2018**, *79*, 463–478. [CrossRef]
8. Adır, V.G.; Stoica, A.M.; Marks, A.; Whidborne, J.F. stabilization and single motor failure recovery of a 4Y octorotor. In Proceedings of the IASTED International Conference on Intelligent Systems and Control (ISC 2011), Cambridge, UK, 11–13 July 2011; pp. 82–87.
9. Michieletto, G.; Ryll, M.; Franchi, A. Control of statically hoverable multi-rotor aerial vehicles and application to rotor-failure robustness for hexarotors. In Proceedings of the 2017 IEEE International Conference on Robotics and Automation (ICRA), Singapore, 29 May–3 June 2017; pp. 2747–2752. [CrossRef]
10. Cutler, M.; Ure, N.K.; Michini, B.; How, J. Comparison of Fixed and Variable Pitch Actuators for Agile Quadrotors. In Proceedings of the AIAA Guidance, Navigation, and Control Conference, Portland, OR, USA, 8–11 August 2011.
11. Oner, K.T.; Cetinsoy, E.; Unel, M.; Aksit, M.F.; Kandemir, I.; Gulez, K. Dynamic model and control of a new quadrotor unmanned aerial vehicle with tilt-wing mechanism. *World Acad. Sci. Eng. Technol.* **2008**, *45*, 58–63.
12. Jiang, G.; Voyles, R. Hexrotor UAV platform enabling dextrous interaction with structures-flight test. In Proceedings of the 2013 IEEE International Symposium on Safety, Security, and Rescue Robotics (SSRR), Linkoping, Sweden, 21–26 October 2013; pp. 1–6. [CrossRef]
13. Gasco, P.S. Development of a Dual Axis Tilt Rotorcraft. Master's Thesis, Cranfield University, Bedford, UK, 2012.
14. Gasco, P.S.; Al-Rihani, Y.; Shin, H.S.; Savvaris, A. A novel actuation concept for a multi rotor UAV. In Proceedings of the International Conference on Unmanned Aircraft Systems (ICUAS), Atlanta, GA, USA, 28–31 May 2013; pp. 373–382. [CrossRef]
15. Nemati, A.; Kumar, M. Modeling and control of a single axis tilting quadcopter. In Proceedings of the American Control Conference, Portland, OR, USA, 4–6 June 2014; pp. 3077–3082. [CrossRef]

16. Scholz, G.; Trommer, G.F. Model based control of a quadrotor with tiltable rotors. *Gyrosc. Navig.* **2016**, *7*, 72–81. [CrossRef]

17. Lim, K.; Shin, J.Y.; Moerder, D.; Cooper, E. A New Approach to Attitude Stability and Control for Low Airspeed Vehicles. In Proceedings of the AIAA Guidance, Navigation, and Control Conference, Providence, RI, USA, 16–19 August 2004. [CrossRef]

18. Gress, G. Using Dual Propellers as Gyroscopes for Tilt-Prop Hover Control. In Proceedings of the Biennial International Powered Lift Conference and Exhibit, Williamsburg, VA, USA, 5–7 November 2002.

19. Gress, G. Lift fans as gyroscopes for controlling compact VTOL air vehicles: Overview and development status of Oblique Active Tilting. In Proceedings of the American Helicopter Society 63th Annual Forum, Virginia Beach, VA, USA, 1–3 May 2007.

20. Sanchez, A.; Escareño, J.; Garcia, O.; Lozano, R. Autonomous Hovering of a Noncyclic Tiltrotor UAV: Modeling, Control and Implementation. *IFAC Proce. Vol.* **2008**, *41*, 803–808. [CrossRef]

21. Klippstein, H.; Diaz De Cerio Sanchez, A.; Hassanin, H.; Zweiri, Y.; Seneviratne, L. Fused Deposition Modeling for Unmanned Aerial Vehicles (UAVs): A Review. *Adv. Eng. Mater.* **2017**, 1700552. [CrossRef]

22. Klippstein, H.; Hassanin, H.; Diaz De Cerio Sanchez, A.; Zweiri, Y.; Seneviratne, L. Additive Manufacturing of Porous Structures for Unmanned Aerial Vehicles Applications. *Adv. Eng. Mater.* **2018**, 1800290. [CrossRef]

23. Pragada, L.K.D.; Katukam, R. 3d printing of quadcopter: A Case Study. *Int. J. Latest Trends Eng. Technol.* **2015**, *5*, 131–140.

24. Hooi, C.G. Design, Rapid Prototyping and Testing of a Ducted Fan Microscale Quadcopter. In Proceedings of the American Helicopter Society 70th Annual Forum, Montreal, QC, Canada, 20–22 May 2014.

25. Alzu'bi, H.; Allateef, I.; Zweiri, Y.; Alkhateeb, B.; Al-Masarwah, I. Quad tilt rotor Vertical Take Off and Landing (VTOL) Unmanned Aerial Vehicle (UAV) with 45 degree rotors. Patent US20130105635, 2 May 2013.

26. Ryll, M.; Bülthoff, H.H.; Giordano, P.R. First flight tests for a quadrotor UAV with tilting propellers. In Proceedings of the 2013 IEEE International Conference on Robotics and Automation, Karlsruhe, Germany, 6–10 May 2013; pp. 295–302. [CrossRef]

27. Ryll, M.; Bülthoff, H.H.; Giordano, P.R. A Novel Overactuated Quadrotor Unmanned Aerial Vehicle: Modeling, Control, and Experimental Validation. *IEEE Trans. Control Syst. Technol.* **2015**, *23*, 540–556. [CrossRef]

28. Zortrax. Zortrax M200. 2016. Available online: https://zortrax.com/printers/zortrax-m200/ (accessed on 28 September 2017).

29. Guzman, M. Modeling and Simulation of an Over-Actuated Dual Tilting Rotors Quadcopter. Master's Thesis, Kingston University, London, UK, 2015.

30. Lim, H.; Park, J.; Lee, D.; Kim, H.J. Build Your Own Quadrotor: Open-Source Projects on Unmanned Aerial Vehicles. *IEEE Robot. Autom. Mag.* **2012**, *19*, 33–45. [CrossRef]

12

Fully Mechatronical Design of an HIL System for Floating Devices

Hermes Giberti [1,*], **Francesco La Mura** [2,*], **Gabriele Resmini** [2] **and Marco Parmeggiani** [2]

[1] Dipartimento di Ingegneria Industriale e dell'Informazione, Università degli Studi di Pavia, Via A. Ferrata 5, 27100 Pavia, Italy

[2] Department of Mechanical Engineering, Politecnico di Milano, 20156 Milano, Italy; gabriele.resmini@mail.polimi.it (G.R.); marco1.parmeggiani@mail.polimi.it (M.P.)

*   Correspondence: hermes.giberti@unipv.it (H.G.); francesco.lamura@polimi.it (F.L.M.);

Abstract: Recent simulation developments in Computational Fluid Dynamics (CFD) have widely increased the knowledge of fluid–structure interaction. This has been particularly effective in the research field of floating bodies such as offshore wind turbines and sailboats, where air and sea are involved. Nevertheless, the models used in the CFD analysis require several experimental parameters in order to be completely calibrated and capable of accurately predicting the physical behaviour of the simulated system. To make up for the lack of experimental data, usually wind tunnel and ocean basin tests are carried out. This paper presents a fully mechatronical design of an Hardware In the Loop (HIL) system capable of simulating the effects of the sea on a physical scaled model positioned in a wind tunnel. This system allows one to obtain all the required information to characterize a model subject, and at the same time to assess the effects of the interaction between wind and sea waves. The focus of this work is on a complete overview of the procedural steps to be followed in order to reach a predefined performance.

Keywords: parallel kinematic machine; kinematic optimisation; hardware in the loop; mechatronic design

1. Introduction

Recent simulation developments in Computational Fluid Dynamics (CFD) have widely increased the knowledge of fluid–structure interaction. This has been particularly effective in the research field of floating bodies such as offshore wind turbines and sailboats, where air and sea are involved. Scale model testing of floating structures is a common practice in this research field and help in the development of new concepts and technological solutions, driving the design choices and rapidly answering the scientific questions. Moreover, this study approach allows one to evaluate related problems such as the definition of the control algorithm structures and quantification of costs and benefits.

Nevertheless, when the effects of wind and wave loads become comparable, the validation of the test models is affected by a scaling issue called Froude–Reynolds conflict, related to the ability to reproduce the effects of wind and wave at the same time in a correct manner. For this reason, hybrid tests both in ocean basins [1] and wind tunnels [2] are the most effective approach used to overcome scaling constraints and exploit separately wind and wave generators.

In this paper, the hardware/software setup developed for hybrid tests in a wind tunnel is presented and fully described. The aim of this work is to provide a complete overview of the design procedures and methodologies required to develop a hardware in the loop device capable of simulating the effects of the sea on a physical scaled model positioned in a wind tunnel. In order to achieve the required performance and in compliance with the dimensional constraints, a fully mechatronical design approach has been used. Moreover, this system is particularly challenging in terms of dynamic response. For these reasons, it can be taken as an example of integrated design not only for this specific application, but also for a generic simulator. In fact, the design scheme followed and the numeric tools to develop the system sizing and select the commercially available components are of general interest.

Figure 1 summarises the design procedure of the HIL device presented in this paper and called Hexafloat. The latter consists of a six-degree-of-freedom (DOF) parallel manipulator custom designed to move a scaled wind turbine (e.g., 1/75 of the 10 MW DTU reference wind turbine [3]) in a wind tunnel. The manipulator end effector is equipped with a six-degrees of freedom weight scale used to measure the constraint reactions between the robot and the scaled wind turbine (i.e., aerodynamic load effects). A real-time mathematical model generates the reference commands to be followed by the robot according to the dynamic of the floating substructure simulated by the system and the measures acquired from the weight scale (i.e., real-time hydro/structure computations). The design procedure, as shown in Figure 1, is made up of four main steps that for descriptive simplicity are presented in sequence in this paper but should be considered as a cyclic procedure that ends asymptotically towards the result. The first step relates to the kinematic topological definition and its optimisation as a function of the workspace dimension and the dimensional constraints. The second and the third steps are related to the sizing procedures necessary to ensure the static and dynamic performance of the device taking into account both the actuation system and structural components. The last step deals with the definition of the hardware and software control issues.

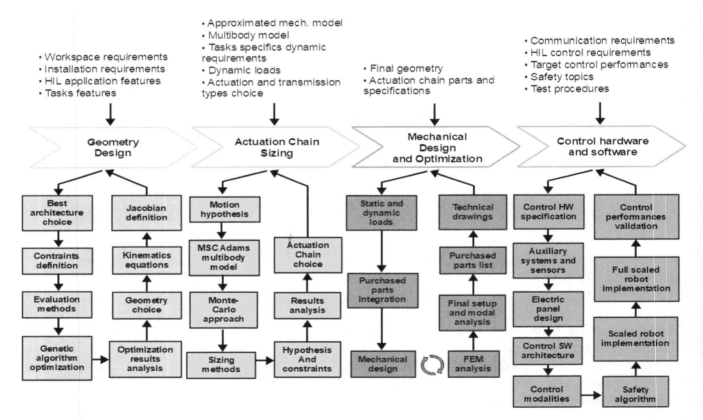

Figure 1. General scheme of the HexaFloat design process.

2. Geometric and Kinematic Design

The starting point of every project is the definition of the operating conditions. Figure 2 summarises the dimensional constraints and shows the position of the desired working volume within the wind tunnel. This position has been chosen in order to optimize the layout of the aerodynamic tests using a scaled turbine with respect to the wind tunnel performance. No commercial solutions suit the requirements due to the restricted space available to install the device so therefore a custom solution must be realized out of necessity.

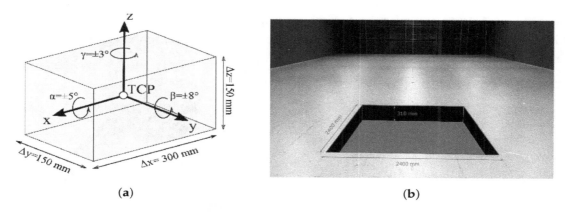

Figure 2. Dimensional constraints: (**a**) Workspace constraints; (**b**) Wind tunnel frontal view and dimensions of the space dedicated to the machine.

Among the possible six-DoF manipulator architectures, a parallel kinematic machine (PKM) has been preferred rather than the serial one, due to its high stiffness and high dynamic performance even if this entails an accurate design of the joints in order to have an adequate working space and the absence of backlash. In the first part of the design procedure, the main focus is on the choice of the kinematic topology and the geometric dimensions [2,4]. The most common and widespread six-DOF PKM architectures among which to choose the right topology for this application can be summed up in three families:

6-UPS: manipulators with kinematic chain characterised by a sequence of a universal joint at the base (U), an actuated prismatic joint (P) that changes the length of the link as well as a spherical joint (S) connected to the mobile platform;

6-RUS: manipulators with fixed length links, moved by actuated revolute joints (R) located at the base. The other joints in the kinematic chain are: an intermediate universal joint and a spherical joint connected to the mobile platform;

6-PUS: manipulators with fixed length links. The actuated prismatic joint is generally composed of a slider moving along a rectilinear guide. The link is connected to the slider by means of a universal joint and to the mobile platform by means of a spherical joint.

The most promising solution within the three families is the six-PUS configuration because the actuation system lays on the floor grounded; moreover, the vertical bulk is limited as opposed to the horizontal one. This topology has been selected in order to guarantee high performance in terms of dynamic response and structural properties with the view to reduce the size vertically. Within this category, the two manipulators shown in Figure 3 have been chosen and evaluated [5]. The first one is called Hexaglide, having parallel linear guides and therefore is characterized by symmetry with respect to the longitudinal median plane. The second is called Hexaslide and is characterized by a radial symmetry with respect to the vertical axis due to the radial distribution of the linear guides in the plane. This characteristic leads to a better isotropy of the workspace.

2.1. Kinetostatic Optimization

An optimization process [5] is required in order to synthesize the geometric parameters of the two chosen robot architectures and evaluate which is the best solution. To properly set up the process, it is first necessary to identify the geometric parameters characterizing the two architectures and thereafter a function that mathematically describes the goal to be achieved. Finally, a set of physical constraints affecting the system have to be identified and defined in mathematical form in order to create the numeric boundaries for the solution. When choosing the number of parameters to be optimized, a trade-off has to be found: if the number of parameters used to characterize a machine is increased, the whole process of optimization would benefit, thus allowing a higher freedom. However, the complexity of the machine increases dramatically. A trade-off needs to be found between global performance and structural modularity, which is critical in order to simplify design and optimization steps.

(a)   (b)

Figure 3. Hexaglide robot (**a**) and Hexaslide robot (**b**) simplified geometry scheme.

The Hexaslide architecture can be analysed taking into consideration the following parameters (Figure 4) for the optimization process:

- s: semi-distance between two parallel guides;
- l_i: length of the links;
- R_p: radius of the circumference on which the platform joint centres are located;
- θ_p: semi-angular aperture between the two segments connecting the origin of the Tool Center Point (TCP) reference frame and a couple of platform joints;
- z_{wsd}: height of the centre of the desired workspace.

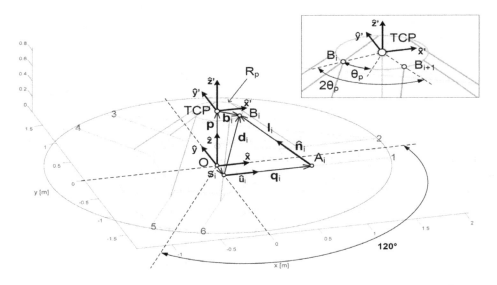

Figure 4. Hexaslide geometric parameters, the Cartesian space is represented in meters.

For this architecture, the same kinematic chain is replicated identically for six times.

The Hexaglide architecture, shown in Figure 5, can be analysed considering a higher number of parameters because three different kinematic chains can be identified in the robot configuration. The parameters to be optimized are:

- $R_{p,1}$: radius of the circumference on which the platform joints 1 and 2 are located;
- $R_{p,2}$: radius of the circumference on which the platform joints 3 and 4 are located;
- $R_{p,3}$: radius of the circumference on which the platform joints 5 and 6 are located;
- $\theta_{p,1}$: semi-angular aperture between the two segments connecting the origin of the TCP reference frame and platform joints 1 and 2;
- $\theta_{p,2}$: semi-angular aperture between the two segments connecting the origin of the TCP reference frame and platform joints 3 and 4;
- $\theta_{p,3}$: semi-angular aperture between the two segments connecting the origin of the TCP reference frame and platform joints 5 and 6;
- t_{01}: vertical distance between TCP and the plane where platform joints 1 and 2 lie;
- t_{02}: vertical distance between TCP and the plane where platform joints 3 and 4 lie;
- t_{03}: vertical distance between TCP and the plane where platform joints 5 and 6 lie;
- l_{01}: length of links 1 and 2;
- l_{02}: length of links 3 and 4;
- l_{03}: length of links 5 and 6;
- s_{01}: semi-distance between parallel guides 1 and 2;
- s_{02}: semi-distance between parallel guides 3 and 4;
- s_{03}: semi-distance between parallel guides 5 and 6;
- s_{h01}, s_{h02}: vertical distance between ground and parallel guides;
- z_{wsd}: height of the center of the desired workspace.

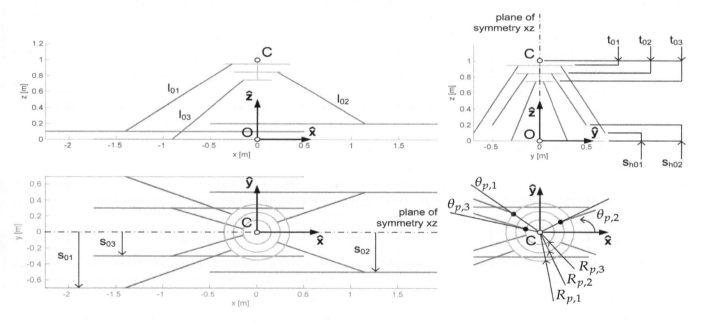

Figure 5. Hexaglide geometric parameters, the Cartesian space is represented in meters.

For this architecture, each kinematic chain is replicated twice, one on both sides of the longitudinal median plain of symmetry.

The goal is to optimize robot kinematic performance while ensuring the desired workspace boundaries. In order to define a proper objective function, the desired workspace is discretized in elementary volumes: for each elementary volume, the specific set of parameters under analysis must be allowed to reach its centre point. Moreover, for each centre point analysed, the resulting machine must be able to orientate the TCP with any possible combination of pitch, roll and jaw angles varying inside a predefined range. If that is not the case, the elementary volume is added to the total volume that the current set of parameters is not able to cover. In order to make the computational cost affordable and restrict the checks to a finite amount of angles combinations, the three angular ranges are discretized. The evaluation of the objective-function is performed firstly by fixing the mobile-platform orientation $\Theta_m = [\alpha_m; \beta_m; \gamma_m]$, and thereafter computing the part of volume that end-effector is unable to reach for these specific orientations, such as:

$$v_{nc}(\Theta_m) = \sum_i \sum_j \sum_k c_{i,j,k}(\Theta_m)\Delta v \qquad \begin{cases} i = 1, \ldots, N_x, \\ j = 1, \ldots, N_y, \\ k = 1, \ldots, N_z, \end{cases} \qquad (1)$$

where the combination of parameters i, j, k unequivocally identify a check-point of the workspace in which N_x, N_y, N_z represent the number of discretization points, respectively, along the x, y, z axes. The variable $c_{i,j,k}(\Theta_m)$ is equal to 1 if the specific parameters prevent the end-effector from reaching the pose defined by i, j, k and Θ_m, and 0, otherwise. The term Δv is the elementary volume resulting from the discretization of the workspace.

This procedure is repeated for all combinations of pitch, roll and jaw angles and the resulting value for the objective-function V_{nc} is computed as:

$$V_{nc} = \sum_m v_{nc}(\Theta_m). \qquad (2)$$

Even if a volume portion is discarded for only one specific orientation, that portion of workspace is excluded for the corresponding geometrical configuration. Using this approach, in the ideal case in which a set of parameters allows the robot to reach any point in the workspace regardless of its orientation, the objective-function would be equal to 0. On the other hand, if the set of parameters does not allow the reaching of a particular point of the workspace, the objective function would assume a value equal to the dimension of the single discrete volume times the number of orientations for which that volume has been discarded.

The kinematic capability of reaching each point with every possible orientation is not in itself enough, but additional constraint definitions are required. The kinematic constraints are defined as follows:

- Distance between the i-th platform joint and the corresponding base joint should not exceed the length of the link for geometric congruence;
- Each actuated joint coordinate has to be comprised within a range defined by the dimension of the machine, since actuators' stroke range have a direct impact on the major bulk direction of the machine, longitudinal for the Hexaglide, and radial for the Hexaslide;
- Each passive joint, both platform and base ones, should respect their respective mobility ranges.

The kinetostatic constraints are defined by the transmission ratio between forces and moments acting on the end-effector and the actuation forces. This transmission ratio for each actuator is computed and the maximum should be lower than a prescribed limit value. As is common for PKMs, this transmission ratio varies considerably within the nominal workspace due to nonlinear kinematics, especially in proximity of the singular configurations. The geometric constraints enforce the respect of the minimum distance between two links and between a link and the mobile platform in order to avoid the problem of self-collision of the component for particular poses.

The solution related to the minimisation of the objective function has been obtained through a single objective genetic algorithm approach [6]. The use of a semi-stochastic search and the evaluation of the performance of different individuals at each iteration makes the process of finding the global minimum of the objective-function to be minimized, easier. The steps characterizing this kind of algorithm are:

- Choice of a sufficiently high number of individuals representing a generation in order to have a significant statistical sample;

- Evaluation of the performance for each individual of the current generation depending on the values assumed by its genes;
- Choice of the group with the better performance that will constitute the elite and will pass unchanged to the next generation;
- Creation of a new generation based on elitary choice, crossover and mutation;
- Computation of the performance of the individuals of the new generation in comparison to the goal desired.

The algorithm would keep modifying parameters trying to cover the desired workspace as much as possible and minimizing the defined objective function. In Tables 1 and 2, the lower and upper bounds imposed at the beginning of the optimization and the optimal values obtained are reported.

Table 1. Limits and results of the Hexaslide optimization.

Symbol	Lower Bound	Upper Bound	Optimal Value	MU
z_{WSd}	400	500	463.6	mm
s	200	300	203.1	mm
l	400	700	686.6	mm
R_p	200	250	238.7	mm
θ_p	0	60	38.5	°

Table 2. Limits and results of the Hexaglide optimization.

Symbol	Lower Bound	Upper Bound	Optimal Value	MU
z_{WSd}	500	700	593.9	mm
s_1	100	980	1050.8	mm
s_2	100	980	277.7	mm
s_3	100	980	330.2	mm
l_1	600	1600	1207.7	mm
l_2	600	1600	873.9	mm
l_3	600	1600	825.7	mm
$R_{p,1}$	200	400	306.2	mm
$R_{p,2}$	200	400	326.7	mm
$R_{p,3}$	200	400	275.9	mm
t_1	−200	0	−183.0	mm
t_2	−200	0	−72.4	mm
t_3	−200	0	22.9	mm
θ_{p1}	10	170	77.3	°
θ_{p2}	10	170	37.9	°
θ_{p3}	10	170	100.6	°

Minimum distance between the links, minimum distance between links and platform, transmission ratio and maximum and minimum strokes of the actuated joint coordinates are mapped throughout the workspace to guide one in the choice of the best architecture. The following conclusions hold respectively for the Hexaslide and Hexaglide.

Hexaslide:

- Topology: both the height and the in plane bulkiness of the manipulator are quite limited. When the robot is in the home position, the links are arranged in such a way that a good compromise is achieved between the capacity to generate velocity in all directions and to bear external forces without too much effort required by the actuators.
- Link-to-link and link-to-platform minimum distances: The link-to-platform minimum distance recorded is above 270 mm throughout the workspace, thus avoiding any risk of collision between the legs.

- Force transmission ratio: the worst case obtained by the computation is closed to the limit value of 5 but restricted to only a few small lower regions of the workspace.
- Actuated joints maximum and minimum stroke: as the joints coordinate distance with respect to the global reference frame is always positive, it is sufficient to check the maximum value in order to evaluate the bulkiness of the robot. This joint coordinate excursion is about 1 m.

Hexaglide

- Topology: The in plane bulkiness of this robot is much higher with respect to the Hexaslide. Moreover, the centre of the desired workspace in the optimized configuration is placed in a higher position compared to the Hexaslide configuration.
- Link-to-link and link-to-platform minimum distances: no risk of self-collision between the elements has been detected.
- Force transmission ratio: the transmission ratio remains limited above 2.5.
- Maximum and minimum stroke: the bulkiness of this solution is higher compared to the Hexaglide one.

It can thus be concluded that Hexaslide architecture is more compact compared to Hexaglide but reaches higher values of force transmission ratio. The Hexaslide architecture has been chosen due to its lower vertical bulkiness, more compact in plane dimensions, better workspace isotropy and lower position of the workspace centre. All of these features make it possible to install the machine under the wind tunnel floor level, reducing the robot influence on the air flow quality and keeping the turbine farther from the wind tunnel ceiling. Moreover, Hexaslide offers two additional advantages: all the elements that make up the links are the same for all six of the kinematic chains and the radial symmetry simplifies the design process. Hereafter, this machine will be referred to as Hexafloat, while the term Hexaslide will refer to the specific architecture of the robot.

2.2. Kinematics Analysis

In order to develop the optimisation problem and design the control algorithm of the robot, it is necessary to solve the inverse and forward kinematics. In this section, for the sake of briefness, only the solution of the inverse and forward kinematic problem of the robot architecture chosen is presented.

2.2.1. Inverse Kinematics (IK)

In order to study *Inverse Kinematics*, two different reference frames have been considered, the global one and the local one with origin in the TCP and built into the robot platform. With reference to Figure 6, two different vector closures for each kinematic chain can be set up in order to compute the joint coordinates vector q. The first vector closure allows one to determine the absolute position vector $\mathbf{d_i}$ of the platform joint B_i with respect to a point that is the intersection of the $\mathbf{q_i}$ direction with its orthogonal plane passing through the global origin. Each $\mathbf{s_i}$ has a fixed length with different orientations in the X_0-Y_0 plane:

$$\mathbf{d_i} = \mathbf{p} + R_{TCP}\mathbf{b_i} - \mathbf{s_i}, \tag{3}$$

where R_{TCP} is the rotation matrix defining the platform orientation. This matrix is used to transform the expression of $\mathbf{b_i}$, constant into a rotating local reference frame, in its equivalent with respect to a zero orientation local frame translating with the platform. Position vector $\mathbf{p}$ and the offset $\mathbf{s_i}$ complete the transformation from the zero orientation local frame to the global one. The second vector closure allows one to determine the position of the i-th slider on the guide:

$$\mathbf{l_i} = \mathbf{d_i} - q_i\hat{\mathbf{u}}_i. \tag{4}$$

The magnitude of $\mathbf{l_i}$ vector corresponds to the length of the robot leg while $\hat{\mathbf{u}}$ represents the guide direction unitary vector. This procedure is identical, except for the orientation of $\mathbf{b_i}$, $\hat{\mathbf{u}}_i$ and $\mathbf{s_i}$, for all six links of the robot.

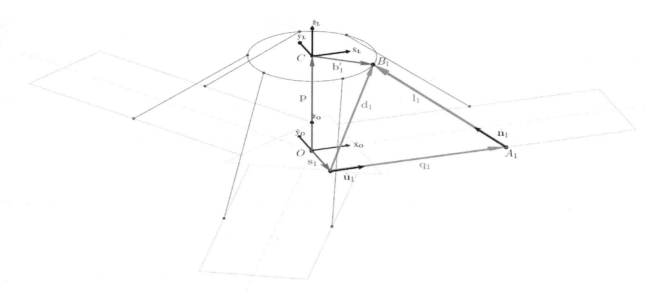

Figure 6. Vector closures used to solve the inverse kinematics (IK) problem: in blue, the 1st closure while in red the 2nd one.

2.2.2. Forward Kinematics (FK)

The Hexafloat robot will not be equipped with sensors able to directly measure the platform position and orientation, therefore a good quality FK could work as a virtual sensor for the pose of the robot. A high frequency estimation of the actual pose of the TCP would be a valuable feedback tool for the other controller in the HIL setup, but a short calculation time and overall stability, in the case of numerical algorithms, is crucial. Extensive research has been conducted to ascertain analytical methods to solve the FK of PKMs, especially for Gough–Steward configuration, but the pose of the robot has not been expressed in explicit form so far.

Considering numerical methods, the Newton–Raphson (NR) algorithm has its numerical stability highly dependent on the accuracy of the initial approximation of the solution vector so a monotonic descent operator can be added obtaining the so-called Modified-Global-Newton–Raphson (MGNR) algorithm, able to estimate the FK solution of six-DoF parallel robots for any initial approximation in the non-singular workspace without divergence. The algorithm requires the definition of a system of six nonlinear equations and the evaluation of a matrix of partial derivatives:

$$\underline{f}_i(\underline{X}) = \underline{0}, \tag{5}$$

$$P_{ij} = \frac{\partial f_i}{\partial x_j}. \tag{6}$$

The evaluation of partial derivatives matrix can be simplified by implementing Jacobian-Free-Monotonic-Descendent (JFMD) algorithm. A first-order Taylor expansion can be used to approximate the partial derivatives matrix in a numerical way. The JFMD method is implemented via the following steps:

- proper initial approximation of x_0 for the solution is chosen and the corresponding $\underline{f}_i(\underline{X}) = \underline{0}$ is calculated;
- the $(k+1)$th solution attempt is calculated according to the following formula:

$$x_{k+1} = x_k - \rho_k [J(x_k)]^{-1} f(x_k), \tag{7}$$

where ρ_k ($0 \leq \rho_k \leq 1$) is the monotonic descent factor. It starts from 1 and during each iteration is calculated as $\rho_k = 2^{-m}$, where m is the number of rechecking times in the corresponding iteration, necessary to obtain the monotonic trend. The approximated Jacobian matrix is evaluated from the first-order Taylor expansion of its partial derivatives and considering a perturbation parameter η of 1×10^{-16}. The i-th row and j-th column of the approximated Jacobian matrix is evaluated as:

$$J_{ij} = \frac{f_i(x + \eta_j \epsilon_j) - f_i(x)}{\eta_j}, \tag{8}$$

• convergence criterion is defined imposing the error to satisfy the following inequality:

$$\| x_k - \rho_k [J(x_k)]^{-1} f(x_k) \|_2 \leq \| f(x_k) \|_2, \tag{9}$$

• the algorithm stops if convergence is achieved or the maximum number of iterations is reached:

$$\| f(x_{k+1}) \|_2 \leq \delta = 1 * 10^{-10} \qquad \text{or} \qquad k \geq k_{max} = 20, \tag{10}$$

where δ is the required computation tolerance and k_{max} is the given maximum number of iterations.

The first initial approximation must be taken close to the Home Position (463.56 mm from the ground reference frame), where the robot is supposed to be when switched on. For the following times, the initial approximation is chosen as the estimated pose at the previous cycle. The logic of the JFMD algorithm is clarified in the flow chart in Figure 7. Thanks to C language implementation, this routine is able to converge to a solution in relative little iteration, with a mean calculation time of around 50 µs in the actual hardware chosen to control the robot. This performance allows one to use FK as a virtual real-time sensor with sufficient accuracy.

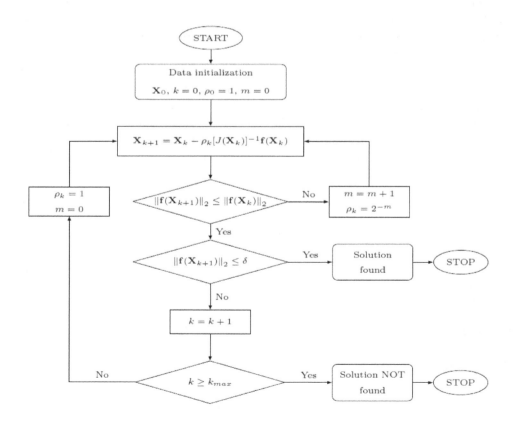

Figure 7. Flow chart of the Jacobian Free Monotonic Descendent (JFMD) algorithm.

2.3. Velocity Analysis and Kinetostatic

In order to determine the relation between the velocity of the TCP and the ones of the joint coordinates, it is necessary to calculate the Jacobian matrix. If all six links are considered together, a compact matrix form coupling the joint velocity vector $\dot{q}$ and the workspace velocity vector w can be defined:

$$\begin{bmatrix} \hat{n}_1^T \hat{u}_1 & \cdots & 0 \\ \vdots & \ddots & \vdots \\ 0 & \cdots & \hat{n}_6^T \hat{u}_6 \end{bmatrix} \dot{q} - \begin{bmatrix} \hat{n}_1^T & \cdots & (b_1 \times \hat{n}_1)^T \\ \vdots & \ddots & \vdots \\ \hat{n}_6^T & \cdots & (b_6 \times \hat{n}_6)^T \end{bmatrix} w = 0,$$

$$[J_q]\dot{q} - [J_{gs}]^{-1}w = 0,$$

and with some algebraic steps:

$$\dot{q} = [J]^{-1}w. \tag{11}$$

The solution of the kinetostatic analysis provides the actuation forces τ_a required to bear the external forces f_{ec} applied to the TCP. Due to the virtual work principle and knowing that the virtual variation of the workspace coordinates is related to the virtual variation of the joint coordinates through the Jacobian matrix $[J]$, the actuation forces can be computed as:

$$\tau_a = -[J]^T f_{ec}. \tag{12}$$

The representation of the way in which forces applied to the robot platform are transmitted to the actuators is obtained considering the unitary hypersphere of forces in the workspace [7]. This unitary hypersphere is transformed into a hyper-ellipsoid in the space of actuation forces. This is a common description used in the robotic field in order to characterise the behaviour of the robot in every point of the working space and it states that, if one of the forces applied to the TCP reaches the maximum value of 1, the other components must be nil. An alternative representation is based on the use of a hyper-cube of unitary semi-side that is transformed into a hyper-polyhedron through the Jacobian matrix, as shown in Figure 8. Moreover, it should be noted that the inverse Jacobian matrix can be split into a translational and a rotational part as presented in [8]. Using this approach, it is possible to note that the rotational components of the inverse Jacobian matrix are dimensional and in fact they correspond to a length. In order to let all the elements of the inverse Jacobian matrix be dimensionless, it is useful to divide the rotational components by a scale factor defined as characteristic length L_c. In this way, a normalized inverse Jacobian matrix is obtained and is used to obtain the maximum actuation force.

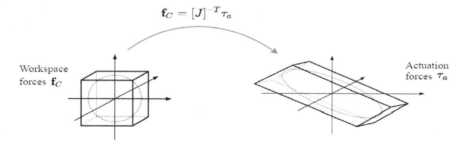

Figure 8. Transformation of the workspace forces unitary hypercube into the actuation forces hyperpolyhedron.

3. Actuation Chain Sizing

A proper sizing of mechanical components and actuating systems requires the knowledge of the most power demanding motion that the machine is required to execute, considering the influence of robot dynamics and carried loads. For the application taken into account, the movement of the end effector is unknown in advance because it depends on the wind effect over the wind turbine and the mathematical model used to define the behaviour of the floating structure. To obtain a scenario of the possible movement of the end effector, it is necessary to consider an approximation of the wave motion of the sea expressed by a combination of six cosinusoidal functions [9,10]. Analysing any combination of a parameter set, which is used to define the sea behaviour, one can obtain a possible end effector motion. The time history of the sea level ε is expressed as:

$$\varepsilon(t) = \rho(t)\cos(\omega t + \varphi(t)),\tag{13}$$

where ρ is the amplitude, ω the pulsation, and φ the phase shift. Both amplitude and phase shift change very slowly in time so it is reasonable to assume they are constant for the whole duration of the simulation and the time history of each DoF is described as:

$$j(t) = A_{0,j} + A_j\cos(2\pi f_j t + \varphi_j) \quad \text{with} \quad j = x, y, z, \alpha, \beta, \gamma,\tag{14}$$

where $A_{0,j}$ is an offset parameter taking into account the initial pose of the robot.

The worst operating conditions may be represented in the case where all frequencies f_j and amplitudes A_j assume their maximum value. However, the intrinsically nonlinear kinematic of the robot may invalidate this assumption. In effect, it is not guaranteed that this is the most demanding case in terms of internal loads, motor torques, velocities and accelerations. Furthermore, a simple co-sinusoidal motion of the end-effector is translated into a periodic motion of the joint coordinates where higher order harmonics with respect to the f_j appear. It is difficult to predict if the energy associated with the higher order harmonics of a specific f_j is bigger than the energy of the harmonics corresponding to a lower f_j.

The four characteristic parameters are thus defined in the following ranges:

- Initial pose $A_{0,j}$: may vary in the range $\pm L_{WSd,j}$;
- Frequency f_j: in the range between f_{min} and f_{max};
- Phase shift φ_j: in the range of $[0, 2\pi]$;
- Amplitude A_j: its maximum range is $[-L_{WSd,j}, L_{WSd,j}]$ and the relation $|A_{0,j} \pm A_j| \leq L_{WSd,j}$ has to be verified to guarantee that the end-effector remains within the boundaries of the desired workspace. The effective range of A_j is obtained by combining the maximum range with the expressed relation, eventually modifying it in the event that $A_{0,i}$ is different from zero.

In order to test any pose in the working volume, a procedure has been designed [11]. Firstly, the workspace has been divided into a finite number of portions. For each of these, an admissible range of motion parameters for each degree of freedom has been chosen. In particular, these ranges have to be discretized in order to obtain a finite number of combinations. Therefore, a set of time histories for the six-DoF end-effector is generated and thereafter the inverse kinematic and dynamic problem solved in order to calculate the motor torques required for each set.

3.1. Multibody Model

The procedure described above needs a mathematical model to be implemented. In particular, a multibody model of the robot has been developed using the commercial software MSC ADAMS 2016 in order to compute the dynamic and kinematic quantities that allow one to properly size mechanical components as well as the actuating system [12]. At the preliminary stage of the project, the inertial properties of the components, that make up the robot, are unknown and must be estimated. The results obtained by the use of the multibody model and the developed procedure are used to refine these values and update the model to subsequently enhance the results accuracy and converge towards the final result.

The developed multibody model is based on a parametric approach in order to be easily integrated into the motor sizing procedure. The first step for the model formulation is the definition of a set of reference frames located in positions that allow one to easily define the inertial properties, joints, applied forces and measurements points. The reference frames created are: a TCP reference frame; six reference frames called B_i in correspondence to the centre of the end-effector joints; six reference frames called A_i in correspondence to the centre of the base universal joints; six reference frames called P_i located at the origin of the guide axes, as shown in Figure 9.

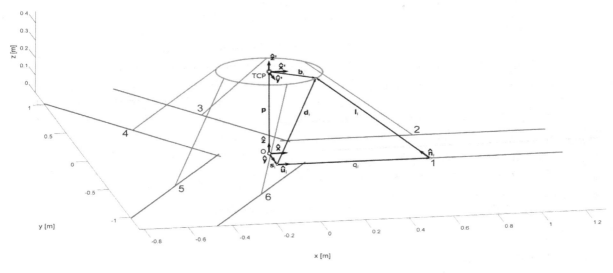

Figure 9. Reference frames configuration, the Cartesian space is represented in meters.

The three main components for each kinematic chain are: platform, links and sliders. The element called platform takes into account the mobile platform, the six spherical joints attached to it, the six axes load cell used to measure the forces and moments exerted by the scaled model of the turbine on the robot. Due to the symmetry, the centre of mass is expected to be located on the z-axis of the TCP reference frame and the principal axes of inertia are expected to be parallel to the axes of the TCP. Links can be represented as cylinders whose centre of mass is located in the middle of the link and whose principal axes of inertia are aligned with those of the A_i reference frame. The elements called sliders take into account the universal joints, joint supports and sliders of the transmission unit. Since these bodies will be subjected to a purely translational motion during simulations, it is sufficient to characterise them with their total mass. This schematization with the equivalent mass and inertial properties of the assembly is possible because the model described in this section is supposed to be rigid. The mechanical and geometrical limits of each joint are represented by locking one or more relative degrees of freedom and giving a limited range of displacement/rotation among the possible ones.

3.2. Monte Carlo Method

Due to the impossibility in the definition of the most demanding task for this application, a statistical approach has been developed, based on the Monte-Carlo Method. This novel approach is implemented as follows:

- Choice of a sufficiently high number M of simulations, each of which has a specific time history for every DoF.
- Definition of the Probability Density Functions (PDF) of the input parameters that define every motion task: it is assumed that, at the initial instant, the TCP may be located with the same probability in any point of the workspace. The PDFs chosen for the parameters describing the motion are respectively: for the initial pose a uniform distribution, for the frequency a uniform distribution, for phase shift a uniform distribution and for the amplitude a Rayleigh distribution.

- Repeated sampling of the chosen PDFs: each parameter is allocated a random value for each simulation. The result is a set of M vectors fully defining a six-DoF motion task to be assigned to the TCP.
- Solution of the IK problem to find the joint coordinates time histories $q(t)$ to be used as inputs for the simulations involving the multibody model.
- Solution of the inverse dynamics problem for each of the M simulations.
- Post-processing: evaluation of a distribution among the M simulations of the parameters of interest.

The results collected are used to size the mechanical components and the actuators. In particular, it is possible to obtain the slider velocity and acceleration, forces exerted by the motors on the sliders, axial forces along the guide and the so-called "Load factor" for the actuator sizing procedure. A summary set of values is required in order to easily compare these quantities with the corresponding limits specified by the manufacturers. These sets are computed through the extraction of the maximum values for each simulation stored in M array; among these, the highest value is extracted. Therefore, the interval from 0 to the highest extracted value is divided into subranges, and, for each of these, the number of occurrences of the M values is assessed. A set of discrete PDFs is obtained and this represents the probability that the maximum value obtained during a simulation is comprised within a certain interval.

3.3. Actuating System Sizing Procedure

The selection of an actuating system requires one to choose both the electric motor and the gearbox unit. In scientific literature, several procedures to size the motor reducer group are available. In this work, the approach proposed in [11,13] is applied. Independently from the procedure used, the motor reducer sizing is based on the checking of the following three relationships:

- Limit on maximum torque:

$$\max |C_m(t)| < C_{m,max} \quad \text{and} \quad d \max |C_m(t)| < C_{t,max},$$

- Limits on nominal torque:

$$C^*_{m,rms} < C_{m,nom},$$

- Limit on maximum speed:

$$\max |\omega_r(t)| < \tau \omega_{m,max} \quad \text{and} \quad \max |\omega_r(t)| < \omega_{t,max}.$$

C_m is the motor torque, $C_{m,max}$ the maximum torque the motor is able to generate, and $C_{t,max}$ is the maximum torque that the transmission unit is able to bear. In addition, ω_r represents the resistant speed computed on the load side, $\omega_{m,max}$ is the maximum speed the motor can reach without damaging its mechanical components, $\omega_{t,max}$ is the maximum speed the slider can achieve without damaging mechanical components. Following the [13] approach, from the power balance and the thermal check inequality, it is possible to define the transmission ratio as follows:

$$\alpha > \beta + [C^*_{r,rms}(\frac{\tau_{rid}}{\sqrt{J_m}}) - \dot{\omega}_{r,rms}(\frac{\sqrt{J_m}}{\tau_{rid}})]^2, \tag{15}$$

where the acceleration factor α and the load factor β are defined as:

$$\alpha = \frac{C^2_{m,nom}}{J_m}; \quad \beta = 2\,[\dot{\omega}_{r,rms}C^*_{r,rms} + (\dot{\omega}_r C^*_r)_{mean}]. \tag{16}$$

The RMS values of torque and acceleration are defined as:

$$C^*_{r,rms} = \sqrt{\frac{1}{t_c} \int_0^{t_c} [C^*_r(t)]^2 \, dt} \quad \text{and} \quad \dot{\omega}_{r,rms} = \sqrt{\frac{1}{t_c} \int_0^{t_c} [\dot{\omega}_r(t)]^2 \, dt}.$$

The acceleration factor α is a motor characteristic that can be evaluated directly from the datasheets, whereas the load factor β depends on the task performed by the actuations system, and, for the case under study, can be calculated using the multibody model. Following the sizing procedure, firstly, the optimal transmission ratio value τ_{opt} is evaluated. Then, this value is compared to the transmission ratio purchasable in the market and the closest is chosen taking into consideration that it is within the available range. In particular, greater attention must be paid to evaluate the maximum velocity required during the task and the relative limit transmission value using the formula:

$$\tau_{lim} = \frac{\max |\omega_r(t)|}{\omega_{m,max}} \quad \max(\tau_{lim}, \tau_{min}) < \tau < \tau_{max}. \tag{17}$$

All the results obtained from the multibody simulations refer to the kinematic and dynamic quantities of the manipulator sliders. In order to transform force into torque and linear velocity or acceleration into angular ones, the following equations have to be used:

$$\dot{\omega}_{r,i} = \frac{\ddot{q}_i(t)}{\tau_{TU}} \quad C_{robot,i}(t) = \tau_{TU} F_{robot,X_i}(t). \tag{18}$$

It is worth noticing that to increase the transmission stiffness and achieve the maximum velocity required, every linear axis of the robot is made up of only a single reduction stage. In particular, the motor is directly connected to the screw by means of a rigid joint and the transmission ratio is the screw lead. Due to the high dynamic performance, stiffness and precision required, recirculating ball screws have been chosen. The double slider configuration for each guide has been selected in order to guarantee a uniform load distribution. The manufacturer's instructions to correctly size these devices have been followed, in particular to the total length of the screw, maximum allowed velocities, forces that the guide can bear and the like. All of this is summarized in Table 3. As concerns the electric motor OMRON (Tokyo, Japan), as well as the heat dissipation check, the maximum torque achievable has been evaluated.

Table 3. Resume of checks done on guides.

Parameter	Value		Limit Value	UM
Length of the guide	1100	>	650	mm
Max velocity	1.67	<	2.66	m/s
Max acceleration	28.5	<	50	m/s^2
F_x	1385	<	12,250	N
F_y	1385	<	69,600	N
F_z	1385	<	69,600	N
M_x	174	<	3028	Nm
M_y	174	<	2290	Nm
M_z	174	<	2290	Nm

Figures 10 and 11 show the maximum torque distribution and maximum rotation speed obtained during simulations. Once fixed a 99% coverage threshold, the corresponding values are extracted and they are used as guidelines in the components' choice. The most restricting limit is imposed by transmission units: the Rollon ballscrew drive (Vimercate (MB), Italy) chosen satisfies the maximum torque test in 94.7% of the cases with Rayleigh approach and in 89.4% with the other one. Even if the coverage is not 99%, the choice has been made to find a suitable trade-off between performances and size. As for maximum torque analysis, the most restricting limit for rotation speed is imposed by the linear transmission unit, but, since the selected transmission satisfies more than 99% of the cases, this test is passed.

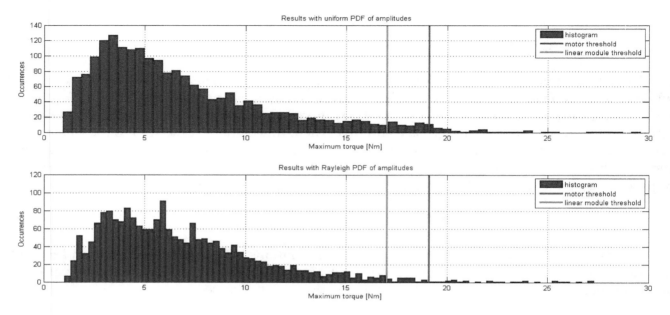

Figure 10. Maximum torque distribution along all the simulations.

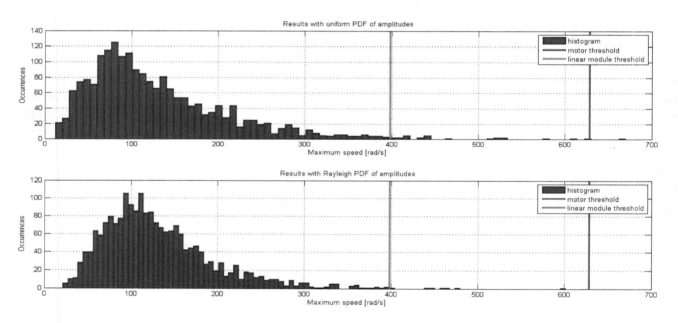

Figure 11. Maximum angular velocity distribution along all the simulations.

4. Mechanical Design and Component Sizing

A design process is always iterative: an initial model, which satisfies the preliminary parameters required, is firstly created, thereafter each part of this model is refined and improved in order to enhance its dynamic characteristics and reduce mass and costs of the whole system. In this process, the multibody model plays an important role: it is used many times in an iterative way to analyse the dynamic behaviour of the system after every single modification in order to check the mechanical components and actuation system. For the development of a simulator, it is necessary to realize two kinds of structural analyses: the first one is a static analysis for which the focus is to determine the stress and strain of any mechanical components. The second analysis is a dynamic one performed to obtain the modal behaviour of the system.

4.1. Robot Description and Overview

In order to clarify the results obtained with static and dynamic analysis, a complete overview of the "as built" project is necessary, Figure 12a. The entire system can be divided into two main parts: the Hexafloat manipulator and the auxiliary systems. The first part is made up of "Grounded elements" (all the fixed supporting structure and the power and actuation units), "Joints", "Links" and "Platform", whereas the auxiliary systems include: "Lifting system" (a tool used to manage the rest task of the system), "Energy chain" (device that houses cables and the like) and "External Sensors" (sensors for monitoring the system during the operating tasks).

Hereafter is a more detailed description of every component and technical solution adopted.

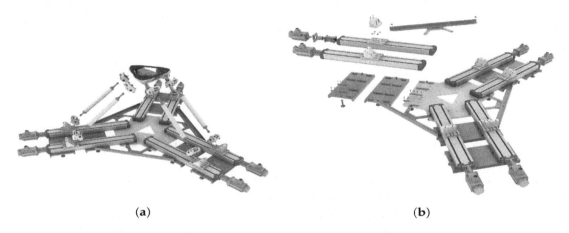

(a)   (b)

Figure 12. Components. (**a**) Hexafloat exploded; (**b**) parts on the ground partially exploded view.

4.1.1. Grounded Components

This is the framework and gives stability to all the moving parts, Figure 12b. The linear actuation systems are fixed onto it. Therefore, this assembly should be stiff enough to bear static and dynamic loads. The *Central plate*, made of aluminium, is used to provide the correct orientation and position of the linear guides through calibrated machining. The linear actuators are grouped into two by two and each couple is oriented with an angle of 120 ° with respect to the other. This leads to a radial symmetry of the machine. Each couple of linear actuators is fixed to the *Central plate*. To provide further stiffness to the system, avoid undesired tilting and relative displacement between the two couples of guides, three "K" links constrain the *Central plate* with the "C" shaped supports over which the actuation system is mounted. *Aluminium joint holders* are used to couple the *Joints* with the linear actuators. Through the optimization process and the solving of the kinematic closure equations, when the robot is in the *Home Position*, the direction of each *Link* is defined and the *Joints holder* shape and inclination designed. The whole framework will be placed in the wind tunnel using 22 levelling elements to distribute the weight uniformly and maintain the guides on the same plane, without misalignments.

4.1.2. Joints

The PUS kinematic chain on the basis of which the manipulator is realized is made up of an actuated prismatic joint followed by a double revolute joint (universal joint) and a spherical one mounted on the *Platform*. Placing a universal joint and a revolute one so that the rotation axis of the latter passes through the intersection of the universal joint rotation axes, it is possible to obtain the same numbers of DoFs that the spherical joint has. Therefore, the same universal joint used in the lower part of the kinematic chain is mounted on the *Platform*, and the rotational degree of freedom is directly realized within the *Link*. In order to obtain the stiffness, precision and mobility required, a custom joint solution has been realized and they are shown in Figure 13a,b. The main feature of such

Joint is the possibility to have a cone shaped motion range with a semi-angular aperture of 45 ° around the normal direction, designed and assembled to have zero backlash.

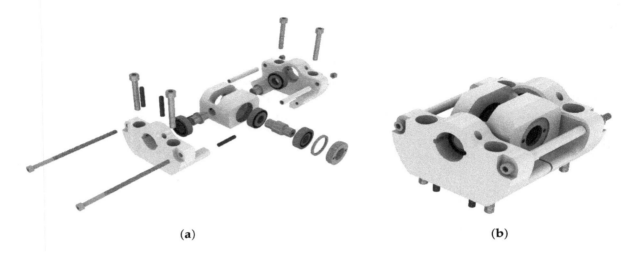

(a)   **(b)**

Figure 13. Custom joint. (**a**) joint exploded view; (**b**) joint assembly.

Each joint is made up of a couple of *Half shells* connected to an *Inner block* through two roller bearings and two *Support shafts*. These components are tightened by two screws and aligned by two calibrated dowel pins. In this way, the *Inner block* has a relative movement with respect to the shell, the further DoF is provided by another two roller bearings that support the *Inner shaft* pinned to the link. All these inner components are packed through a *Distance ring* and a *Closing ring*. *Half shells* and *Inner block* are made of aluminium, while other components are made of steel. The *Joint* is fixed to the *Joints holder* through four screws and aligned with two calibrated dowel pins.

4.1.3. Links

The *Links*, Figure 14, connect the *mobile platform* to the fixed one. These bear the pay load made up of the mobile platform, *RUAG (RUAG Aviation - Aerodynamic, Schiltwald, Switzerland) load cell*, scaled wind turbine and sensors. The RUAG six-DOF weight scale, model W192–6I, is dimensioned for the following forces and moments: $F_x = \pm 1500$ N, $F_y = \pm 1000$ N, $F_z = \pm 5000$ N, $M_x = \pm 500$ Nm, $M_y = \pm 1000$ Nm and $M_z = \pm 600$ Nm. Due to the complexity of this device, it is necessary to use a calibration matrix capable of taking into account not only the weight scale deformation but also the working temperature. This calibration matrix, provided by the manufacturer, follows quality procedures in use at the wind tunnel and it is available at the internet address of the Life50+ project. The *Link* is composed of:

- Lower rod, made of steel, at one side, directly connected to the *Inner shaft* of the *Joint* assembled on the *grounded components*, the other side is used to pack up the bearing group between a mechanical stop and nuts;
- Bearing case, made of steel, which houses the bearings that provide the rotational DoF along the *Link* axis, and is connected to the *Leg pipe* by six screws;
- Leg pipe, a hollow cylinder made of aluminium;
- Distance washer, made of steel, it allows the regulation of the total *Link* length through a threaded connection with the *Upper rod*. It is also connected to the *Leg pipe* by six screws;
- Upper rod, made of steel, the final component connected to the *Inner shaft* of the *Joint* assembled on the *Platform*;

Figure 14. Link exploded view.

4.1.4. Platform

The *Platform*, Figure 15a, is realised in order to define the right position of the joints and the six-DOF weight scale. It is made up of the following elements:

- Bottom plate, made of aluminium, necessary to sustain and distribute the load and it is the frame of the *Platform*;
- Three Angular joints holder, made of aluminium, needed to guarantee the correct orientation angle of the *Joints*;
- Top plate, made of carbon fibres, to give added rigidity to the structure;
- RUAG 6-axis load cell, to measure loads exchanged between the machine and the wind turbine.

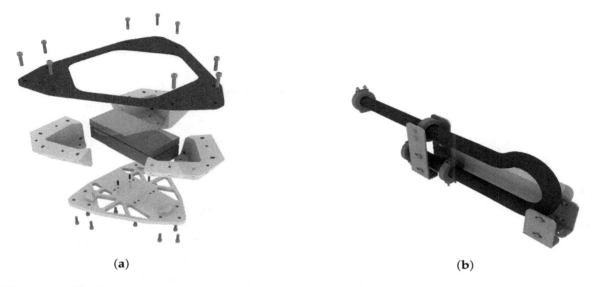

(a)   (b)

Figure 15. Platform and energy chain. (**a**) platform exploded view; (**b**) energy chain assembly.

4.1.5. Auxiliary System

Two auxiliary systems have been designed: an *Energy chain* (Figure 15b) and a *Lifting system* (Figure 16b). The Hexafloat machine is designed to stay below the floor of the wind tunnel when not in operation. In this configuration, the machine has to overcome a singularity to reach Home Position and a *Lifting system* helps the robot to pass this critical point in the rise and return phases. This system can be schematically modelled as in Figure 16a as an isosceles three-hinged arc, in which hinge A is placed on ground, B on the *Moving carriage* and C at the central part of the *Three points beam*, thus allowing a vertical movement of the *Three points beam* top end (point D). Furthermore, a four-link mechanism is coupled to this system in order to keep it parallel to the ground along the whole stroke of the *Coupling platform*, mounted on top of the lifting mechanism. The *Moving carriage* is driven by the motor through a trapezoidal screw, which converts motion from angular to linear.

Measuring and actuation devices will be mounted on the *Platform* and on-board the scaled wind turbine, and these instruments need to be fed by electric energy and have to be connected to a controller in order to guarantee a real-time data exchange. Therefore, a cable housing system is needed in order to protect cables, maintain a good flexibility while following the *Platform* movements, without interfering with these. The system is made up of the following elements: four "L" plates supports, made of steel;

two lateral connecting plates, made of aluminium; two mounting brackets with strain relief mounted, respectively, on the *Platform* and on *Parts on ground*; an intermediate mounting bracket; a mounting sliding bracket and a flexible chain.

(a)   (b)

Figure 16. Lifting system. (**a**) lifting system scheme; (**b**) lifting system assembly.

4.2. Static Analysis

Through the results obtained from the multibody analysis, one can determine the most critical load acting on the system. In particular, the interest is on the force that must be borne by the links because, through this, the remaining parts of the system are loaded. For safety reasons, an overestimated value of 2500 N is assumed both in traction and compression for every link. The static analysis has been performed using a commercial FEM solution and several models have been studied in order to take into consideration both single elements and their subassembly. Taking into account the final configuration, one can state:

- Grounded components: The symmetry of the assembly allows one to take into account only one third of the frame. Moreover, only the aluminium profile of the linear actuator is taken into account because it bears the vertical loads. A "worst" vertical load (i.e., 2500 N) is applied on each guide and positioned with different combinations of the total actual stroke of the sliders (i.e., 0%, 25%, 50%, 75%, 100%). Local stresses and strains never exceed the limit values, Figure 17a.
- Joint: bearings have been substituted with rigid components to simplify the assessment. Different configurations of angle are tested both as regards compression and traction with an applied load of 2500 N, Figure 18a,b. Only the rod shows visible deformations and stresses. In order to better investigate the behaviour of internal components, other analyses have been carried out and no problems have been reported by the results:

  - Inner Block: a load of 2500 N is split into two equal loads, each one acting on a set of bearings. Maximum stress registered is 85 MPa, which is far below the admissible stress of 250 MPa, Figure 19a.
  - Support shaft: a load of 1250 N is applied for simulating the presence of two Support blocks per Joint. The test is effected by loading one end of the shaft and maintaining the other one fixed, Figure 19b.
  - Inner shaft: a load of 2500 N is applied in the midspan and both ends are pinned in order to simulate the presence of the two bearings, Figure 20a.

- Link: an axial load of 2500 N is applied and only traction and compression have been tested. Bearings are substituted with rigid parts, Figure 20b. Considerable stress has been noted on the upper and lower rods, but their values is well below the critical one, Figure 21a,b.
- Platform: two horizontal forces are applied at different heights on the TCP. The first of 200 N located at 1.0 m and the second of 100 N at 1.5 m corresponding, respectively, to equivalent

inertial load and aerodynamic thrust due to the pay load. These loads are borne very well by the structure both in terms of stress and displacement, Figure 17b.

- Lifting system: this device has to generate a total lift of 150 mm in a time span of about 15 s with a cycloidal motion curve and must also bear the weight of all the moving parts of the robot. This system has been developed in order to help the robot to overcome the singularity configuration during the start and stop procedures. The analysis reveals maximum values occurrence at the initial phase of the rise and FEM analysis ensures that stress remains below the critical values.

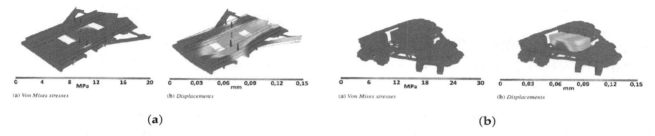

Figure 17. Movable and fixed platform analysis. (**a**) parts on ground static analysis; (**b**) platform static analysis.

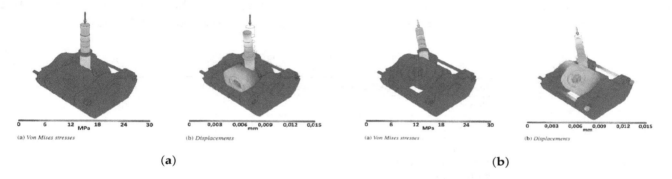

Figure 18. Joint analysis. (**a**) joint static analysis; (**b**) joint static analysis, inclined configuration.

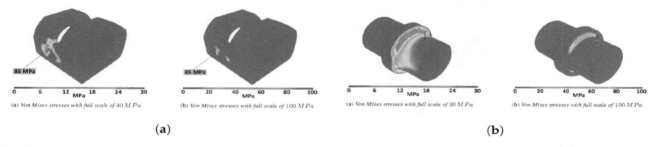

Figure 19. Inner block and shaft analysis. (**a**) inner block static analysis; (**b**) support shaft static analysis.

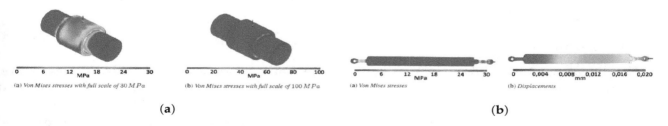

Figure 20. Inner shaft and link analysis. (**a**) inner shaft static analysis; (**b**) link static analysis.

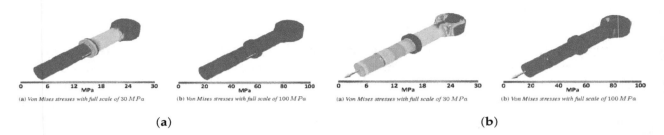

(a)   (b)

Figure 21. Lower rod analysis. (**a**) lower rod compression analysis; (**b**) lower rod traction analysis.

4.3. Purchased Components' Sizing

Final choices regarding motors, linear actuators and the energy chain are listed.

4.3.1. Motors

The selected model is the OMRON R88M-K2K030F-BS2, whose main characteristics are collected in Table 4(a). The *Lifting system* motor, required to bear the torque increment given by the low efficiency of the trapezoidal screw adopted, is the OMRON R88M-K20030T-BS2, the main characteristics of which are reported in Table 4(b). Both motor models have an auxiliary brake for safety reasons and an encoder to allow one to have a control of position and velocity. For the 2 KW motor of the robot, the encoder is a quadrature incremental encoder with a maximum resolution of 4,194,304 cnt/rev.

Table 4. Motors' characteristics.

(a) OMRON R88M-K2K030F-BS2 Main Characteristics		
Characteristic	Value	MU
Tension	400	V
Nominal Power	2000	W
Nominal Torque	6.37	Nm
Maximum Torque	19.1	Nm
Nominal velocity	3000	rpm
Maximum velocity	5000	rpm
(b) OMRON R88M-K20030T-BS2 Main Characteristicss		
Characteristic	Value	MU
Tension	230	V
Nominal Power	200	W
Nominal Torque	0.64	Nm
Maximum Torque	1.91	Nm
Nominal velocity	3000	rpm
Maximum velocity	6000	rpm

The encoder signal is processed in order to obtain a lower resolution of 131,072 cnt/rev. The resulting linear resolution of 3,276,800 cnt/m is obtained considering a lead of 0.04 m/rev for the ball-screw linear axis. For the lifting system motor, the maximum encoder resolution is 131,072 cnt/rev. The modified used resolution is of 13,107.2 cnt/rev, thus obtaining the same linear resolution of the robots axis due to a smaller lead of 0.004 m/rev.

4.3.2. Linear Actuator

The linear actuator chosen is the model TH145 SP4, produced by ROLLON and whose characteristics are reported in Table 5(a). The following customizations are required: screw lead of 40 mm per revolute, to allow the respecting of required performances; two calibrated centring holes on its lower side to correctly assemble the central plate; two calibrated centring holes on the external carriage to allow the correct positioning of the Joint holder.

Table 5. Other components' characteristics.

(a) ROLLON TH145 SP4 Main Characteristics

Characteristic	Value	MU
Total length	1100	mm
Width	145	mm
Thickness	85	mm
Screw diameter	20	mm
Number of charts	2	-
Total system mass	30	kg

(b) Igus Triflex TRE.40 Main Characteristics

Characteristic	Value	MU
Total length	1600	mm
Number of chain links	116	-
External diameter	43	mm
Curvature radius	58	mm
Maximum single cable diameter	13	mm
Maximum relative rotation	±10	°

4.3.3. Other Components

The selected energy chain is Igus (Colonia, Germany) Riflex TRE.40 whose main characteristics are reported in Table 5(b). Tapered roller bearings chosen both for the links and joints are the INA FAG 30202-A. These support both axial and radial loads, thus avoiding the usage of more than one bearing. KTR-TOOLFLEX20M motor coupling is chosen. Due to the reduced space available for the *Lifting system*, a more compact model of bearing INA FAG 3000-B-2RS-TVH has been chosen. It is a double row angular contact ball bearing, which can bear both axial and radial loads. KTR Rotex 19/92Sh-A/2.1-ϕ11/2.0-ϕ8 motor coupling has been chosen for the *Lifting system* and it provides the required performance and encumbrance constraints.

4.4. Modal Analysis

In this section, the approach followed to assess the correctness of the overall design, in terms of dynamic behaviour, is reported. More specifically, the goal is to verify that the first normal modes of the coupled structure coincide with those of the turbine only, in order to minimize the dynamic effects of the robot. Nevertheless, the modal behaviour of a robot strongly depends on the specific pose of the end effector, since the mass and stiffness distribution change according to the pose.

The normal modes and their associated frequencies are computed in order to verify that the frequencies in the first normal mode were high enough. In particular, they must be higher than the well defined frequency range for all the robot poses in to the working volume. The frequency range depends on the dynamic behaviour of the scaled turbine.

To analyse the dynamic response of the manipulator, the entire workspace was discretized, and the trend of the frequency corresponding to the first normal mode over the entire workspace was mapped on specific planes that intersect the robot workspace. The procedure is listed below:

- Identification of planes that intersect the workspace.
- Identification of a grid of equally spaced points on each plane.
- Discretization of pitch, roll and yaw angles describing the end-effector orientation:

  - Three roll angles: $-5°, 0°, +5°$,
  - Five pitch angles: $-8°, -4°, 0°, +4°, +8°$,
  - Three yaw angles: $-3°, 0°, +3°$.

- Modification of the pose of the robot in order to have the end-effector placed in correspondence to each point of the grid and exploring all the possible orientations.
- In correspondence to each pose, linearization of the flexible virtual model and computation of the frequency associated to the first normal mode.
- For each point of the grid, recording the lowest value of frequency among the ones obtained by changing the orientation angles.

The final result is a set of maps, one for each intersecting plane, which show the trend of the lowest frequencies regardless of the orientation of the robot, as shown in Figure 22. To perform these sets of simulations, a simplified design of the robot was created. Lower simplified joints are tied to the ground applying spherical pins and the simulation provides the first three eigenmode frequencies. Links can be regarded as the main cause for a possible deterioration of the behaviour of the coupled system. The main deformations are concentrated on links and an advanced multibody model has been developed in the "Adams" environment, Figure 23, whose links are made of two extremity elements and an intermediate one characterised by aluminium properties. The results given by the two numerical environments are reported in Table 6.

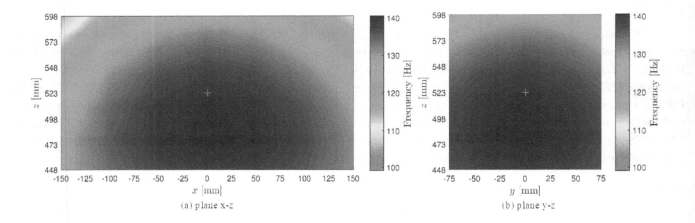

Figure 22. Modal frequencies maps.

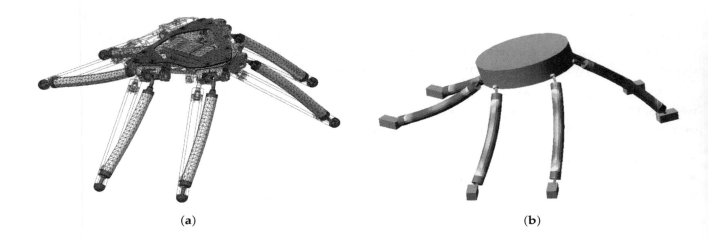

Figure 23. Numerical modal analysis. (a) Inventor 1st mode of vibration; (b) Adams 1st mode of vibration.

Table 6. Eigenmodes frequencies in Home Position.

(a) Eigenmodes Frequencies Calculated through Inventor	
Eigenmode index	Frequency
I	151 Hz
II	154 Hz
III	210 Hz

(b) Eigenmodes Frequencies Calculated through Simplified Adams Model	
Eigenmode index	Frequency
I	169 Hz
II	171 Hz
III	222 Hz

A frequency response analysis has been performed on the wind turbine and on the coupled system made by the wind turbine and the Hexaslide. The aim of this simulation is to check the Hexaslide is not influencing the frequencies of the turbine's mode of vibrating, thus making the presence of Hexaslide negligible at least from a mechanical point of view.

A swept sine load of 100 N amplitude between 0.1 Hz and 250 Hz has been applied to the rotor centre in wind (x) direction and the displacements and rotation have been measured. This Force is able to well excite the first and most relevant eigenmodes of the turbine.

The results are reported in Figure 24. The first two wind turbine eigenfrequencies are preserved even in the coupled system, while two peaks appear between 75 Hz and 100 Hz.

It can be concluded that, in the range of 0–60 Hz, a robot's dynamics does not influence the turbine eigenfrequencies.

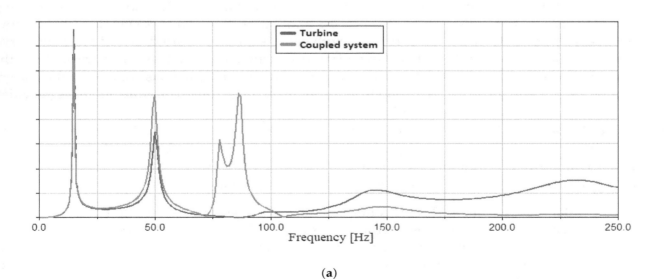

(a)

Figure 24. *Cont.*

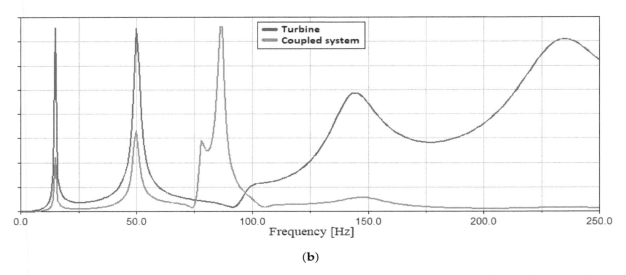

(b)

Figure 24. Frequency Response analysis of Turbine and Hexaslide with turbine. (**a**) displacement along $\hat{x}_O$ direction; (**b**) rotation around the $\hat{y}_O$ axis.

5. Control Architecture and Electronics

A simplified scheme of the electric panel configuration is reported in Figure 25. The core of the electrical panel is represented by Power PMAC, the property of Delta Tau Data Systems (Chatsworth, CA, USA). There are essentially two zones: one with AC voltage, represented by solid black and blue lines, powered by a 380 V AC line and brings it to the Power PMAC fed at 230 V and to the power circuits of the servo amplifiers fed at 380 V, and the other one with 24 V DC voltage, downstream of the 24 V power supply, which is represented by the solid and dashed green lines. This circuit provides power to the safety relay, limit switches and proximity sensors and to the Beckhoff modules for Ether-CAT communications. Red lines represent the transmission of data between different components.

- **EtherCAT modules**: the Beckhoff EtherCAT module EK1100 is connected to two EL1008 modules, each of them providing eight digital inputs, and to one EL2008 module that makes available eight digital outputs.
- **Servo amplifiers**: their main function is to power the motors according to the signals coming from the motion controller. They also process and gather the feedback signals of the motors' encoders to bring them to Power PMAC. Both signals are transported to and from the servo amplifier into a single cable that is then split into an actual terminal board to be brought to the correct connectors of Power PMAC. Each servo amplifier has the main task of closing the current and phase commutation feedback loop for the motor, starting from the torque reference provided by the Power PMAC motion controller.
- **Power PMAC**: this is the core of the electrical architecture, it is a general-purpose embedded computer with a built-in motion and machine-control application. It also provides a wide variety of hardware machine interface circuits that allow the connection to common servo and stepper drives, feedback sensors, and analogue and digital I/O points.

The modular rack is the most flexible configuration, since it permits the user to choose which CPU card, digital or analogue I/O card, axis interface cards and the like to use in the system. Power PMAC can handle all of the tasks required for machine control, constantly switching back and forth between the different tasks thousands of times per second. On this powerful controller, the main control software of the robot has been designed and implemented. Figure 26 shows the general overview of the software architecture. In particular, it is structured so that a primary and a secondary states machine manages the principal functionality of the system, among which the management of the logic state (start, stop, homing, jog, and the like) and the functioning of the machine (managing exceptions, motion programs, debug towards the user). The Position-Based-Admittance-Control (PBAC) control scheme used to close the position loop is highlighted in light blue. Further details in this regard are set out in [14].

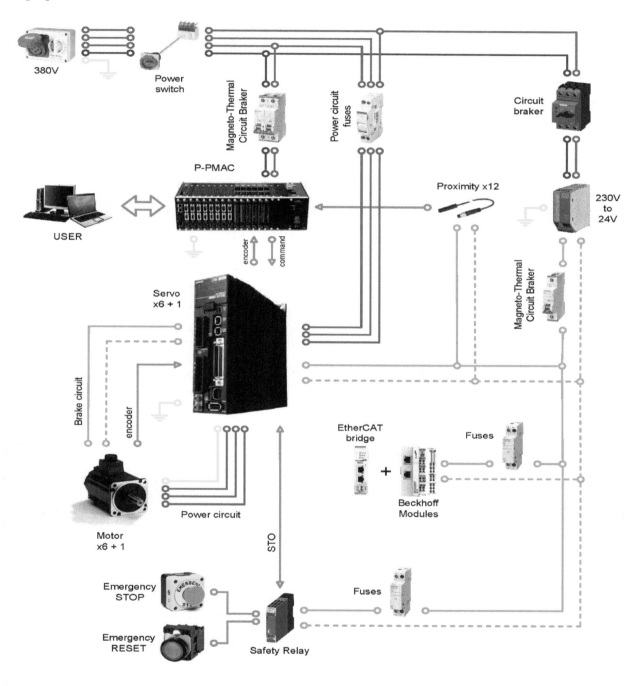

Figure 25. Simplified scheme of an electrical layout.

The lowest level of the control is constituted by the position and velocity servo loops, giving the analogic torque output reference for the seven servo actuator. Each high level control modality developed has been designed to respect real-time performances desired as well as safety, using advanced tools such as position based admittance control, buffering with time-based control, motion look-ahead for smooth blending, fast C written nonlinear FK and demanding algebra, workspace boundaries check with controlled dynamics on the limits [15], and acceleration saturation with workspace reference tracking check. For more details about control architecture and algorithms, refer to [14,15].

The Human–Machine Interface (HMI) designed allows full control by the user of every significant functioning parameter, simplifying state change and enhancing safety. The HMI communicates with Power PMAC by means of Telnet communication protocol and is completely written in C# language.

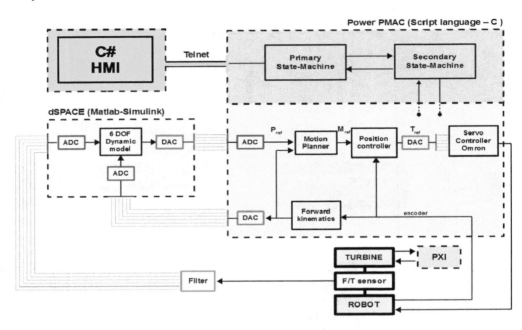

Figure 26. Complete control scheme.

6. Conclusions

The experimental verification of the dynamic response of the robot is currently being finalized. The main control features have been tested on a scaled model of the manipulator with optimum results [14,15]. The full scale is now fully operative and its behaviour is under test. For the sake of completeness in Figure 27, the full scale system equipped with accelerometers during the modal analysis campaign is reported. From the first broad results, it appears confirmed that the robot will not interfere, from a dynamic point of view, with the wind turbine scale model [3].

Further experimental modal analysis is staged for the next months, to fully explore the performances not only of the complete mechanical system, but of the whole machine under control influence. First verifications of control performances have already given satisfying results, but further refinement is done every day to an optimized controlled system behaviour in every required operative condition.

This paper shows the design methodology of the HexaFloat system, a six-DoF robot for wind tunnel hybrid testing of floating offshore wind turbines. This setup consists of a parallel kinematic robot, "HexaFloat", designed and developed to test the dynamics of floating offshore wind turbine concepts, selected within LIFES50+ project, at the Politecnico di Milano wind tunnel, through a hybrid methodology which combines HIL, in real-time measurements (i.e., aerodynamic forces on the wind turbine scale model) and computations (i.e., hydrodynamic forces on platform). This represents the complementary test approach, with respect to the one developed at the Sintef Ocean basin. The final test campaign for LIFES50+ project is staged for July 2018. The complete HIL setup is currently under testing for performance verification, disturbance influence analysis, control of final refinements and tuning and controllers' communication optimization.

Figure 27. Hexaslide experimental setup.

Author Contributions: H.G. and F.L.M. conceived the devices and the sizing tools used; G.R. and M.P. designed the system; F.L.M. developed the control software of the devices; F.L.M., tested the system. H.G. and F.L.M. wrote wrote the paper.

References

1. Sauder, T.; Chabaud, V.; Thys, M.; Bachynski, E.E.; Sæther, L.O. Real-time hybrid model testing of a braceless semi-submersible wind turbine: Part I—The hybrid approach. In Proceedings of the 2016 35th International Conference on Ocean, Offshore and Arctic Engineering, Madrid, Spain, 17–22 June 2016; American Society of Mechanical Engineers: New York, NY, USA, 2016; pp. V006T09A039.

2. Bayati, I.; Belloli, M.; Ferrari, D.; Fossati, F.; Giberti, H. Design of a 6-DoF robotic platform for wind tunnel tests of floating wind turbines. *Energy Procedia* **2014**, *53*, 313–323. [CrossRef]

3. Bayati, I.; Belloli, M.; Bernini, L.; Giberti, H.; Zasso, A. Scale model technology for floating offshore wind turbines. *IET Renew. Power Gener.* **2017**, *11*, 1120–1126. [CrossRef]

4. Giberti, H.; Ferrari, D. A novel hardware-in-the-loop device for floating offshore wind turbines and sailing boats. *Mech. Mach. Theory* **2015**, *85*, 82–105. [CrossRef]

5. Fiore, E.; Giberti, H. Optimization and comparison between two 6-DoF parallel kinematic machines for HIL simulations in wind tunnel. *MATEC Web Conf.* **2016**, *45*, 04012. [CrossRef]

6. Ferrari, D.; Giberti, H. A genetic algorithm approach to the kinematic synthesis of a 6-DoF parallel manipulator. In Proceedings of the 2014 IEEE Conference on Control Applications (CCA), Juan Les Antibes, France, 8–10 October 2014; pp. 222–227.

7. Merlet, J. Jacobian, manipulability, condition number, and accuracy of parallel robots. *J. Mech. Des.* **2006**, *128*, 199–206, doi:10.1115/1.2121740. [CrossRef]

8. Legnani, G.; Tosi, D.; Fassi, I.; Giberti, H.; Cinquemani, S. The "point of isotropy" and other properties of serial and parallel manipulators. *Mech. Mach. Theory* **2010**, *45*, 1407–1423, doi:10.1016/j.mechmachtheory. 2010.05.007. [CrossRef]

9. Vinje, T. The statistical distribution of wave heights in a random seaway. *Appl. Ocean Res.* **1989**, *11*, 143–152, doi:10.1016/0141-1187(89)90024-2. [CrossRef]

10. Xu, D.; Li, X.; Zhang, L.; Xu, N.; Lu, H. On the distributions of wave periods, wavelengths, and amplitudes in a random wave field. *J. Geophys. Res. C* **2004**, *109*.10.1029/2003JC002073. [CrossRef]

11. Fiore, E.; Giberti, H.; Bonomi, G. An innovative method for sizing actuating systems of manipulators with generic tasks. *Mech. Mach. Sci.* **2017**, *47*, 297–305, doi:10.1007/978-3-319-48375-7-32. [CrossRef]

12. Fiore, E.; Giberti, H.; Ferrari, D. Dynamics modeling and accuracy evaluation of a 6-DoF Hexaslide robot. In *Nonlinear Dynamics*; Springer: Berlin, Germany, 2016; Volume 1, pp. 473–479.

13. Giberti, H.; Cinquemani, S.; Legnani, G. A practical approach to the selection of the motor-reducer unit in electric drive systems. *Mech. Based Des. Struct. Mach.* **2011**, *39*, 303–319, doi:10.1080/15397734.2011.543048. [CrossRef]

14. La Mura, F.; Todeschini, G.; Giberti, H. High Performance Motion-Planner Architecture for Hardware-In-the-Loop System Based on Position-Based-Admittance-Control. *Robotics* **2018**, *7*, 8. [CrossRef]

15. La Mura, F.; Romanó, P.; Fiore, E.; Giberti, H. Workspace Limiting Strategy for 6 DOF Force Controlled PKMs Manipulating High Inertia Objects. *Robotics* **2018**, *7*, 10. [CrossRef]

13

Hexapods with Plane-Symmetric Self-Motions

Georg Nawratil

Institute of Discrete Mathematics and Geometry, Vienna University of Technology,
Wiedner Hauptstrasse 8-10/104, 1040 Vienna, Austria; nawratil@geometrie.tuwien.ac.at;

Abstract: A hexapod is a parallel manipulator where the platform is linked with the base by six legs, which are anchored via spherical joints. In general, such a mechanical device is rigid for fixed leg lengths, but, under particular conditions, it can perform a so-called self-motion. In this paper, we determine all hexapods possessing self-motions of a special type. The motions under consideration are so-called plane-symmetric ones, which are the straight forward spatial counterpart of planar/spherical symmetric rollings. The full classification of hexapods with plane-symmetric self-motions is achieved by formulating the problem in terms of algebraic geometry by means of Study parameters. It turns out that besides the planar/spherical symmetric rollings with circular paths and two trivial cases (butterfly self-motion and two-dimensional spherical self-motion), only one further solution exists, which is the so-called Duporcq hexapod. This manipulator, which is studied in detail in the last part of the paper, may be of interest for the design of deployable structures due to its kinematotropic behavior and total flat branching singularities.

Keywords: hexapod; self-motion; spatial symmetric rolling; plane-symmetric motion; Duporcq manipulator

---

1. Introduction

In planar kinematics, the instantaneous pole P traces the so-called fixed/moving polode in the fixed/moving system during the constrained motion of a given mechanism. It is well known that this motion can also be generated by the rolling of the moving polode ϕ along the fixed polode ϕ_0 without sliding. If the polodes are symmetric with respect to the pole tangent t, then the motion is called planar symmetric rolling (cf. Figure 1, left). In 1826, this motion was first (with the exception of the already known symmetric circle rolling yielding the limacons of Pascal) studied by Quetelet [1], who pointed out the following property (cf. [2]): *The path x of a point X under this special planar motion can be generated by the reflexion of a point X_0 of the fixed system on each tangent of ϕ_0.* This can also be reformulated as follows: *x can be obtained by a central dilation with center X_0 and scale factor 2 (i.e., central doubling) of X_0's pedal-curve f with respect to ϕ_0.* A detailed study of the planar symmetric rolling was done by Bereis [3], Bottema [4] and Tölke (cf. [2] and the references given therein).

The spherical counterpart of this motion is called spherical symmetric rolling and was extensively studied by Tölke in a series of papers, which are summarized and referenced in [2]. The spherical version of the above given characterization also holds true for the spherical symmetric rolling (cf. Figure 1, right).

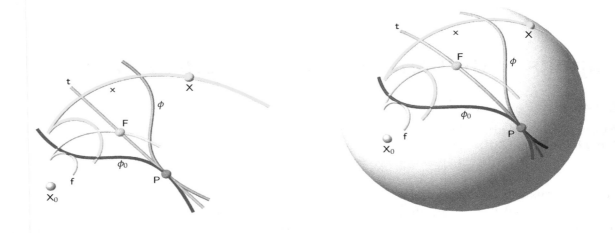

Figure 1. Sketch of the planar symmetric rolling (**left**) and the spherical symmetric rolling (**right**). The pedal-point of the fixed point X_0 with respect to the pol-tangent t is denoted by F.

From another perspective, a planar/spherical symmetric rolling can also be generated by reflecting the fixed system in a 1-parametric continuous set of lines/great circles. This point of view is of importance for the spatial generalization of symmetric rollings, which can be done in multiple ways:

1. Darboux noted in [5] (No. 61) a 2-parametric spatial motion, which is generated by the rolling of a moving surface Φ on an indirect congruent fixed surface Φ_0. It also holds that the path-surface of a point X can be generated by the reflexion of a point X_0 of the fixed system on each tangent-plane of Φ_0; for example, the path-surface can be obtained by a central doubling of X_0's pedal-surface with respect to Φ_0's tangent-planes.

2. Krames [6] considered the so-called line-symmetric motion as the 1-parametric spatial analogue of the planar/spherical symmetric rolling. These motions are obtained by reflecting the moving system in a 1-parametric continuous set of lines, which form the so-called *basic surface* Γ (cf. Figure 2, left). Krames reasoned this by the fact that the path x of a point X under a line-symmetric motion can be generated by the reflexion of a point X_0 on each generator g of Γ; for example, x can be obtained by a central doubling of X_0's pedal-curve f with respect to Γ's rulings. However, it should be pointed out that Γ differs from the fixed axode Φ_0 (generated by the central tangents of Γ). However, Φ_0 and the moving axode Φ are at each time instant symmetric with respect to the axis p of the instantaneous screw, which is in general not an instantaneous rotation. For further details and references on this motion type, please see [7,8] (§7 of Ch. 4) and [9].

3. It is astonishing that neither Tölke [2] (Section 3.1) nor Krames [6] (p. 394) mentioned the more apparent generalization by reflecting the fixed system in a 1-parametric continuous set of planes. Less attention was paid to these so-called *plane-symmetric motions* in the literature until now. We summarize the known results in the next section.

Remark 1. *Note that the term plane-symmetric motion was also used in [10] (§3.3) for a superset of the above described motions, which is characterized by the sole property that "the same equation describes the motion and its inverse, but with respect to reference systems that are a reflection of each other". In order to avoid confusions, we point out that we do not mean this superset by using this wording.*

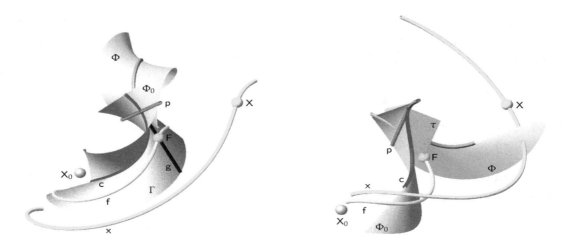

Figure 2. Sketch of the line-symmetric motion (**left**) and the spatial symmetric rolling (**right**). For the illustrations, the basic surface Γ of the line-symmetric motion and the fixed axode Φ_0 of the spatial symmetric rolling have been chosen as tangent-surfaces of a straight cubic circle c. F denotes the pedal-point of the fixed point X_0 with respect to (left) the generator g of Γ and (right) the tangent-plane τ along the instantaneous axis p of rotation, respectively.

1.1. Review on Plane-Symmetric Motions

The basic properties of this motion type are reported in [8] (§8 of Ch. 4). Given is a 1-parametric continuous set of planes $\tau(t)$, where the parameter t can be seen as time. By reflecting the fixed frame $\mathfrak{F}_0$ on the plane $\tau(t)$, we obtain the pose $\mathfrak{F}_0^t$ of the plane-symmetric motion.

Let us consider to infinitesimal neighboring poses $\mathfrak{F}_0^t$ and $\mathfrak{F}_0^{t+\Delta t}$ of the plane-symmetric motion. Now, one can transform $\mathfrak{F}_0^t$ into $\mathfrak{F}_0^{t+\Delta t}$ by a reflexion on $\tau(t)$ followed by a further reflexion on $\tau(t + \Delta t)$. It is well known that this is a pure rotation about the line of intersection of $\tau(t)$ and $\tau(t + \Delta t)$. Moreover, this is exactly a torsal ruling of the developable surface enveloped by the given 1-parametric set of planes. As a consequence, the fixed axode Φ_0 is a developable surface (It is well known (e.g., [11] (Thms. 5.1.7 and 6.1.3)) that every developable surface is composed of cylindrical, conical or tangent-surfaces) and the corresponding moving axode Φ is obtained by reflecting Φ_0 in Φ_0's tangent-plane τ along the instantaneous axis p of rotation (cf. Figure 2, right). Now, the path x of a point X under a plane-symmetric motion can be generated by the reflexion of a point X_0 on each tangent-plane τ of Φ_0; i.e., x can be obtained by a central doubling of X_0's pedal-curve f with respect to Φ's tangent-planes.

Due to all these properties, the plane-symmetric motion seems to be the straightforward spatial counterpart of the planar/spherical symmetric rolling. Therefore, we call a plane-symmetric motion also a *spatial symmetric rolling*.

As far as the author knows, these spatial symmetric rollings are only explicitly mentioned in a practical example by Kunze and Stachel [12], who pointed out that the relative motion of opposite systems of a threefold-symmetric Bricard linkage (e.g., the *invertible cube* of Schatz) is a plane-symmetric one. Clearly, this also holds for the more general class of plane-symmetric Bricard linkages [13], where the two opposite systems not containing a rotation-axis spanning the plane of symmetry also possess a plane-symmetric relative motion during the overconstrained motion of the closed 6R-chain.

1.2. Motivation and Outline

One of the author's main research interests are hexapods with self-motions, i.e., overconstrained parallel manipulators where the platform is linked with the base by six legs, which are anchored via spherical red joints (Due to the spherical joints at the platform and the base, each leg can rotate about its carrier line without changing the pose of the platform. These uncontrolled leg-movements are not meant by the term *self-motion*). All these mechanical devices are solutions to the still unsolved problem posed by the French Academy of Science for the *Prix Vaillant* of the year 1904, which is also known as Borel–Bricard problem and reads as follows [14]: *"Determine and study all displacements of a rigid body in which distinct points of the body move on spherical paths."* In order to avoid trivial solutions of the problem, the following assumption should hold for the remainder of the article.

Assumption 1. *The platform anchor points* $m_1, \ldots, m_6$ *of the hexapod as well as the corresponding base anchor points* $M_1, \ldots, M_6$ *should span in each case at least a plane.*

It is well known that so-called architecturally singular hexapods (A hexapod is called architecturally singular if the six legs belong in each relative pose of the platform with respect to the base to a linear line complex) possess self-motions in each pose (over $\mathbb{C}$). These special solutions to the Borel–Bricard problem are already well studied (A review on this topic is given in [15] (Section 3.1)). The approaches for the determination of non-architecturally singular hexapods recorded in the literature (Note that we do not claim that the following list of given references is complete), can roughly be divided into the following two groups:

1.   Assumptions on the geometry of the platform and base; e.g.,

   (a) linear mapping between platform and base [16–22],
   (b) symmetry properties of platform and base [20–24],
   (c) special topology (e.g., octahedral structure [25]),

   or a combination of these assumptions (e.g., [20–22])

2.   Assumptions on the self-motion; e.g.,

   (a) line-symmetric self-motion [9],
   (b) type II Darboux–Mannheim self-motion [26],
   (c) Schoenflies self-motion [27],
   (d) translational self-motion [28],
   (e) self-motion of maximal degree [29],

   or more generally characterizations like *linear relations between direction cosines* [30–33].

Note that these assumptions are done in order to reduce the complexity of the problem, as one has to deal with 30 design parameters (24 for the geometry and six leg lengths, whereby the number of 30 can be reduced by one due to the freedom of scaling) and six degrees of freedom.

We want to follow the second approach by assuming that the self-motions are symmetric rollings. Therefore, this paper closes a gap as line-symmetric self-motions and point-symmetric (Point-symmetric motions are obtained by reflecting the fixed system in a 1-parametric continuous set of points and according to [7] (Section 8), these motions are pure translations) self-motions are already well-studied [9,28]. In addition, this motion-type seems to be a good candidate for self-motions, due to the following property implied by the symmetry of the motion:

Theorem 1. *If a point* A *of the moving system traces a spherical curve with center* B_0 *during a plane-symmetric motion, then also the point* B *of the moving system has a spherical trajectory about the point* A_0, *where* A *and* A_0 *as well as* B *and* B_0, *are plane-symmetric points of the moving and fixed frame with respect to the tangent-plane* τ *along the instantaneous axis* p *of rotation. As a consequence, the set of points with spherical trajectories is indirectly congruent to the set of corresponding sphere centers.*

In the remainder of the paper, we call the replacement of the point pair (A, B_0) by (B, A_0) the *"symmetric leg-replacement"*.

Remark 2. *Clearly, the lower dimensional version of Theorem 1 is also true for the planar/spherical symmetric rolling. Moreover, Theorem 1 also holds for point-symmetric motions, if "plane-reflection" is substituted by "point-reflection". A similar result holds for line-symmetric motions; one only has to replace "plane-reflection" by "line-reflection" and "indirectly congruent" by "directly congruent" (see e.g., [9]).*

The paper is structured as follows: We start with the discussion of planar/spherical symmetric rolling motions with circular paths in Section 2.1. In Section 2.2, we formulate the problem of determining hexapods with plane-symmetric self-motions in terms of algebraic geometry by means of Study parameters. Based on this description, the problem is solved in Section 3. One of the obtained solutions is the so-called *Duporcq hexapod*, which is discussed in more detail in Section 4. The paper is closed by a conclusion (cf. Section 5).

2. Preliminary Considerations and Preparatory Work

As far as the author knows, no hexapods with plane-symmetric self-motions are reported in the literature so far. From known results in planar/spherical kinematics, which are reviewed in the next subsection, we can immediately construct such hexapods.

2.1. Planar/Spherical Symmetric Rollings with Circular Paths

Clearly, a pure rotation is a planar/spherical symmetric rolling where every point of the moving system traces a circle. Besides this trivial case, which we meet again under the notation of a so-called butterfly self-motions (cf. later given Theorem 4), the following planar/spherical symmetric rollings with circular trajectories exist:

- The planar symmetric rolling motions with points running on circular paths are well known due to the study of Bereis [3]. In this case, the polodes are either ellipses or hyperbolas and the focals (two real, two complex) of the moving ellipse/hyperbola are running on circles. They are the Burmester points of this motion. These motion can be realized by the mechanisms illustrated in Figure 3.

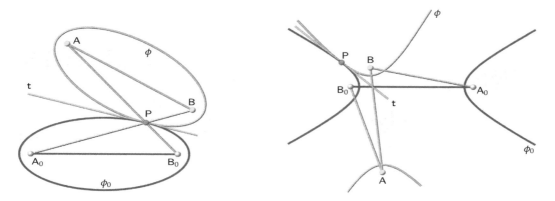

Figure 3. Twin-crank mechanisms with non-counter-rotating cranks (**left**): In this case, the polodes are ellipses. Twin-crank mechanism with counter-rotating cranks (**right**): In this case, the polodes are hyperbolas.

- Unfortunately, the considerations of Bereis cannot be generalized straightforward to the sphere (cf. [2] (p. 195)), as in spherical kinematics six Burmester points exist (e.g., [8] (p. 216)). However, we can do the reasoning in a different way. Due to [28] (Theorem 6), one can assume without loss of generality that only two points of a moving body can have spherical trajectories. According to the spherical version of Theorem 1, a second point is also running on a circle due to the symmetric leg-replacement (With the exceptional case that the first leg is orthogonal to the pole tangent, but this will not yield a closed loop; i.e., a spherical parallel manipulator). Thus, we can only end up with a spherical isogram illustrated in Figure 4, which is studied in more detail in [34].

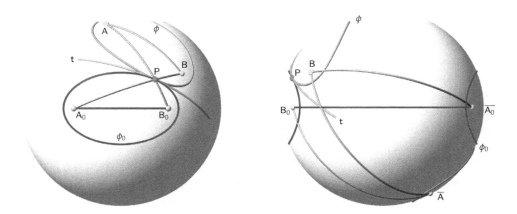

Figure 4. As on the sphere points can be replaced by their antipodes, it can easily be seen that every spherical conic can be interpreted as a spherical ellipse (e.g., [35] (Section 10.1)). The left and the right figure show the same symmetric rolling motion. If we replace A and A_0 by their antipodal points $\overline{A}$ and $\overline{A_0}$, respectively, and look on the sphere from the right side, then we get the figure illustrated on the right-hand side.

From the discussed planar and spherical case, one can easily construct hexapods with plane-symmetric self-motions (see Figure 5).

Remark 3. *Note that the hexapods of Figure 5 do not only possess the illustrated plane-symmetric self-motions, but also the already mentioned butterfly self-motions (cf. later given Theorem 4).*

Figure 5. Hexapods with plane-symmetric self-motions, where the platform (green) and the base (blue) are both planar. The axodes of the self-motions are cylinders (**left**) and cones (**right**), respectively, but we only illustrated the planar/spherical directrices of these singular quadrics to see better their connection to the planar/spherical symmetric rolling displayed in Figures 3 and 4, respectively.

2.2. Mathematical Framework

For the algebraic formulation of our problem, we want to use Study parameters $(e_0 : e_1 : e_2 : e_3 : f_0 : f_1 : f_2 : f_3)$, which are nothing else than homogenized dual unit-quaternions $\mathfrak{E} + \varepsilon\mathfrak{F}$ with

$$\mathfrak{E} = e_0 + e_1\mathbf{i} + e_2\mathbf{j} + e_3\mathbf{k} \quad \text{and} \quad \mathfrak{F} = f_0 + f_1\mathbf{i} + f_2\mathbf{j} + f_3\mathbf{k}, \tag{1}$$

where $\mathbf{i}, \mathbf{j}, \mathbf{k}$ are the well-known quaternionic units and ε the dual unit with the property $\varepsilon^2 = 0$.

Now, all real points of the seven-dimensional Study parameter space P^7, which are located on the so-called Study quadric $\Psi : \sum_{i=0}^{3} e_i f_i = 0$, correspond to a Euclidean displacement with exception of the three-dimensional subspace E of Ψ given by $e_0 = e_1 = e_2 = e_3 = 0$, as its points cannot fulfill the condition $N \neq 0$ with $N = e_0^2 + e_1^2 + e_2^2 + e_3^2$. The translation vector $\mathbf{t} := (t_1, t_2, t_3)^T$ and the rotation matrix $\mathbf{R} := (r_{ij})$ of the corresponding Euclidean displacement $\mathbf{x} \mapsto \mathbf{R}\mathbf{x} + \mathbf{t}$ are given by:

$$t_1 = 2(e_0 f_1 - e_1 f_0 + e_2 f_3 - e_3 f_2), \quad t_2 = 2(e_0 f_2 - e_2 f_0 + e_3 f_1 - e_1 f_3), \quad t_3 = 2(e_0 f_3 - e_3 f_0 + e_1 f_2 - e_2 f_1),$$

and

$$\mathbf{R} = \begin{pmatrix} e_0^2 + e_1^2 - e_2^2 - e_3^2 & 2(e_1 e_2 - e_0 e_3) & 2(e_1 e_3 + e_0 e_2) \\ 2(e_1 e_2 + e_0 e_3) & e_0^2 - e_1^2 + e_2^2 - e_3^2 & 2(e_2 e_3 - e_0 e_1) \\ 2(e_1 e_3 - e_0 e_2) & 2(e_2 e_3 + e_0 e_1) & e_0^2 - e_1^2 - e_2^2 + e_3^2 \end{pmatrix},$$

if the normalizing condition $N = 1$ is fulfilled.

Clearly, the reflection on a plane is an orientation-reversing congruence transformation, which cannot be described directly by the Study parameters. Therefore, we follow the approach of Selig and Husty [7] (Section 8), which is as follows: We start with a reflexion on a fixed plane; say the xy-plane of the fixed frame $\mathfrak{F}_0$. By this plane-reflection of $\mathfrak{F}_0$, we obtain $\overline{\mathfrak{F}}_0$. In addition, we apply the reflexion on the plane $\tau(t)$, which finally yields the pose $\overline{\mathfrak{F}}_0^t$. As the composition of two plane-reflexions is again a direct congruence transformation, we can describe the plane-symmetric motions in this way. If $\tau(t)$ and the xy-plane of $\mathfrak{F}_0$ are

- not parallel, then the composition is a rotation about the line of intersection,
- parallel, then the composition is a translation orthogonal to these planes.

This yields that the plane-symmetric motions are given by $e_3 = f_0 = f_1 = f_2 = 0$. Moreover, it should be noted that the Study condition is fulfilled identically, thus the set of plane-symmetric motions corresponds to a three-dimensional generator space P of Ψ which intersects E in a line. Based on this description, we analyze the relation between plane-symmetric motions and line-symmetric ones in the next theorem:

Theorem 2. *A plane-symmetric motion is also a line-symmetric one if and only if there exists a linear relation $\alpha e_0 + \beta e_1 + \gamma e_2 + \delta f_3 = 0$ with $(\alpha, \beta, \gamma, \delta) \neq (0,0,0,0)$ between the remaining Study parameters.*

Proof. For the proof, we need an algebraic characterization of line-symmetric motions in terms of Study parameters. It is well-known that there always exist, a Cartesian frame in the moving system in a way that $e_0 = f_0 = 0$ holds for a line-symmetric motion. Then, $(e_1 : e_2 : e_3 : f_1 : f_2 : f_3)$ are the Plücker coordinates of the generators of the basic surface with respect to the fixed frame.

A change of the moving system can be achieved by a so-called right multiplication; i.e., $(\mathfrak{E} + \varepsilon \mathfrak{F}) \circ (\mathfrak{R} + \varepsilon \mathfrak{S})$ where $\circ$ stands for the quaternionic multiplication. If we denote this product by $\mathfrak{G} + \varepsilon \mathfrak{H}$, the corresponding entries g_0 and h_0 read as follows (under consideration of $e_3 = f_0 = f_1 = f_2 = 0$):

$$g_0 := r_0 e_0 - r_1 e_1 - r_2 e_2, \quad h_0 := s_0 e_0 - s_1 e_1 - s_2 e_2 - r_3 f_3. \tag{2}$$

If $\delta = 0$ holds, then we set $r_0 = \alpha, r_1 = -\beta, r_2 = -\gamma$ and $s_0 = s_1 = s_2 = r_3 = 0$. For $\delta \neq 0$, we set $s_0 = \alpha, s_1 = -\beta, s_2 = -\gamma, r_3 = -\delta$ and $r_0 = r_1 = r_2 = 0$. For both cases, we get $g_0 = h_0 = 0$, which finishes the sufficiency of the linear relation between e_0, e_1, e_2, f_3.

Its necessity can also be seen from Equation (2), as without such a linear relation, the condition $g_0 = h_0 = 0$ can only be fulfilled for $r_0 = r_1 = r_2 = r_3 = 0$, which yields a contradiction as $\mathfrak{R}$ has to differ from the zero-quaternion. $\square$

A further important theorem in this context is the following:

Theorem 3. *A plane-symmetric motion is also a line-symmetric one if and only if it is a planar motion or a spherical motion.*

Proof. If the linear relation equals $f_3 = \alpha e_0 + \beta e_1 + \gamma e_2$, then it can easily be checked by direct computations that the point $(\gamma, -\beta, -\alpha)$ is mapped to the point $(\gamma, -\beta, \alpha)$ for all e_0, e_1, e_2 fulfilling $N = 1$. Therefore, $(\gamma, -\beta, \alpha)$ is the center of the spherical motion.

If the linear relation equals $\alpha e_0 + \beta e_1 + \gamma e_2 = 0$, then it can easily be checked by direct computations that the direction $(\gamma, -\beta, -\alpha)$ is mapped to the direction $(\gamma, -\beta, \alpha)$ for all e_0, e_1, e_2 fulfilling $N = 1$. Therefore, the direction $(\gamma, -\beta, \alpha)$ remains fixed under the motion. Moreover, the translation vector (t_1, t_2, t_3) is orthogonal to this direction, which already proves that the motion is planar. $\square$

These two theorems imply the following statement:

Corollary 1. *If we embed the planar and spherical symmetric rollings into SE(3), then they can also be seen as line-symmetric motions.*

Therefore, the self-motions of the hexapods illustrated in Figure 5 are plane-symmetric and line-symmetric at the same time. This raises also the question of whether self-motions exist, which are plane-symmetric but not line-symmetric. The answer is given within the next section.

3. Plane-Symmetric Self-Motions

The coordinate vector of the base point M_i with respect to the fixed system is given by $\mathbf{M}_i = (A_i, B_i, C_i)^T$. The position of the corresponding platform anchor point $m_i(t)$ is obtained by reflecting a point $m_{i,0}$ with fixed coordinates $\mathbf{m}_i = (a_i, b_i, c_i)^T$ in a 1-parametric continuous set of planes $\tau(t)$. Instead of these reflexions, we use direct isometries based on the Study representation described in Section 2.2 (i.e., $e_3 = f_0 = f_1 = f_2 = 0$). Therefore, the locus of the corresponding platform anchor point m_i with respect to the fixed frame can be parametrized as $\mathbf{R}\overline{\mathbf{m}}_i + \mathbf{t}$ with $\overline{\mathbf{m}}_i = (a_i, b_i, -c_i)^T$.

The condition that the point m_i is located on a sphere centered in M_i with radius d_i is a quadratic homogeneous equation in the Study parameters according to Husty [36]. For our setup, this so-called sphere condition Λ_i has the following form:

$$\Lambda_i: \quad (a_i^2 + b_i^2 + c_i^2 + A_i^2 + B_i^2 + C_i^2 - d_i^2)N - 4(c_i + C_i)e_0 f_3 + 4(b_i + B_i)e_1 f_3 - 4(a_i + A_i)e_2 f_3$$
$$- 2(a_i A_i + b_i B_i - c_i C_i)e_0^2 - 2(a_i A_i - b_i B_i + c_i C_i)e_1^2 + 2(a_i A_i - b_i B_i - c_i C_i)e_2^2 \qquad (3)$$
$$- 4(c_i B_i + b_i C_i)e_0 e_1 + 4(c_i A_i + a_i C_i)e_0 e_2 - 4(b_i A_i + a_i B_i)e_1 e_2 + 4f_3^2 = 0.$$

It corresponds to a quadric in the three-dimensional projective space P^3 with homogenous coordinates $(e_0 : e_1 : e_2 : f_3)$. The symmetric leg-replacement (cf. Theorem 1) can also easily be seen within this formula, as it is invariant under the following permutations: $A_i \leftrightarrow a_i$, $B_i \leftrightarrow b_i$, $C_i \leftrightarrow c_i$. Due to this symmetry, we only have to find spatial rolling motions where three points have a spherical trajectory. This means that the corresponding three quardrics Λ_1, Λ_2 and Λ_3 of P^3 have to have a curve in common, which can be a

1. straight line,
2. conic section,
3. cubic curve,
4. quartic curve.

In the following subsections these cases are discussed separately.

3.1. Intersection Curve Is a Straight Line

It is well-known that straight lines in the Study quadric correspond with either rotations about a line or straight translations. As the second option is not possible due to the sphere condition, we are only left with the rotation case. In the first step, we ask under which conditions two quadrics Λ_1 and Λ_2 have a straight line in common.

- **General Case:** Let us assume that $M_1 \neq M_2$ and $m_1(t) \neq m_2(t)$ hold. Clearly, the straight line in P^3 has to correspond with a rotation about the line G spanned by M_1 and M_2. Therefore, the line $g(t)$ spanned by $m_1(t)$ and $m_2(t)$ generates either a hyperboloid, cone or cylinder of revolution with axis G. Moreover, all these poses of the platform points have to be obtained by plane-reflexions of the points $m_{1,0}$ and $m_{2,0}$, respectively. This already implies that the 1-parametric set of planes $\tau(t)$ has to be a pencil of planes with axis G. Therefore, the leg lengths d_1 and d_2 are given by

$$d_i = dist(M_i, m_{i,0}) = \sqrt{(A_i - a_i)^2 + (B_i - b_i)^2 + (C_i - c_i)^2}, \qquad (4)$$

which is already the necessary and sufficient condition for the two quadrics Λ_1 and Λ_2 to have a straight line in common.

- **Special Case:** As the case $M_1 = M_2$ and $m_1(t) = m_2(t)$ cannot arise (legs are identical), we only have to discuss one further case due to the symmetric leg-replacement. Without loss of generality, we can assume $M_1 \neq M_2$ and $m_1(t) = m_2(t)$. Now, $m_1(t) = m_2(t)$ has to trace a circle about the line G, which in fact implies the same condition given in Equation (4) for $i = 1, 2$.

Under consideration of the notation that (M_i, m_i) and (M_{i+3}, m_{i+3}) are coupled by the symmetric leg-replacement (for $i = 1, 2, 3$), we can immediately formulate the following theorem.

Theorem 4. *Up to symmetric leg-replacements, the three quadrics Λ_1, Λ_2 and Λ_3 have a line in common if and only if Equation (4) holds for $i = 1, 2, 3$ and M_1, M_2, M_3 are collinear. The corresponding self-motion of the hexapod is a butterfly self-motion about the line spanned by M_1, M_2, M_3, where $M_i = m_{i+3}$ holds for $i = 1, 2, 3$.*

As these butterfly self-motions (cf. Figure 6, left) are trivial, they are not of further interest.

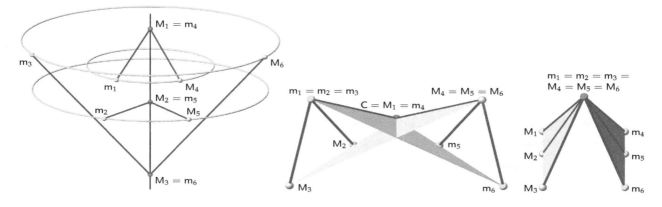

Figure 6. In all three illustrations, the plane of symmetry is always a vertical projecting plane; **left**: butterfly self-motion of a hexapod. Note that not necessarily the three legs obtained by the symmetric leg-replacements have to be added, but any legs where neither M_i or $m_i(t)$ is collinear with M_1, M_2, M_3 for $i = 4, 5, 6$; **center**: situation after performing the Δ-transform; **right**: two-dimensional spherical self-motion.

3.2. Intersection Curve Is a Conic

As the conic is a planar curve, there has to exist a linear relation between the homogenous coordinates $(e_0 : e_1 : e_2 : f_3)$ of P^3. Therefore, we can apply the Theorems 2 and 3, which imply that we can only end up with planar/spherical symmetric rollings already discussed in Section 2.1.

3.3. Intersection Curve Is Cubic

A necessary condition that Λ_1, Λ_2 and Λ_3 have a cubic curve in common is that the intersection of two quadrics split up into a line and this cubic. Therefore, condition Equation (4) has to hold. It can easily be checked that Λ_i splits up into two planes:

$$\Lambda_i : \quad 4(C_i e_0 - B_i e_1 + A_i e_2 - f_0)(c_i e_0 - b_i e_1 + a_i e_2 - f_0) = 0 \tag{5}$$

under consideration of Equation (4). Therefore, the cubic has to split up into three lines, which all correspond to plane-symmetric butterfly self-motions already described in Theorem 4. As a consequence, no further discussion of this case is necessary.

3.4. Intersection Curve Is Quartic

We start with the following lemma, which helps to exclude the discussion of special cases arising.

Lemma 1. *If* M_1, M_2, M_3 *are collinear and* $m_1 = m_2 = m_3$ *holds (under consideration of symmetric leg-replacements), then the hexapod can only have the following plane-symmetric self-motions:*

1.   *butterfly self-motion,*
2.   *two-dimensional spherical self-motion,*
3.   *planar/spherical symmetric rollings of Section 2.1.*

Proof. If the carrier line G of M_1, M_2, M_3 is always identical with the reflected carrier line g of m_4, m_5, m_6, then it is clear that the motion can only be a butterfly motion (cf. Figure 6, left).

Moreover, it is trivial, that the motion can only be a planar one if G is always parallel to g ($\Rightarrow$ planar symmetric rolling of Section 2.1).

Now, we discuss the remaining case that, during the plane-symmetric self-motion, one configuration exists, where G and g intersect in one point C. As the first three legs are always in a pencil of lines, one can make a so-called Δ-transform [37] (without changing the self-motion) such that $M_1 = C$ holds. This results in the following relations (cf. Figure 6, center):

$$\overline{M_1 M_4} = \overline{m_4 M_4} \quad \text{and} \quad \overline{m_1 m_4} = \overline{M_1 m_4}. \tag{6}$$

Under consideration of the plane-symmetric setup, these conditions can only be fulfilled if

• $M_1 = m_4$ holds, which yields the spherical symmetric rolling (with center $M_1 = m_4$) of Section 2.1,
• $M_4 = m_1$ holds, which implies a two-dimensional spherical self-motion (with center $M_4 = m_1$; cf. Figure 6, right).

This finishes the proof. $\square$

Remark 4. *Note that, for the two-dimensional spherical self-motion, the collinearity condition of* M_1, M_2, M_3 *is not necessary. For the leg lengths of Equation (4) and* $m_1 = m_2 = m_3$, *the three quadrics* Λ_1, Λ_2 *and* Λ_3 *have already a plane in common due to Equation (5).*

If Λ_1, Λ_2 and Λ_3 have a quadric curve in common, they are contained within a pencil of quadrics, which is already spanned by two of them. Therefore, we make the following ansatz:

$$\Sigma : \quad \lambda_1 \Lambda_1 + \lambda_2 \Lambda_2 + \Lambda_3 = 0 \quad \text{with} \quad \lambda_1 \lambda_2 \neq 0. \tag{7}$$

In order to simplify the resulting direct computations, we can select the fixed frame in a clever way based on the following lemma:

Lemma 2. *By applying symmetric leg-replacements, we can assume that* M_1, M_2, M_3 *span a plane (under consideration of Assumption 1).*

Proof. If M_1, M_2, M_3 are collinear (span the line G), we apply the symmetric leg-replacement to the ith leg for $i \in \{1, 2, 3\}$. Due to Assumption 1, at least one of the M_{i+3} are not located on G, thus, after a renumeration of anchor points, the lemma holds. $\square$

Due to Lemma 2, we can assume without loss of generality that the origin of the fixed frame equals M_i, that M_j is located on the positive x-axis of the fixed frame and that M_k is located in the xy-plane of the fixed frame for pairwise distinct $i, j, k \in \{1, 2, 3\}$. As M_1, M_2, M_3 is a triangle there always exist at least four (This number results from the fact that each triangle has at least two acute angles, whose two vertices can be used as M_i) choices for i, j, k in a way that M_k is located in the 1st quadrant of the xy-plane. After a may necessary renumeration, we can assume:

$$\mathbf{M}_1 = (0,0,0)^T, \quad \mathbf{M}_2 = (A_2, 0, 0)^T, \quad \mathbf{M}_3 = (A_3, B_3, 0)^T, \tag{8}$$

with $A_2 > 0$, $A_3 > 0$ and $B_3 > 0$. Moreover, by selecting the unit-length in a suitable way, we can achieve $A_2 = 1$.

Based on this choice of the fixed frame, we inspect the coefficients of the linear combination Σ given in Equation (7) with respect to the Study parameters. We denote the coefficient of $e_0^i e_1^j e_2^k f_3^l$ by Σ_{ijkl}. From $\Sigma_{1100} = -4c_3 B_3$, we get $c_3 = 0$. Moreover, we can compute d_3^2 from Σ_{2000}. Then, Σ_{0200} equals $4b_3 B_3$, which implies $b_3 = 0$. From $\Sigma_{1100} = 4\lambda_2 c_2$, we get $c_2 = 0$. Now, $\Sigma_{1001} = 4\lambda_1 c_1$ yields $c_1 = 0$. Then, we express A_3 and B_3 from Σ_{0101} and Σ_{0011}, which results in

$$A_3 = -a_3 - \lambda_1 a_1 - \lambda_2 a_2 - \lambda_2, \qquad B_3 = -\lambda_1 b_1 - \lambda_2 b_2. \tag{9}$$

Moreover, we can set $\lambda_1 = -1 - \lambda_2$ due to Σ_{0002}. Then, we are only left with the following two conditions arising from Σ_{0110} and Σ_{0020}, respectively:

$$-a_3 b_1 - a_3 b_1 \lambda_2 + a_3 \lambda_2 b_2 - \lambda_2 b_2 = 0, \quad -a_3^2 + a_3 a_1 + a_3 a_1 \lambda_2 - a_3 \lambda_2 a_2 - a_3 \lambda_2 + \lambda_2 a_2 = 0. \tag{10}$$

Eliminating λ_2 out of these equations by resultant method yields:

$$a_3(a_3 - 1)(a_3 b_1 - a_3 b_2 - b_1 a_2 + a_1 b_2). \tag{11}$$

Therefore, we distinguish the following cases:

1. For $a_3 = 0$, Equation (10) imply $\lambda_2 a_2 = 0$ and $\lambda_2 b_2 = 0$, respectively. $a_2 = b_2 = 0$ imply the conditions of Lemma 1.

2. For $a_3 = 1$, Equation (10) imply $\lambda_2 = -1$. Then, the second and third leg are identical under consideration of symmetric leg-replacement.

3. For $a_3 b_1 - a_3 b_2 - b_1 a_2 + a_1 b_2 = 0$, we have to distinguish two cases:

    (a) $b_1 = b_2$: now, the condition simplifies to $b_1(a_1 - a_2) = 0$. As $b_1 = 0$ implies $B_3 = 0$ a contradiction, we set $a_1 = a_2$. Then, Equation (10) imply $\lambda_2 = -a_3$, which results in the conditions of Lemma 1.

    (b) $b_1 \neq b_2$: Under this assumption, we can solve this equation for a_3. A further two cases have to be distinguished:

i. $b_2a_1 - b_1a_2 - b_2 = 0$: If one solves this equation for a_i, then Equation (10) implies $b_j = 0$ for distinct $i, j \in \{1, 2\}$. In both cases, we end up with the conditions of Lemma 1.

ii. $b_2a_1 - b_1a_2 - b_2 \neq 0$: Under this assumption, we can solve the condition implied by Equation (10) for λ_2, which yields:

$$\lambda_2 = \frac{(b_1a_2 - b_2a_1)b_1}{(b_1 - b_2)(b_2a_1 - b_1a_2 - b_2)}. \tag{12}$$

It can easily be checked that the obtained solution corresponds to the hexapod's platform and base illustrated in Figure 7, which are also known as Duporcq's complete quadrilaterals [38]. In the remainder of the paper, this interesting solution, which is discussed/studied in more detail in the next section, is called *Duporcq hexapod*. Based on this notation, we can formulate the following theorem.

Theorem 5. *Besides the trivial cases mentioned in Lemma 1, the quadrics Λ_1, Λ_2 and Λ_3 belong to a pencil if and only if they correspond to sphere conditions of three legs of a Duporcq hexapod (which are not identical under symmetric leg-replacements).*

Figure 7. Illustration of Duporcq's complete quadrilaterals: The base (**left**) is congruent with the platform (**right**).

4. Duporcq Hexapod

Due to the results obtained in Section 3 and Theorems 2 and 3, we can conclude that only the Duporcq hexapod of Theorem 5 possesses plane-symmetric self-motions, which are neither planar nor spherical motions. Therefore, we discuss this hexapod in more detail in this section.

In [38], Duporcq describes the following remarkable motion: *Let $M_1, \dots, M_6$ and $m_1, \dots, m_6$ be the vertices of two complete quadrilaterals, which are congruent. Moreover, the vertices are labeled in a way that m_i is the opposite vertex of M_i for $i \in \{1, \dots, 6\}$ (cf. Figure 7). Then, there exists a 2-parametric line-symmetric motion where each m_i is running on spheres centered in M_i.*

It is well known [39] (Section 1) that this is an architecturally singular hexapod and that one can remove any leg without changing the direct kinematics of the mechanism. The resulting pentapod is called *Duporcq pentapod* and its line-symmetric self-motions were also studied in [39]. For the coordinatisation of the platform points and base points used in Section 3.4, the 2-parametric line-symmetric self-motion fulfills $e_0 = f_0 = 0$ (cf. [39] (Section 4)).

Remark 5. *Note that the theoretic results of Section 4 are visualized on the basis of the following example:*

$$a_1 = \frac{3}{2}, \quad a_2 = b_1 = 3, \quad b_2 = \frac{9}{4}, \quad d_1^2 = \frac{17}{2}, \quad d_2^2 = \frac{33}{2}. \tag{13}$$

This input data implies

$$a_3 = \frac{15}{2}, \quad A_3 = \frac{8}{7}, \quad B_3 = \frac{6}{7}, \quad d_3^2 = \frac{13231}{196}, \tag{14}$$

with respect to the coordinatisation of the platform points and base points used in Section 3.4.

4.1. Plane-Symmetric Self-Motions of the Duporcq Hexapod

First of all, it should be pointed out that the plane-symmetric self-motions of the Duporcq manipulator were not known until now. They can be computed as follows: We express f_3 from the condition $\Lambda_2 - \Lambda_1$ (which is linear in f_3). Plugging the resulting expression into Λ_1 implies a homogenous quartic equation Y in e_0, e_1, e_2, which already represents the plane-symmetric self-motion (cf. Figure 8, left).

In the following, we are interested in the transition poses between this one-dimensional plane-symmetric self-motion and the above-mentioned two-dimensional line-symmetric one. Therefore, we only have to intersect the quartic curve Y with $e_0 = 0$, which yields four of these so-called *branching singularities* [40]. These four transition poses are totally flat configurations of the Duporcq hexapod (cf. Table 1, Figure 8, red left and Figure 9).

Remark 6. *Note that a further prominent example of a hexapod, which possesses flat poses during its self-motion, is Bricard's flexible octahedron of type 3 (cf. [41]).*

Moreover, it should be mentioned that the Duporcq hexapod is a kinematotropic mechanism (according to the notation of Wohlhart [42]). To the best of the author's knowledge, only one further hexapod with this property is known so far, which is the so-called Wren platform (see [42] (Section 3) and [21] (Section 2.2)).

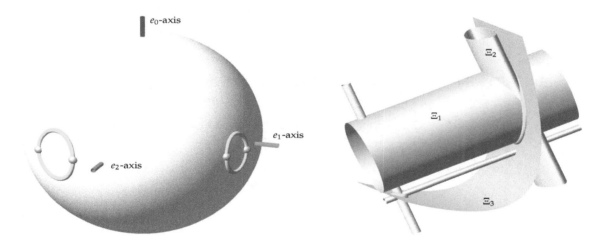

Figure 8. (**left**:) the quartic Y is displayed under consideration of the normalization condition $N = 1$. For the example at hand, it consists of two components (as antipodal points yield the same displacement). Intersection points of the displayed spherical curve with the equator plane yield the branching singularities between plane-symmetric and line-symmetric self-motions. They are numbered from left to right by 1 to 4. (**right**): visualization of the surfaces Ξ_i under the assumption that $u_0 = 0$ corresponds to the ideal plane. The surface Ξ_i is a cylinder in direction of the u_i-axis (for $i = 1, 2, 3$).

Table 1. The Study parameters of the four flat transition poses illustrated in Figure 9. As they result as roots of a polynomial of degree 4, they can be computed explicitly, but, in order to avoid too long expressions, they are displayed numerically.

Flat Pose	e_1	e_2	f_3
1	−0.63171148011492395006	0.77520358996267041460	0.24434973773984142590
2	−0.26236530678800600560	0.96496862425367773706	0.36840718493416854565
3	0.89932040897259076870	0.43729029489044469464	−1.6168042368274940498
4	0.98317707611585865513	0.18265496708349071532	−2.3876030965525136289

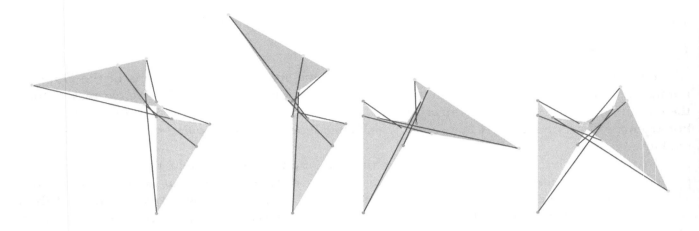

Figure 9. The four flat transition poses numbered from left to right by 1 to 4.

For the example at hand, the fixed axode can be described in the dual representation (If $ax + by + cz + d = 0$ is the equation of the plane of symmetry, then its dual representation is given by the homogenous quadruplet $(u_0 : u_1 : u_2 : u_3) = (d : a : b : c)$ according to [11] (Section 6.2)) as the intersection of the following three surfaces displayed in Figure 8, right:

$$
\begin{aligned}
\Xi_1: \quad & 178596u_3^2u_2^2 + 45924u_0^4 + 69696u_3^4 + 464508u_2^3u_0 + 293436u_2u_3^2u_0 \\
& + 573049u_2^2u_0^2 + 124921u_3^2u_0^2 + 108900u_2^4 + 276336u_2u_0^3 = 0, \\
\Xi_2: \quad & 108900u_1^4 + 592944u_1^3u_0 + 39204u_3^2u_1^2 + 688345u_1^2u_0^2 + 361152u_3^2u_1u_0 \\
& - 300932u_1u_0^3 + 18724u_0^4 + 831744u_3^2u_0^2 = 0, \\
\Xi_3: \quad & 264u_1^2 + 709u_1u_0 + 198u_2u_1 + 912u_2u_0 + 326u_0^2 = 0.
\end{aligned}
\tag{15}
$$

Based on these surfaces, it can be checked (e.g., by computing the Hilbert-polynomial) that the fixed axode corresponds to an algebraic curve of degree 4 in the dual representation. This curve can easily be parametrized as follows (two branches):

$$
u_0 = 1, \quad u_2 = -\frac{264u_1^2 + 709u_1 + 326}{6(33u_1 + 152)}, \quad u_3 = \pm\frac{\sqrt{w}}{6(33u_1 + 152)},
\tag{16}
$$

with $w = 300932u_1 - 18724 - 688345u_1^2 - 592944u_1^3 - 108900u_1^4$. Moreover it can be seen (cf. Figure 8, right) that the curve has two components. The left one is obtained for $u_1 \in [0.07650139; 0.27046582]$ and the right one for $u_1 \in [-3.17251656; -2.61929914]$. Note that the borders of the two intervals are the roots of w, which can be computed explicitly, but, in order to avoid too long expressions, we displayed them numerically.

Based on the parametrization given in Equation (16), one can easily calculate (cf. [11] (Equation (6.8))) the curve of regression of the fixed axode, which is also displayed in Figure 10.

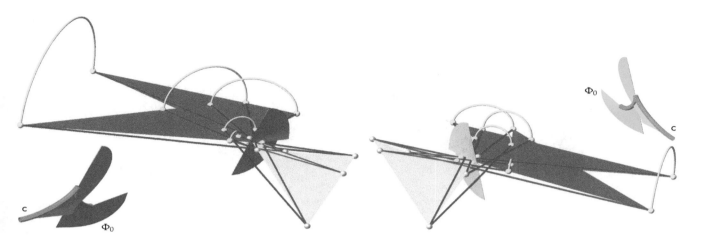

Figure 10. Trajectories of the platform points during the plane-symmetric self-motion between the flat poses 1 and 2 (**left**) and the flat poses 3 and 4 (**right**). Moreover, the fixed axodes Φ_0 are displayed, which look like cones upon the first viewing. However, a blow up (in the lower left corner and upper right corner, respectively) of the region of the supposed vertex shows the line of regression c of Φ_0. For the illustrated self-motion, the tangents of c in the two end points span the carrier plane (xy-plane) of the flat poses. If one considers the complete self-motion, then c has four cusps (obtained by reflecting the illustrated curve at the xy-plane).

4.2. Point-Symmetric Self-Motions of the Duporcq Hexapod

Finally, we want to correct a statement given in [39] (Remark 4), where it is stated that

1. the Duporcq manipulator also has pure translational one-dimensional self-motions,
2. each two-dimensional line-symmetric self-motion of a Duporcq manipulator contains a pure translational one-dimensional sub-self-motion.

The first statement is true in contrast to the second one. In fact, the pure translational self-motion (which can be considered as point-symmetric self-motion) has two branching singularities, where they can switch into a 2-parametric line-symmetric self-motion. This can easily be seen as follows:

For the coordinatisation of the platform points and base points used in Section 3.4, the 1-parametric point-symmetric motion fulfills $e_0 = e_1 = e_2 = f_3 = 0$. It can be computed by expressing f_1 from $\Lambda_2 - \Lambda_1$ (which is linear in f_1). Plugging the resulting expression into Λ_1 implies a homogenous quadratic equation in e_3, f_0, f_2, which already represents the point-symmetric self-motion. By the additional condition $f_0 = 0$, we obtain the two mentioned branching singularities, which are again totally flat configurations of the Duporcq hexapod (cf. Table 2 and Figure 11).

Table 2. The Study parameters of the two flat transition poses illustrated in Figure 11. As they result as roots of a polynomial of degree 2, they can be computed explicitly, but in order to avoid too long expressions they are again displayed numerically.

Flat Pose	e_3	f_1	f_2
1	1	0.1406805116103807682	−0.2234541534831142304
2	1	2.9246864608666834522	−1.0586559382600050357

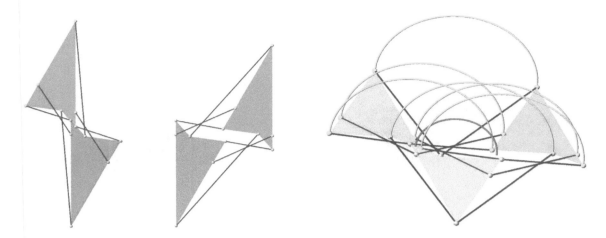

Figure 11. The first and second flat pose (**left** and **center**, respectively) and the translational self-motion between them (**right**). This circular translation can also be seen as a point-symmetric motion, where the corresponding curve (half-circle) is illustrated in red.

Finally it should be noted that there is no branching singularity between plane-symmetric self-motions and point-symmetric self-motions as $e_0 = e_1 = e_2 = e_3 = 0$ has to hold, which contradicts the normalizing condition $N = 1$. Summed up one can say, that the Duporcq hexapod is a twofold kinematotropic mechanism, as there are branching singularities between the two-dimensional line-symmetric self-motion and the one-dimensional

- point-symmetric self-motion,
- plane-symmetric self-motion.

Due to its kinematotropic behavior and its total flat branching singularities the Duporcq manipulator is possibly of interest for the design of deployable structures.

5. Conclusions

This paper gives a complete classification of hexapods with plane-symmetric self-motions. It turns out that besides the planar/spherical symmetric rollings with circular paths and two trivial cases (butterfly self-motion and two-dimensional spherical self-motion), only one further solution exists, which is the so-called Duporcq hexapod. This is the only manipulator possessing plane-symmetric self-motions, which are neither planar nor spherical motions (and therefore also no line-symmetric motions). Moreover, the Duporcq hexapod is may be of interest for the design of deployable structures due to its kinematotropic behavior and total flat branching singularities.

References

1. Quetelet, L.A.J. Memoire sur une nouvelle maniere de considerer les caustiques, produites soit par reflexion soit par refraction. *Brux. Nouv. Mem.* **1826**, *3*, 89–140. (In French)
2. Tölke, J. Ebene euklidische und sphärische symmetrische Rollungen. *Mech. Mach. Theory* **1978**, *13*, 187–198. (In German) [CrossRef]
3. Bereis, R. Über die symmetrische Rollung. *Österr. Ing. Arch.* **1953**, *7*, 243–246. (In German)
4. Bottema, O. Characteristic properties of the symmetric plane motion. *Proc. K. Ned. Akad. Wet. B* **1972**, *75*, 145–151.
5. Darboux, G. *Lecons sur la Theorie Generale des Surfaces*; Gauthier-Villars: Paris, France, 1887. (In French)

6. Krames, J. Über Fußpunktkurven von Regelflächen und eine besondere Klasse von Raumbewegungen (Über symmetrische Schrotungen I). *Monatsh. Math. Phys.* **1937**, *45*, 394–406. (In German) [CrossRef]
7. Selig, J.M.; Husty, M. Half-turns and line symmetric motions. *Mech. Mach. Theory* **2011**, *46*, 156–167. [CrossRef]
8. Bottema, O.; Roth, B. *Theoretical Kinematics*; North-Holland: Amsterdam, The Netherlands, 1979.
9. Gallet, M.; Nawratil, G.; Schicho, J.; Selig, J.M. Mobile Icosapods. *Adv. Appl. Math.* **2017**, *88*, 1–25. [CrossRef]
10. Hernandez-Gutierrez, I. Screw Surfaces in the Analysis and Synthesis of Mechanisms. Ph.D. Thesis, Stanford University, Stanford, CA, USA, 1989.
11. Pottmann, H.; Wallner, J. *Computational Line Geometry*; Springer: Berlin/Heidelberg, Germany, 2001.
12. Kunze, S.; Stachel, H. Über ein sechsgliedriges räumliches Getriebe. *Elem. Math.* **1974**, *29*, 25–32. (In German)
13. Baker, J.E. The single screw reciprocal to the general plane-symmetric six-screw linkage. *J. Geom. Graph.* **1997**, *1*, 5–12.
14. Husty, M.E. Borel's and R. Bricard's Papers on Displacements with Spherical Paths and their Relevance to Self-Motions of Parallel Manipulators. In *Proceedings of the International Symposium on History of Machines and Mechanisms–HMM 2000*; Ceccarelli, M., Ed.; Kluwer: Dordrecht, The Netherlands, 2000; pp. 163–171.
15. Nawratil, G. Correcting Duporcq's theorem. *Mech. Mach. Theory* **2014**, *73*, 282–295. [CrossRef] [PubMed]
16. Karger, A. Singularities and self-motions of equiform platforms. *Mech. Mach. Theory* **2001**, *36*, 801–815. [CrossRef]
17. Karger, A. Singularities and self-motions of a special type of platforms. In *Advances in Robot Kinematics: Theory and Applications*; Lenarcic, J., Thomas, F., Eds.; Kluwer: Dordrecht, The Netherlands, 2002; pp. 155–164.
18. Karger, A. Parallel manipulators with simple geometrical structure. In *Proceedings of the EUCOMES 08: The Second European Conference on Mechanism Science*; Ceccarelli, M., Ed.; Springer: Dordrecht, The Netherlands, 2008; pp. 463–470.
19. Nawratil, G. Non-existence of planar projective Stewart Gough platforms with elliptic self-motions. In *Computational Kinematics: Proceedings of the 6th International Workshop on Computational Kinematics (CK2013)*; Thomas, F., Perez Garcia, A., Eds.; Springer: Dordrecht, The Netherlands, 2013; pp. 49–57.
20. Nawratil, G. On equiform Stewart Gough platforms with self-motions. *J. Geom. Graph.* **2013**, *17*, 163–175.
21. Nawratil, G. Congruent Stewart Gough platforms with non-translational self-motions. In *Proceedings of the 16th International Conference on Geometry and Graphics*; Schröcker, H.-P., Husty, M., Eds.; Innsbruck University Press: Innsbruck, Austria, 2014; pp. 204–215.
22. Nawratil, G. On the Self-Mobility of Point-Symmetric Hexapods. *Symmetry* **2014**, *6*, 954–974. [CrossRef]
23. Dietmaier, P. Forward kinematics and mobility criteria of one type of symmetric Stewart-Gough platforms. In *Recent Advances in Robot Kinematics*; Lenarcic, J., Parenti-Castelli, V., Eds.; Kluwer: Dordrecht, The Netherlands, 1996; pp. 379–388.
24. Karger, A.; Husty, M. Classification of all self-motions of the original Stewart-Gough platform. *Comput. Aided Des.* **1998**, *30*, 205–215. [CrossRef]
25. Nawratil, G. Self-motions of parallel manipulators associated with flexible octahedra. In *Proceedings of the Austrian Robotics Workshop*; Hofbaur, M., Husty, M., Eds.; Umit Lecture Notes: Hall in Tyrol, Austria, 2011; pp. 232–248.
26. Nawratil, G. Planar Stewart Gough platforms with a type II DM self-motion. *J. Geom.* **2011**, *102*, 149–169. [CrossRef]
27. Husty, M.L.; Karger, A. Self motions of Stewart-Gough platforms, an overview. In Proceedings of the Workshop on Fundamental Issues and Future Research Directions for Parallel Mechanisms and Manipulators, Quebec City, QC, Canada, 2–3 October 2002; Gosselin, C.M., Ebert-Uphoff, I., Eds.; pp. 131–141.
28. Nawratil, G. Introducing the theory of bonds for Stewart Gough platforms with self-motions. *ASME J. Mech. Rob.* **2014**, *6*. [CrossRef]
29. Gallet, M.; Nawratil, G.; Schicho, J. Liaison Linkages. *J. Symb. Comput.* **2017**, *79*, 65–98. [CrossRef]
30. Borel, E. Mémoire sur les déplacements à trajectoires sphériques. In *Mém. Présent. Var. Sci. Acad. Sci. Natl. Inst. Fr. TOME XXXIII*; Imprimerie Nationale: Paris, France, 1908; pp. 1–128. (In French)
31. Bricard, R. Mémoire sur les déplacements à trajectoires sphériques. *J. École Polytech.* **1906**, *11*, 1–96. (In French)
32. Karger, A. New Self-Motions of Parallel Manipulators. In *Advances in Robot Kinematics: Analysis and Design*; Lenarcic, J., Wenger, P., Eds.; Springer: Dordrecht, The Netherlands, 2008; pp. 275–282.
33. Karger, A. Self-motions of Stewart-Gough platforms. *Comput. Aided Geom. Des.* **2008**, *25*, 775–783. [CrossRef]

34. Figliolini, G.; Angeles, J. The Spherical Equivalent of Bresse's Circles: The Case of Crossed Double-Crank Linkages. *ASME J. Mech. Rob.* **2014**, *9*. [CrossRef]

35. Glaeser, G.; Stachel, H.; Odehnal, B. *The Universe of Conics*; Springer Spektrum: Berlin/Heidelberg, Germany, 2016.

36. Husty, M.L. An algorithm for solving the direct kinematics of general Stewart-Gough platforms. *Mech. Mach. Theory* **1996**, *31*, 365–380. [CrossRef]

37. Borras, J.; Thomas, F.; Torras, C. On Delta Transforms. *IEEE Trans. Robot. Autom.* **2009**, *25*, 1225–1236. [CrossRef]

38. Duporcq, E. Sur un remarquable déplacement à deux paramétres. *Bull. Soc. Math. Fr.* **1901**, *29*, 1–4. (In French)

39. Nawratil, G.; Schicho, J. Duporcq Pentapods. *ASME J. Mech. Rob.* **2017**, *9*. [CrossRef]

40. Gogu, G. Branching singularities in kinematotropic parallel mechanisms. In *Computational Kinematics: Proceedings of the 5th International Workshop on Computational Kinematics*; Kecskemethy, A., Müller, A., Eds.; Springer: Berlin/Heidelberg, Germany, 2009; pp. 341–348.

41. Stachel, H. Flexible Polyhedral Surfaces With Two Flat Poses. *Symmetry* **2015**, *7*, 774–787. [CrossRef]

42. Wohlhart, K. Kinematotropic linkages. In *Recent Advances in Robot Kinematics*; Lenarcic, J., Parenti-Castelli, V., Eds.; Kluwer: Dordrecht, The Netherlands, 1996; pp. 359–368.

Permissions

All chapters in this book were first published in MDPI; hereby published with permission under the Creative Commons Attribution License or equivalent. Every chapter published in this book has been scrutinized by our experts. Their significance has been extensively debated. The topics covered herein carry significant findings which will fuel the growth of the discipline. They may even be implemented as practical applications or may be referred to as a beginning point for another development.

The contributors of this book come from diverse backgrounds, making this book a truly international effort. This book will bring forth new frontiers with its revolutionizing research information and detailed analysis of the nascent developments around the world.

We would like to thank all the contributing authors for lending their expertise to make the book truly unique. They have played a crucial role in the development of this book. Without their invaluable contributions this book wouldn't have been possible. They have made vital efforts to compile up to date information on the varied aspects of this subject to make this book a valuable addition to the collection of many professionals and students.

This book was conceptualized with the vision of imparting up-to-date information and advanced data in this field. To ensure the same, a matchless editorial board was set up. Every individual on the board went through rigorous rounds of assessment to prove their worth. After which they invested a large part of their time researching and compiling the most relevant data for our readers.

The editorial board has been involved in producing this book since its inception. They have spent rigorous hours researching and exploring the diverse topics which have resulted in the successful publishing of this book. They have passed on their knowledge of decades through this book. To expedite this challenging task, the publisher supported the team at every step. A small team of assistant editors was also appointed to further simplify the editing procedure and attain best results for the readers.

Apart from the editorial board, the designing team has also invested a significant amount of their time in understanding the subject and creating the most relevant covers. They scrutinized every image to scout for the most suitable representation of the subject and create an appropriate cover for the book.

The publishing team has been an ardent support to the editorial, designing and production team. Their endless efforts to recruit the best for this project, has resulted in the accomplishment of this book. They are a veteran in the field of academics and their pool of knowledge is as vast as their experience in printing. Their expertise and guidance has proved useful at every step. Their uncompromising quality standards have made this book an exceptional effort. Their encouragement from time to time has been an inspiration for everyone.

The publisher and the editorial board hope that this book will prove to be a valuable piece of knowledge for researchers, students, practitioners and scholars across the globe.

List of Contributors

Jianhua Li, Jianfeng Sun and Guolong Chen
School of Mechanical & Electrical Engineering, Lanzhou University of Technology, Lanzhou 730050, China

Yunwang Li
School of Mechatronic Engineering, China University of Mining and Technology, Xuzhou 221116, China
Department of Mechanical Engineering, Stevens Institute of Technology, Hoboken, NJ 07030, USA

Shirong Ge, Lala Zhao, Xucong Yan and Yuwei Zheng
School of Mechatronic Engineering, China University of Mining and Technology, Xuzhou 221116, China

Yong Shi
Department of Mechanical Engineering, Stevens Institute of Technology, Hoboken, NJ 07030, USA

Sumei Dai
Department of Mechanical Engineering, Stevens Institute of Technology, Hoboken, NJ 07030, USA
School of Mechanical and Electrical Engineering, Xuzhou University of Technology, Xuzhou 221018, China

Ivan Giorgio and Dionisio Del Vescovo
Department of Mechanical and Aerospace Engineering, SAPIENZA Università di Roma, via Eudossiana 18, 00184 Rome, Italy
Research center on Mathematics and Mechanics of Complex Systems, Università degli studi dell'Aquila, 67100 L'Aquila, Italy

Raffaele Di Gregorio and Mattia Cattai
Department of Engineering, University of Ferrara, 44100 Ferrara, Italy

Henrique Simas
Raul Guenther laboratory of Applied Robotics, Department of Mechanical Engineering, Federal University of Santa Catarina, Florianópolis, SC 88040-900, Brazil

Naim Md Lutful Huq, Md Raisuddin Khan, Amir Akramin Shafie, Md Masum Billah and Syed Masrur Ahmmad
Department of Mechatronics Engineering, International Islamic University Malaysia, Selangor 53100, Malaysia

Alfredo Valverde and Panagiotis Tsiotras
School of Aerospace Engineering, Georgia Institute of Technology, Atlanta, GA 30313, USA

Guojun Liu
School of Mechanical Engineering, Hunan Institute of Science and Technology, Yueyang 414006, China

Bozun Wang, James T. Allison and Albert E. Patterson
Department of Industrial and Enterprise Systems Engineering, University of Illinois at Urbana-Champaign, 117 Transportation Building, 104 South Mathews Avenue, Urbana, IL 61801, USA

Yefei Si
Department of Mechanical Science and Engineering, University of Illinois at Urbana-Champaign, 144 Mechanical Engineering Building, 1206 West Green Street, Urbana, IL 61801, USA

Charul Chadha
Department of Aerospace Engineering, University of Illinois at Urbana-Champaign, 306 Talbot Laboratory, 104 South Wright Street, Urbana, IL 61801, USA

Giovanni Legnani
Department of Mechanical and Industrial Engineering, University of Brescia, 25123 Brescia, Italy
Institute of Intelligent Industrial Technologies and Systems for Advanced Manufacturing, National Research Council, 20133 Milan, Italy

Irene Fassi
Institute of Intelligent Industrial Technologies and Systems for Advanced Manufacturing, National Research Council, 20133 Milan, Italy

Marco Faroni and Antonio Visioli
Dipartimento di Ingegneria Meccanica e Industriale, University of Brescia, Via Branze 38, 25123 Brescia, Italy

Manuel Beschi and Nicola Pedrocchi
Istituto di Sistemi e Tecnologie Industriali Intelligenti per il Manifatturiero Avanzato, National Research Council, Via Alfonso Corti 12, 20133 Milano, Italy

Ali Bin Junaid
Department of Mechanical Engineering, KU Leuven, 3000 Leuven, Belgium

Alejandro Diaz De Cerio Sanchez and Javier Betancor Bosch
Faculty of Science, Engineering and Computing, Kingston University London, London SW15 3DW, UK

Nikolaos Vitzilaios
Department of Mechanical Engineering, University of South Carolina, Columbia, SC 29208, USA

Yahya Zweiri
Faculty of Science, Engineering and Computing, Kingston University London, London SW15 3DW, UK
Khalifa University Center for Autonomous Robotic Systems, Department of Aerospace Engineering, Khalifa University of Science and Technology, Abu Dhabi, UAE

Hermes Giberti
Dipartimento di Ingegneria Industriale e dell'Informazione, Università degli Studi di Pavia, Via A. Ferrata 5, 27100 Pavia, Italy

Francesco La Mura, Gabriele Resmini and Marco Parmeggiani
Department of Mechanical Engineering, Politecnico di Milano, 20156 Milano, Italy

Georg Nawratil
Institute of Discrete Mathematics and Geometry, Vienna University of Technology, Wiedner Hauptstrasse 8-10/104, 1040 Vienna, Austria

Index

Printed in the USA
CPSIA information can be obtained
at www.ICGtesting.com
JSHW051342041223
53221JS00006B/61

9 781647 266769